Conference Proceedings of the Society for Experimental Mechanics Series

Series Editor

Kristin B. Zimmerman, Ph.D.
Society for Experimental Mechanics, Inc.
Bethel, CT, USA

More information about this series at http://www.springer.com/series/8922

Chad S. Korach • Srinivasan Arjun Tekalur • Pablo Zavattieri
Editors

Mechanics of Biological Systems and Materials, Volume 6

Proceedings of the 2016 Annual Conference on Experimental and Applied Mechanics

Springer

Editors
Chad S. Korach
Department of Mechanical Engineering
State University of New York
at Stony Brook
Stony Brook, NY, USA

Srinivasan Arjun Tekalur
Michigan State University
East Lansing, MI, USA

Pablo Zavattieri
Purdue University
West Lafayette, IN, USA

ISSN 2191-5644 ISSN 2191-5652 (electronic)
Conference Proceedings of the Society for Experimental Mechanics Series
ISBN 978-3-319-41350-1 ISBN 978-3-319-41351-8 (eBook)
DOI 10.1007/978-3-319-41351-8

Library of Congress Control Number: 2016949413

Printed on acid-free paper

This Springer imprint is published by Springer Nature
The registered company is Springer International Publishing AG Switzerland

Preface

Mechanics of Biological Systems and Materials represents one of ten volumes of technical papers presented at the SEM 2016 Annual Conference and Exposition on Experimental and Applied Mechanics organized by the Society for Experimental Mechanics and held in Orlando, FL, June 6–9, 2016. The complete proceedings also includes volumes on *Dynamic Behavior of Materials*; *Challenges in Mechanics of Time-Dependent Materials*; *Advancement of Optical Methods in Experimental Mechanics*; *Experimental and Applied Mechanics*; *Micro-and Nanomechanics*; *Mechanics of Composite and Multifunctional Materials*; *Fracture, Fatigue, Failure and Damage Evolution*; *Residual Stress, Thermomechanics and Infrared Imaging, Hybrid Techniques and Inverse Problems*; and *Joining Technologies for Composites and Dissimilar Materials.*

Each collection presents early findings from experimental and computational investigations on an important area within experimental mechanics, the *mechanics of biological systems and materials* being one of these areas.

This volume summarizes the exchange of ideas and information among scientists and engineers involved in the research and analysis of how mechanical loads interact with the structure, properties, and function of living organisms and their tissues. The scope includes experimental, imaging, numerical, and mathematical techniques and tools spanning various lengths and time scales. Establishing this symposium at the Annual Meeting of the Society for Experimental Mechanics provides a venue where state-of-the-art experimental methods can be leveraged in the study of biomechanics. A major goal of the symposium was for participants to collaborate in the asking of fundamental questions and the development of new techniques to address bio-inspired problems in society, human health, and the natural world. The organizers would like to thank all the speakers and staff at SEM for enabling a successful program.

Stony Brook, NY, USA — Chad S. Korach
East Lansing, MI, USA — Srinivasan Arjun Tekalur
West Lafayette, IN, USA — Pablo Zavattieri

Contents

Chapter 1
Mechanic Adaptability of Metastatic Cells in Colon Cancer

V. Palmieri, D. Lucchetti, M. Papi, F. Calapà, G. Ciasca, A. Sgambato, and M. De Spirito

Abstract Tumour microenvironment contributes importantly to the phenotype development of cancer cells, which receive and adapt to chemical and physical signals. It is hypothesized that remodelling of the actin cytoskeleton may enable tumor cells to evade normal apoptotic signalling and permit the acquisition of metastatic properties such as anchorage-independent growth and enhanced cell migration. Among different cancer cell lines, the SW480 and SW620 colon cells, derived from the same patient at different stages of tumour progression, represent a unique model to study the changes in mechanical properties during metastases. The SW480 derive from a primary colon tumour, while the SW620 originate from a population of malignant cells appeared in a lymph node metastases months after the initial surgery. In this work we measured by AFM, combined with Electron and Confocal Microscopy, the mechanic adaptability of cancer cell which reached in vivo the lymph node environment. We demonstrate how to resist to apoptotic stimuli in the lymphatic circulation, where the flow shear stress reduce cell survival, the metastatic cell increase their non-specific adhesion while retaining a flexible and soft cytoskeleton, related to uncontrolled growth properties. These features allow accomplishing the spreading in the circulation aim of metastatic cells.

Keywords Cancer mechanics • Colon • AFM • SEM

1.1 Introduction

Cancer mechanics is the study of forces involved in the intricate interplay between cells and extracellular environment. The emerging idea in this research field is that forces acting on cells can control biochemical signals responsible for cell proliferation, migration and apoptosis [1]. Cells sense and respond to the mechanical properties of the environment, such as the stiffness of the extracellular matrix (ECM) or the compression exerted by neighbouring tissues, by balancing external forces with changes in cytoskeleton organization and shape remodelling. This response leads to the activation of signalling pathways of cells spreading, growth, motility and death [2].

The understanding of the forces that link environment and tumour cells can help to understand the metastatic process, leading cause of mortality among cancer patients. During this process, cells detach from neighbours of the primary tumor, remodel the external matrix and migrate towards the vasculature and lymphatics. These processes require dynamic modulation of cell shape and cytoskeleton together with changes in gene expression [2]. For example metastatic cells can detach by altering surface proteins expression and reducing adhesion or by altering cytoskeleton polymerization in order to acquire a higher plasticity to squeeze easily through ECM [1].

Specific cellular and extra-cellular mechanical properties have recently been exploited as possible biomarkers of disease [3]. In particular, in cancer progression, mechanical properties would allow identifying physical attributes of cells that are more likely to metastasize [3, 4].

In this study we compared the mechanical and morphological features of colon cancer cell populations isolated from a primary tumor and from its lymph-node metastases. We used the SW480 and SW620 colon cancer cell lines that offer the unique advantage of having been isolated from the same patient at different stages of cancer progression. In particular SW480 cell line originates from a Dukes' stage B colon carcinoma while the SW620 cell line has been derived from a lymph

V. Palmieri (✉) • M. Papi • G. Ciasca • M. De Spirito
Institute of Physics, Università Cattolica del Sacro Cuore, L.go Francesco Vito 1, 00168 Rome, Italy
e-mail: vplabcemi@gmail.com

D. Lucchetti (✉) • F. Calapà • A. Sgambato
Institute of Pathology, Università Cattolica del Sacro Cuore, L.go Francesco Vito 1, 00168 Rome, Italy
e-mail: dnlucchetti@gmail.com

C.S. Korach et al. (eds.), *Mechanics of Biological Systems and Materials, Volume 6*,
Conference Proceedings of the Society for Experimental Mechanics Series, DOI 10.1007/978-3-319-41351-8_1

node metastasis developed in the same patient few months later the surgical removal of the tumor [5]. For their limited genetic variability, the SW480 and SW620 lines represent a unique model to analyse biophysical changes occurring during metastatic progression in vivo [5].

The mechanical and morphological properties of both cell lines have been studied by Scanning Electron Microscopy (SEM), Atomic Force Microscopy (AFM) and Confocal Microscopy and a set of morphological parameters concerning cell shape, membrane roughness and protrusions appearance has been analysed. We elucidated the cytoskeleton organization associated with metastasis and provided a further understanding of the mechanical aspects implied in malignant transformation of cells that can represent possible targets for the prediction, treatment and even prevention of cancer.

1.2 Results and Discussion

It has been previously reported that SW480 and SW620 cells have different appearance in culture: most SW480 cells have a spreading, epithelial-type morphology (E-type), while a small fraction displays a rounded morphology (R-type). In contrast, SW620 cells are known to display an ovoid morphology and form small aggregates [6]. We observed clear morphological differences in SEM micrographs (Fig. 1.1) with a prevailing rounded shape for SW620 cells (Fig. 1.1b) and a mixed elongated and rounded morphology in the SW480 cultures (Fig. 1.1a). E-type and R-type cells in the SW480 line were analysed separately as previously described [7]: cells with an Aspect Ratio higher than 1.5 were considered E-type cells (Fig. 1.1c).

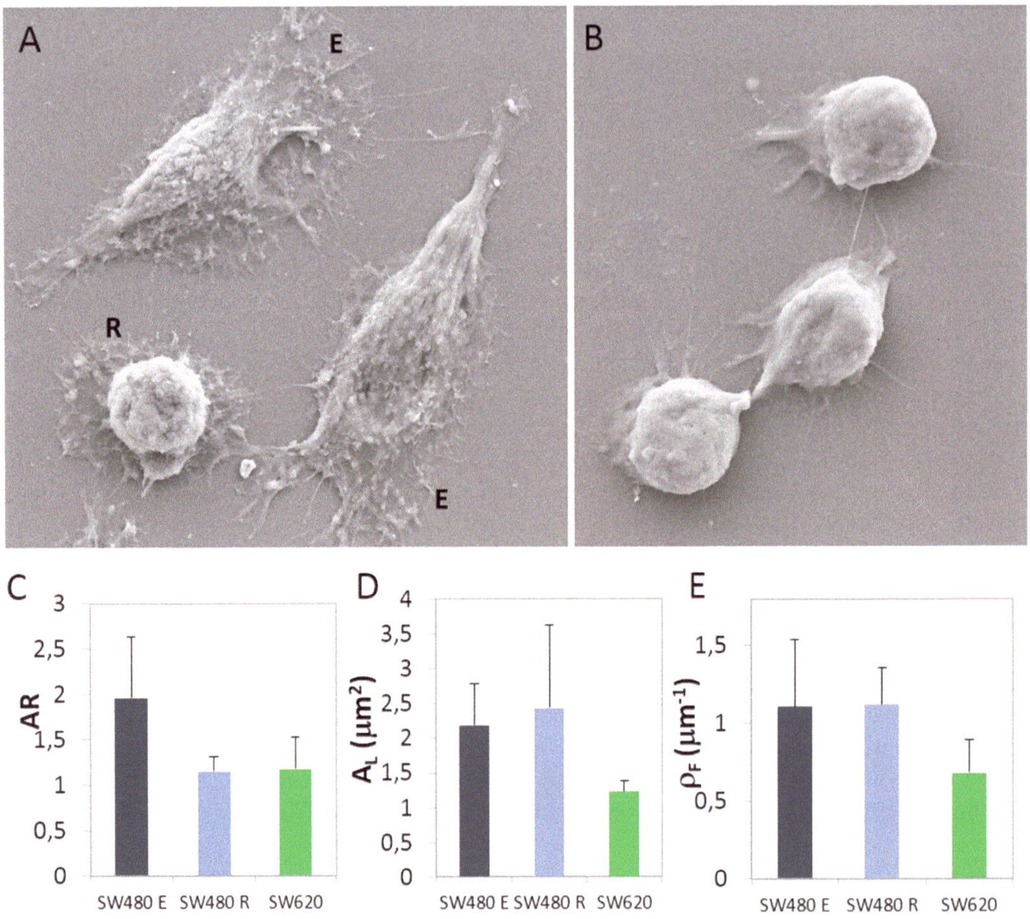

Fig. 1.1 SEM Morphological characterization of SW480 (E-type or R type) and SW620 cells. Cells have different morphologies: SW480 cells comprise an elongated (E) and a rounded (R) (**a**) population both showing a large lamellipodia area and fine protrusions emanating from it. On the contrary, SW620 cells have a rounded morphology with a reduced number of protrusions and a small lamellipodia area (**b**). (**c**) The SW480 line can be divided in two distinct populations using a threshold based on Aspect Ratio of the cells. Indeed elongated cells (E-type) have a mean AR around two while rounded cells (R-type) of one. The SW620 cells have all an aspect ratio always less than 1.5 so they were not divided in sub-populations. Both E-type and R-type cells display a larger lamellipodia area (**d**) and a higher density of filopodia (**e**) compared to SW620 cells. Results are statistically significant ($p < 0.001$ one-way ANOVA and Tukey's multiple comparison tests)

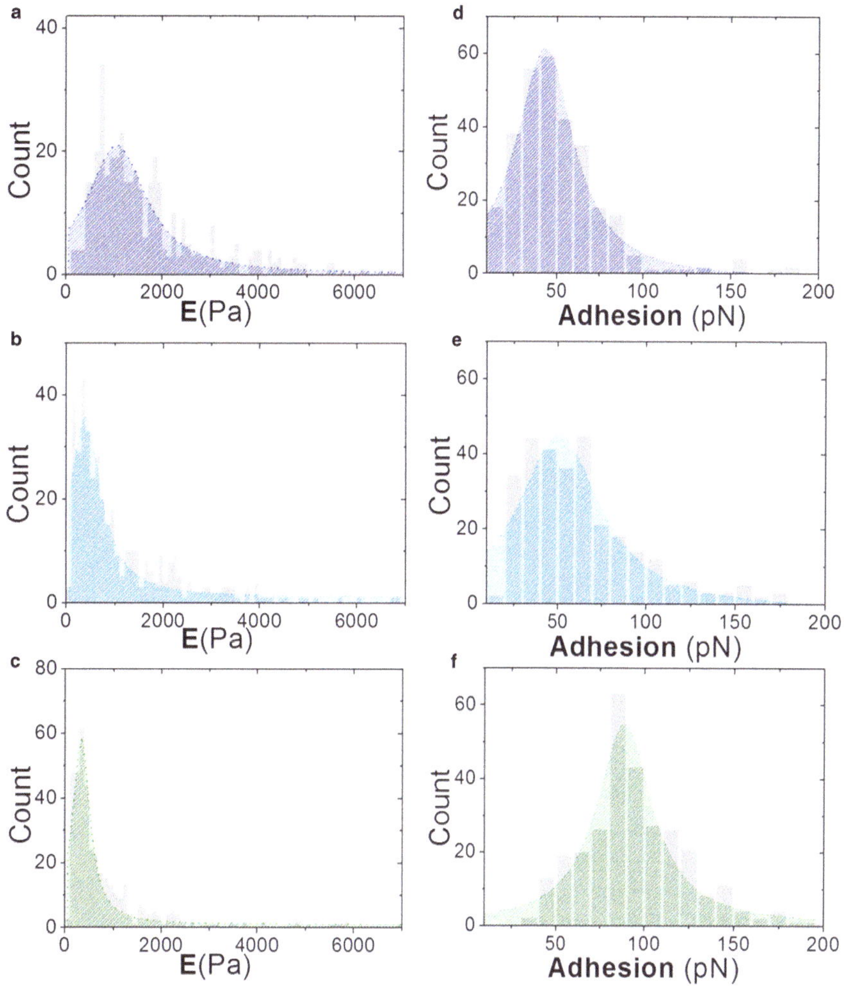

Fig. 1.2 Young Moduli of SW480 cells (E-type in (**a**) and R-type in (**b**)) compared to SW620 cells (**c**). Cells with the highest Young modulus were SW480 E-type, which exhibited a larger distribution peaked at 1200 Pa. Both R-type SW480 and SW620 cells exhibited a narrower distribution peaked at 500 Pa, hence displaying a "softer" cytoskeleton. Non-specific cell adhesion of SW620 (**f**) was significatively higher than in E-type (**d**) and R-type (**e**) SW480 cells

In SW620 cells the AR never exceeded 1.5 (Fig. 1.1c). This criterion has been used also for measurements performed with AFM and Confocal Microscopy.

The membrane protrusions of SW480 and SW620 cells appeared markedly different in SEM micrographs. In Fig. 1.1a, a large area around the cell body of SW480 cells is visible with thin filaments protruding from it. These structures are respectively the lamellipodia and filopodia, highly dynamic thin cell wall extensions of actin filaments. On the other hand, SW620 cells displayed a less extended lamellipodia area (Fig. 1.1b).

Differences observed in membrane protrusions were quantified by means of two parameters: density of filopodia along the cell perimeter (ρ_F) and Lamellipodia Area (A_L). The SW480 cells displayed a higher ρ_F and a larger A_L in respect to SW620 cells (Fig. 1.1d, e). It is important to note also a sharp difference in cell surface roughness in SEM micrographs. Indeed SW620 display a much smoother membrane compared to both populations of SW480 line.

Cell mechanics was analysed by means of Atomic Force Microscopy (AFM), recording strain-stress characteristics (force deformation curves) on cell surface and Hertz Model to calculate cell elasticity. From the obtained curves, cell stiffness (Young Modulus) and non-specific cell adhesion were derived (Fig. 1.2).

Young Moduli of SW480 E-type or R-type and SW620 cells are shown in Fig. 1.2a–c respectively. SW480 E-type cells displayed the highest values of Young modulus, with a wide distribution peaked at 1060 Pa. Both R-type SW480 and SW620 cells exhibited a narrower distribution peaked at 500 Pa, hence displaying a "softer" cytoskeleton. The low values of Young moduli measured (below 1 kPa) for all cell lines considered in this study, indicate a high deformability and compliance, typical of malignant phenotypes [8].

The adhesion distributions of SW480 E-type, SW480 R-type and SW620 cells are shown in Fig. 1.2d–f, respectively. The metastatic line SW620 displayed the highest adhesion with a distribution peaked at 95 pN. The two populations of SW480 cell line showed comparable adhesion values with averages around 50 pN, in accordance to the similar surface roughness observed in SEM micrographs which is known to be strongly related to cell adhesion properties at the nanoscale [9].

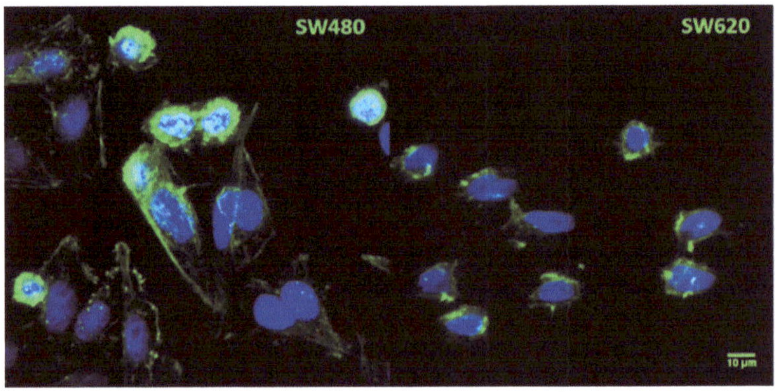

Fig. 1.3 Representative confocal images of Phalloidin labeled actin in SW480 and SW620 cells. Both elongated and rounded cells are visible in the microscope field in SW480 sample (*left*), while only rounded cells are visible in SW620 micrograph (*right*). Cells nuclei are labelled in *blue* with DAPI

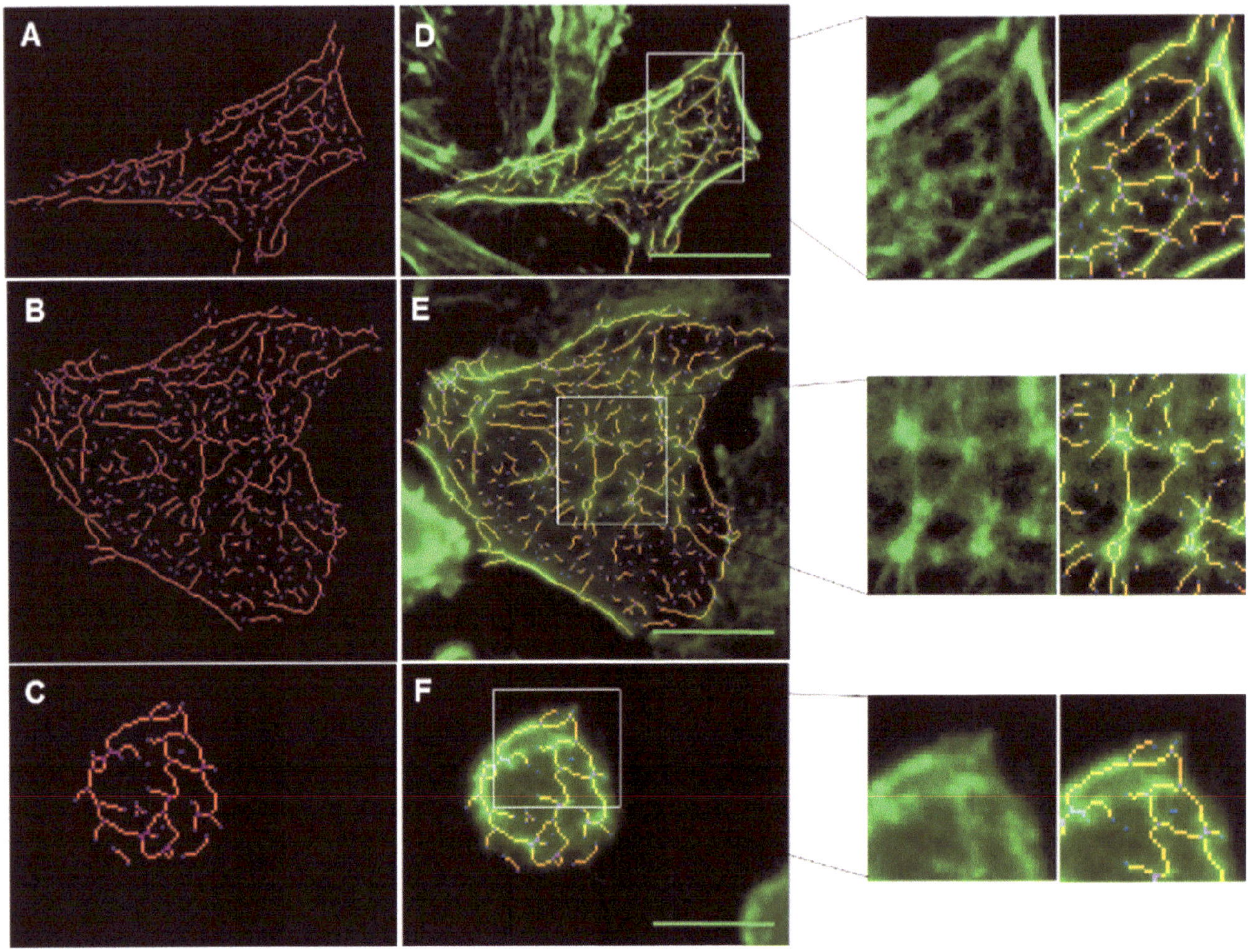

Fig. 1.4 Representative confocal images of skeletonized actin in SW480 E-type (**a**), SW480 R-type (**b**) and SW620 cells (**c**). Skeleton branches are labelled in *orange*, junctions in *purple* and end points in *blue*. On the *right*, skeletons are superimposed on the original confocal images of SW480 E-type (**d**), SW480 R-type (**e**) and SW620 (**f**) cells. Scale Bar is 22 µm in (**d**, **e**) and 10 µm in (**f**). In the *insets* detail of cytoskeleton of each cell are shown

The high values of adhesion, the smooth surface topography and the decreased number of protrusions observed in the SW620 cells suggested an altered actin organization, which has been then analysed by confocal imaging (Fig. 1.3). In Fig. 1.3 representative confocal micrographs of SW480 and SW620 cells are shown with actin in green and cell nuclei in blue.

In order to establish cytoskeleton organization differences among different cell lines, two parameters were considered: actin fibers Coherency and density of junctions of actin network (ρ_J).

Coherency is calculated from the structure tensor of each pixel in the image and is bounded between 0 (isotropic areas) and 1 (highly oriented structures) [10]. In Fig. 1.4a–c three representative confocal images of skeletonized actins filaments

Table 1.1 Analysis of actin organization from confocal images

	ρ_J (μm^{-2})	Coherency
SW480-E	1.59 ± 0.52	0.4 ± 0.14
SW480-R	1.82 ± 1.07	0.19 ± 0.13
SW620	0.82 ± 0.54	0.17 ± 0.11

Average density of cytoskeleton junctions, normalized to cell area, obtained after skeletonization of actin network on confocal images ρ_J. The number of junctions is highly reduced only in the SW620 cells. Coherency of cytoskeleton obtained with OrientationJ plugin. Coherency is bounded between 0 and 1, with 1 indicating highly oriented structures and 0 indicating isotropic areas. The coherency is markedly reduced in SW480-R and SW620 cells

acquired on SW480 E-type (a), SW480 R-type (b) and SW620 cells (c) are shown. The same images were superimposed to the corresponding original confocal images in Fig. 1.4d–e. Analysis of skeletonized images allowed to highlight actin branches (orange), junctions (purple) and isolated points (blue) (inset in Fig. 1.4c). Analysis of Coherency was also calculated on the original confocal images (Fig. 1.4d–f) according to the procedure described in the Experimental section.

Image analysis revealed that, as expected from the lower stiffness values, the rounded cells have a decreased Coherency (Table 1.1). Indeed we measured a mean value of Coherency of 0.4 for E-type SW480 cells and around 0.2 for R-type SW480 cells and SW620 cells. On the other hand, the increased adhesion together with the reduced number of junctions are markers of an actin network destructuration in SW620 cells (Table 1.1).

To control their migratory and invasive capabilities, cancer cells are able to reorganize both their membrane protrusions and cytoskeleton [11, 12]. Filopodia and lamellipodia protrusions drive cell migration by attaching to the substrate and generating forces to pull the cell body forward, while cytoskeleton organization influences cell shape and mechanics as well as cell response to external forces [13]. Protrusions and, more importantly, cytoskeleton subversion in cancer cells can lead to changes in cell growth, stiffness, movement and invasiveness [14].

In order to analyse cytoskeleton of metastatic and non-metastatic cells, different high resolution techniques were used in this study. Filopodia and lamellipodia protrusions and cell surface morphology were analyzed by Scanning Electron Microscopy. Mechanical cell parameters (elasticity and cell adhesion) were obtained recording Force-Distance Curves on cell surface by Atomic Force Microscopy and actin cytoskeleton organization was quantified by measuring fibers anisotropy (Coherency) and networking (Junctions density) by image analysis of Confocal Microscopy measurements.

The specialized functions of SW480 populations, i.e. proliferate at the primary site (SW480 R-type) or metastasize (SW480 E-type), are reflected in cell shape and mechanical properties. We observed that SW480 R-type display a rounded morphology with a decreased stiffness and actin anisotropy (Coherency) compared to E-type cells.

The decrease of cell stiffness, that can be a consequence of a destructuration of actin bundles (reduced Coherency), can represent a mechanistic pathway that drives uncontrolled growth and evasion of apoptosis [1]. Indeed, the actin cytoskeleton is a major determinant of cell polarity and cell-to-cell junctions, essential for normal tissue homeostasis. When cell polarity is lost, tissue integrity is compromised, resulting in overgrowth, aberrant invasive behaviour and promotion of tumours [14].

The higher cell deformability and actin isotropy in rounded cells are also markers of their ability to grow in culture without anchorage [6, 15]. Anchorage-independent growth is a condition where cell proliferation does not depend on culture substrate or cell-to-cell contact and has been observed for SW480 R-type and SW620 cells but not for SW480 E-type cells [6]. It has been recently demonstrated that the HCT-8 colon cancer cells become rounded and "soft" when, losing the mechano-sensitivity to the surrounding environment, gain this ability to grow "independently" [16]. Similarly, SW480 R-type cells have lost sensitiveness to the environment and grow uncontrollably while SW480 E-type cells, with a higher stiffness, are still capable of sensing neighbouring tissues and invade them. Cells derived from lymph node metastasis (SW620 cells) have a rounded morphology similarly to SW480 R-type cells and share with them the ability to grow without anchorage which is reflected in the mean cell elasticity and a reduced fibers Coherency which are comparable to the values obtained for SW480 R-type cells.

Conversely, cells surface appeared markedly altered in the metastatic SW620 cells, which displayed a smoother surface and a decreased filopodia density and lamellipodia area in comparison to SW480 primary tumor cells.

SW620 cells are known to have a limited migration capacity compared to both populations of SW480 cells line [15]. This feature was reflected in a decreased number of protrusions on their surface.

The smoothness of cells surface can be a consequence of the reduction of filopodia structures and/or a marker of an impairment of organization of cortical layer of cytoskeleton underlying the membrane of SW620 cells [17].

SW620 cells are representative of cancer cells that have detached from primary tumor and have invaded lymph nodes, therefore they must have acquired specific properties to survive in the lymphatic stream. Indeed, once in the vasculature, cancer cells are exposed to fluid shear forces of the stream. The majority of circulating cancer cells are rapidly and lethally damaged in the microvasculature. To survive, circulating cancer cells must adhere to the vessel walls and eventually penetrate it [18]. We observed an increase in cell adhesion of metastatic SW620 cells which can be consequence of an altered arrangement of the cortical actin network underlying membrane, as demonstrated by Mescola et al. [19]. Indeed, actin network of SW620 cells, imaged with Confocal microscopy, displayed a reduction in the number of junctions compared to SW480 cells. This actin network "weakening" linearly correlates with the AFM values of adhesion. Therefore, we suggest that modification of the actin organization observed in SW620 cells is necessary to become more "sticky" to vasculature walls and be more stable in the lymphatic and blood vessels.

1.3 Experimental

1.3.1 Cell Cultures

The SW480 and SW620 human colon carcinoma-derived cell lines were purchased from the American Type Culture Collection (ATCC; Manassas, VA) and were maintained in RPMI supplemented with 10 % fetal bovine serum, penicillin-streptomycin (100 U/ml), and 2 mM L-glutamine at 37 °C, in a humid 5 % CO_2 atmosphere, as previously reported [20]. All image analysis measurements were performed on 60 cells per sample on three independent samples.

1.3.2 Scanning Electron Microscopy (SEM)

For SEM imaging, cells on coverslips were washed three times with 0.1 M Sodium Cacodylate buffer (pH 7.4) and then incubated with 0.1 M Sodium Cacodylate buffer and 2.5 % glutaraldehyde for 4 h. Then cells were dehydrated serially in 30, 60, 80 and 100 % ethanol. Finally samples were fixed for 30 min with 2 % osmium tetra-oxide (OsO_4). The SEM procedures were completed by drying of the samples, and sputtering of 8 nm gold layer as reported previously [21]. Micrographs were acquired with a Zeiss Supra 25 microscope (Germany) with a secondary electron detector.

In order to quantify cell morphology, cell axes were measured by means of ImageJ software. From these values, the ellipse fitting each single cell was drawn and perimeter (P) and area (A) were calculated as follows

$$P = 2\pi\sqrt{\frac{a^2 + b^2}{2}} \tag{1.1}$$

$$A = \pi ab \tag{1.2}$$

where a and b are the major and minor semi axis of the cell.

The number of filopodia per cell (N_F) was extracted from images and normalized by the cell perimeter to obtain the density of filopodia (ρ_F) along the cell membrane:

$$\rho_F = \frac{N_F}{P} \tag{1.3}$$

In order to quantify the lamellipodia area, cell major (a_B) and minor (b_B) axis, were measured without taking into account the lamellipodia extensions. From these parameters, the cell "body" area was calculated and subtracted from the total Area (A) to recover the Lamellipodia Area (A_L):

$$A_L = A - \pi a_B b_B \tag{1.4}$$

In the SW480 cell line, cells belonging to the R-type or E-type were distinguished on the basis of the Aspect Ratio (AR) defined as:

$$AR = \frac{a_B}{b_B} \tag{1.5}$$

Cells with AR higher than 1.5 were considered E-type cells.

1.3.3 Atomic Force Microscopy (AFM)

Cells were kept in the cell culture medium at a constant temperature (37 °C) throughout data acquisition. Cantilevers with a silica conical tip characterized by an end radius of ~10 nm and a half conical angle of 20° have been used (CSC16-MikroMasch). All these cantilevers, with a nominal spring constant of k = ~0.02 N/m, were accurately calibrated by using thermal method. Force-distance (F-D) curves were obtained using a fixed force set point and keeping a constant speed of 4.0 μm/s [22]. The total vertical displacement was set to 12 μm. F–D curves were analysed using the data processing supplied with the JPK Nanowizard AFM system by applying baseline subtraction, conversion to tip-sample separation, identification and fitting of each jump in the retraction curve to determine the quantitative parameters described in the text. To recover the local cell Young Modulus (E) all the reaction forces F(δ) obtained from the approaching curves were fitted to the Hertz curve:

$$F(\delta) = \frac{2E\,tan\,(\alpha)}{\pi(1 - \nu^2)}\delta^2 \tag{1.6}$$

where δ is the indentation depth, $\nu = 0.5$ is the cells Poisson's ratio and $\alpha = 20°$ the half opening angle of the AFM tip apex.

Local cell adhesion has been extracted directly from the absolute value of the force minimum in the retract curve [19, 23, 24].

1.3.4 Confocal Microscopy

For immunofluorescence cells were stained with FITC-labeled Phalloidin (Life Technologies) and DAPI (fluoromount G with DAPI, Electron Microscopy Sciences) as reported previously [24, 25]. Immunofluorescence images were obtained with a multichannel white light source with DAPI or FITC filter settings on a CARV II spinning-disk microscope (Crisel Instruments, Rome, Italy) by using a 60× oil immersion objective (NA 1.4). Z-stacks were acquired for each sample. Background values (defined as intensities below 7 % of the maximum intensity) were set to zero and coloured black as previously reported [26]. Image processing and analysis was performed with ImageJ software [27]. After having measured the Aspect Ratio of each cell in order to distinguish between elongated and rounded cells (AR threshold = 1.5), actin organization was analysed with different ImageJ plugins. Firstly, the OrientationJ plugin has been used to obtain the Coherency parameter as described previously [10]. The analysis of Coherency was performed on maximum intensity projections of the acquired Z stacks. OrientationJ evaluates the local orientation of every pixel of an image by calculating the structure tensor for each pixel. The Coherency (C) parameter is the ratio between the difference and the sum of the structure tensor eigenvalues and is bounded between 0 (isotropic areas) and 1 (highly oriented structures) [10]. To analyse the number of knots of actin network the Skeletonize tool of ImageJ has been used in order to segment 2D Z-projections. For each cell the number of junctions was calculated and normalized to cell area to obtain the density of Junctions (ρ_J).

1.3.5 Statistical Analysis

Data were analysed by one-way ANOVA followed by Tukey's multiple comparison test. A value of $p < 0.001$ was considered statistically significant.

1.4 Conclusions

A better understanding of cells mechanics can help to predict growth and spreading ability of cancer cells. In order to investigate structural differences in cellular architecture between non-metastatic and metastatic cancer cells, we analysed the colon cancer SW480 and SW620 cell lines, which offer the unique advantage of representing different stages of disease progression of the same cancer. In fact, SW480 and SW620 cell lines were derived, respectively, from the primary tumour and a lymph-node metastasis from the same patient. Two distinct sub-populations have been described within the SW480 cell line, the E-type and R-type cells, whose name reflects their Elongated and Rounded morphology, respectively [7, 28]. Though being both derived from a primary tumour, these populations display opposite invasive properties when inoculated in nude mice: while the E-type cells have the ability to metastasize but form spontaneously regressive primary tumours, R-type cells form large primary tumors without invasion or nodal metastases [28]. On the other hand, the SW620 cell line comprises a single population of rounded metastatic cells [5].

In this article we demonstrated that cell mechanics can be related to actin organization and showed that a decrease in cell elasticity is accompanied by a high isotropy of actin fibers (SW480 R-type and SW620 cells) while a decreased number of actin network junctions is related to an increase in cell adhesion (SW620 cells) and to a smooth cell surface. We also demonstrated that mechanical stiffness and cell adhesion are modulated by cancer cells during the metastatic process.

We hypothesize that regulation of cell mechanics can allow cancer cells to acquire specialized functions essential for their ability to grow, invade surrounding tissues and metastasize. In particular soft rounded cells, insensitive to the environment, are responsible for the increase of the tumour volume thanks to their uncontrolled growth without anchorage On the other hand, elongated and highly motile cells are more prone to invade neighbouring tissues due to their high plasticity coupled to sensitiveness to the ECM. In order to cope with shear forces of the lymphatic and hematic flow, cells organize a more "flexible" cortical actin network to increase their adhesion to vessels: then once a metastatic cell reaches a lymph node, again it modulates its mechanical properties to survive in a new environment [18].

Acknowledgements This research has been supported by Università Cattolica del Sacro Cuore of Rome. Measurements were performed at the Laboratorio Centralizzato di Microscopia ottica ed elettronica facility (LABCEMI) of Università Cattolica del S. Cuore (Rome, Italy). We are extremely thankful to Mario Amici for the technical support in experiments. The authors declare no commercial or financial conflict of interest.

References

1. Katira, P., Bonnecaze, R.T., Zaman, M.H.P.: Modeling the mechanics of cancer: effect of changes in cellular and extra-cellular mechanical properties. Front. Oncol. **3**, 145 (2013)
2. Lopez, J.I., Mouw, J.K., Weaver, V.M.: Biomechanical regulation of cell orientation and fate. Oncogene **27**(55), 6981–6993 (2008)
3. Ciasca, G., Papi, M., Di Claudio, S., Chiarpotto, M., Palmieri, V., Maulucci, G., Nocca, G., Rossi, C., De Spirito, M.: Mapping viscoelastic properties of healthy and pathological red blood cells at the nanoscale level. Nanoscale **7**(40), 17030–17037 (2015)
4. Suresh, S.: Biomechanics and biophysics of cancer cells. Acta Biomater. **3**(4), 413–438 (2007)
5. Hewitt, R.E., McMarlin, A., Kleiner, D., Wersto, R., Martin, P., Tsokos, M., Stamp, G.W., Stetler-Stevenson, W.G.: Validation of a model of colon cancer progression. J. Pathol. **192**, 446–454 (2000)
6. Palmieri, V., Lucchetti, D., Maiorana, A., Papi, M., Maulucci, G., Calapà, F., Ciasca, G., Giordano, R., Sgambato, A., De Spirito, M.: Mechanical and structural comparison between primary tumor and lymph node metastasis cells in colorectal cancer. Soft Matter **11**, 5719–5726 (2015)
7. Palmieri, V., Lucchetti, D., Maiorana, A., Papi, M., Maulucci, G., Ciasca, G., Svelto, M., Sgambato, A., De Spirito, M.: Biomechanical investigation of colorectal cancer cells. Appl. Phys. Lett. **105**(12), 123701–123704 (2014)
8. Cross, S.E., Jin, Y.S., Rao, J., Gimzewski, J.K.: Nanomechanical analysis of cells from cancer patients. Nat. Nanotechnol. **2**(12), 780–783 (2007)
9. Butta, H.J., Cappella, B., Kappla, M.: Force measurements with the atomic force microscope: technique, interpretation and applications. Surf. Sci. Rep. **59**(1–6), 1–152 (2005)
10. Rezakhaniha, R., Agianniotis, A., Schrauwen, J.T., Griffa, A., Sage, D., Bouten, C.V., van de Vosse, F.N., Unser, M., Stergiopulos, N.: Experimental investigation of collagen waviness and orientation in the arterial adventitia using confocal laser scanning microscopy. Biomech. Model. Mechanobiol. **11**(3–4), 461–473 (2012)
11. Yilmaz, M., Christofori, G.: Mechanisms of motility in metastasizing cells. Mol. Cancer Res. **8**(5), 629–642 (2010)
12. Angelucci, C., Maulucci, G., Colabianchi, A., Iacopino, F., D'Alessio, A., Maiorana, A., Palmieri, V., Papi, M., De Spirito, M., Di Leone, A., Masetti, R., Sica, G.: Stearoyl-CoA desaturase 1 and paracrine diffusible signals have a major role in the promotion of breast cancer cell migration induced by cancer-associated fibroblasts. Br. J. Cancer **112**(10), 1675–1686 (2015)
13. Yamaguchi, H., Condeelis, J.: Regulation of the actin cytoskeleton in cancer cell migration and invasion. Biochim. Biophys. Acta. **1773**(5), 642–652 (2007)

14. Stevenson, R.P., Veltman, D., Machesky, L.M.: Actin-bundling proteins in cancer progression at a glance. J. Cell Sci. **125**(Pt 5), 1073–1079 (2012)
15. Kubens, B.S., Zänker, K.S.: Differences in the migration capacity of primary human colon carcinoma cells (SW480) and their lymph node metastatic derivatives (SW620). Cancer Lett. **131**, 55–64 (1998)
16. Tang, X., Wen, Q., Kuhlenschmidt, T.B., Kuhlenschmidt, M.S., Janmey, P.A., Saif, T.A.: Attenuation of cell mechanosensitivity in colon cancer cells during in vitro metastasis. PLoS One **7**(11), e50443 (2012)
17. Dokukin, M.E., Guz, N.V., Gaikwad, R.M., Woodworth, C.D., Sokolov, I.: Cell surface as a fractal: normal and cancerous cervical cells demonstrate different fractal behavior of surface adhesion maps at the nanoscale. Phys. Rev. Lett. **107**(2), 028101 (2011)
18. Korb, T., Schlüter, K., Enns, A., Spiegel, H.U., Senninger, N., Nicolson, G.L., Haier, J.: Integrity of actin fibers and microtubules influences metastatic tumor cell adhesion. Exp. Cell Res. **299**(1), 236–247 (2004)
19. Mescola, A., Vella, S., Scotto, M., Gavazzo, P., Canale, C., Diaspro, A., Pagano, A., Vassalli, M.: Probing cytoskeleton organisation of neuroblastoma cells with single-cell force spectroscopy. J. Mol. Recognit. **25**(5), 270–277 (2012)
20. Palmieri, V., Lucchetti, D., Gatto, I., Maiorana, A., Marcantoni, M., Maulucci, G., Papi, M., Pola, R., De Spirito, M., Sgambato, A.: Dynamic light scattering for the characterization and counting of extracellular vesicles: a powerful noninvasive tool. J. Nanopart. Res. **16**(9), 1–8 (2014)
21. Ciasca, G., Businaro, L., Papi, M., Notargiacomo, A., Chiarpotto, M., De Ninno, A., Palmieri, V., Carta, S., Giovine, E., Gerardino, A., De Spirito, M.: Self-assembling of large ordered DNA arrays using superhydrophobic patterned surfaces. Nanotechnology **24**(49), 495302 (2013)
22. Papi, M., Brunelli, R., Familiari, G., Frassanito, M.C., Lamberti, L., Maulucci, G., Monaci, M., Pappalettere, C., Parasassi, T., Relucenti, M., Sylla, L., Ursini, F., De Spirito, M.: Whole-depth change in bovine zona pellucida biomechanics after fertilization: how relevant in hindering polyspermy? PLoS One **7**(9), e45696 (2012)
23. Bugli, F., Posteraro, B., Papi, M., Torelli, R., Maiorana, A., Paroni Sterbini, F., Posteraro, P., Sanguinetti, M., De Spirito, M.: In vitro interaction between alginate lyase and amphotericin B against *Aspergillus fumigatus* biofilm determined by different methods. Antimicrob. Agents Chemother. **57**(3), 1275–1282 (2013)
24. Scaglione, G.L., Lancellotti, S., Papi, M., De Spirito, M., Maiorana, A., Baronciani, L., Pagliari, M.T., Arcovito, A., Di Stasio, E., Peyvandi, F., De Cristofaro, R.: The type 2B p.R1306W natural mutation of von Willebrand factor dramatically enhances the multimer sensitivity to shear stress. J. Thromb. Haemost. **11**(9), 1688–1698 (2013)
25. Di Simone, N., De Spirito, M., Di Nicuolo, F., Tersigni, C., Castellani, R., Silano, M., Maulucci, G., Papi, M., Marana, R., Scambia, G.: Gasbarrini Potential new mechanisms of placental damage in celiac disease: anti-transglutaminase antibodies impair human endometrial angiogenesis. Biol. Reprod. **89**(4), 88 (2013)
26. Maulucci, G., Troiani, D., Eramo, S.L., Paciello, F., Podda, M.V., Paludetti, G., Papi, M., Maiorana, A., Palmieri, V., De Spirito, M., Fetoni, A. R.: Time evolution of noise induced oxidation in outer hair cells: role of NAD(P)H and plasma membrane fluidity. Biochim. Biophys. Acta **1840**(7), 2192–2202 (2014)
27. Schindelin, J., Arganda-Carreras, I., Frise, E., Kaynig, V., Longair, M., Pietzsch, T., Preibisch, S., Rueden, C., Saalfeld, S., Schmid, B., Tinevez, J.Y., White, D.J., Hartenstein, V., Eliceiri, K., Tomancak, P., Cardona, A.: Fiji: an open-source platform for biological-image analysis. Nat. Methods **9**(7), 676–682 (2012)
28. Yoon, W.H., Lee, S.K., Song, K.S., Kim, J.S., Kim, T.D., Li, G., Yun, E.J., Heo, J.Y., Jung, Y.J., Park, J.I., Kweon, G.R., Koo, S.H., Park, H. D., Hwang, B.D., Lim, K.: The tumorigenic, invasive and metastatic potential of epithelial and round subpopulations of the SW480 human colon cancer cell line. Mol. Med. Rep. **1**, 5763–5768 (2008)

Chapter 2
Nano-Mechanical Response of Red Blood Cells

Massimiliano Papi, Gabriele Ciasca, Valentina Palmieri, Giuseppe Maulucci, Cristina Rossi, Eleonora Minelli, and Marco De Spirito

Abstract In their physiological function, red blood cells (RBCs) need to undergo large deformations in order to pass through capillaries and small vessels. In several pathological conditions, including diabetes mellitus and Alzheimer's disease, this extreme deformability appears to be deeply impaired and an increase in the RBCs stiffness is usually detected.

Given the key role played by the mechanical properties of RBCs, we investigated their viscous-elastic response by AFM nano-mapping. High-resolution maps demonstrate that healthy erythrocytes are stiffer in their canter and softer at the periphery. The RBC stiffness profile shows a cylindrical symmetry that appears to be strongly correlated with their typical biconcave shape.

Our measurements show that the Young's modulus is strongly depending on the indentation rate, demonstrating that viscous forces have a key role in determining their mechanical response. The importance of viscous forces is further stressed by the comparison between healthy and pathological erythrocyte. Our data show that pathological RBCs are not simply stiffer than healthy ones. Conversely they display a different dependence on the indentation rate that leads to an apparent increase in stiffness. Taken together our results show that both the local stiffness distribution and the viscoelastic response provide important information on RBC biomechanics.

Keywords Atomic force microscopy • Stiffness • Dissipation • Biomechanics • Red blood cells

2.1 Introduction

Red blood cells (RBC) have a typical biconcave shape with dimensions of approximately 8 μm in diameter and 2 μm in thickness and contain an interior viscous liquid enclosed by a viscoelastic membrane [1–7]. Such membrane consists of a nearly incompressible lipid bilayer attached to a spectrin protein network, held together by short actin filaments [7].

The peculiar RBC membrane structure ensures the integrity of the cell in narrow capillaries whose cross-section is smaller than the size of the cells. Under physiological conditions, indeed, RBCs must undergo repeated and severe deformations when travelling through small capillaries with diameter not more than 3–5 μm. Moreover, when passing through the spleen RBCs are required to traverse extremely narrow slits with sizes less than 1 μm [8].

There is a growing evidence that this extreme deformability is significantly altered in pathological conditions, such as diabetes mellitus, essential hypertension, arteriosclerosis and coronary artery disease [5, 9, 10]. On the other hand, an altered red blood cell deformability may contribute to the onset and the development of many pathologies. For example, membranopathies and hemoglobinopathies are known to alter the RBC deformability, affecting the blood flow in large and small vessels [11–13].

On the one hand, RBC modifications occur at the whole cell level, on the other hand they are closely related to changes in the molecular composition and organization of the cell that, in their turn, occur at the nanoscale.

This has made it necessary to develop quantitative tools able to probe RBC changes at nanometer and piconewton scales. In this regard, Atomic Force Microscopy (AFM) is an extremely powerful technique as it permits to probe cells, tissues and molecule at the nanoscale in nearly physiological conditions [6, 14–25].

One of the key characteristics to look at RBC biomechanical properties by AFM is the Young's modulus (E) that provides information on cell stiffness. The nanoscale mapping of E has been proven to be effective in distinguishing between normal

M. Papi (✉) • G. Ciasca • V. Palmieri • G. Maulucci • E. Minelli • M. De Spirito
Institute of Physics, Catholic University of Sacred Heart, Largo F. Vito 1, 00168 Rome, Italy
e-mail: massimiliano.papi@unicatt.it

C. Rossi
Institute of Biochemistry and Clinical Biochemistry, Catholic University of Sacred Heart, Largo F. Vito 1, 00168 Rome, Italy

C.S. Korach et al. (eds.), *Mechanics of Biological Systems and Materials, Volume 6*,
Conference Proceedings of the Society for Experimental Mechanics Series, DOI 10.1007/978-3-319-41351-8_2

and pathological erythrocytes in several disease conditions [6–9, 17–19], opening the way to the development of novel diagnostic tools.

Many biomechanics AFM studies rely on the basic assumption that RBCs behaves like an elastic body: dissipative forces are neglected and Young's modulus is treated as unaffected by probe dynamics. As recently demonstrated in [6], this assumption cannot be considered strictly valid and dissipative forces play a key role in the biomechanical response of red blood cells at the nanoscale level. This result is in close agreement with the peculiar structure of the RBC membrane that can be modelled as a network of viscoelastic springs mediating elastic and viscous response.

One of the major properties to evaluate quantitatively the contribution of dissipative forces is to measure the percentage of energy dissipated during the AFM indentation process or Hysteresis (H). Hysteresis is rarely considered in AFM experiments on the biomechanical behaviour of red blood cells. Nevertheless, the nanoscale mapping of H can provide valuable information on the RCB modifications occurring at the molecular level at the onset of many pathologies.

In this work we describe a novel scanning probe-based nanoscale mapping methodology that generates Hysteresis maps of cells and tissues, which have been subjected to AFM indentation.

To test this method we used red blood cells extracted from an healthy donors and patients with iron overload and hyperferritinemia, showing that H may provide highly valuable information with potential application in the clinical practice.

2.2 Material and Methods

The blood was obtained from healthy donor volunteers and patients with iron overload and hyperferritinemia, anticoagulated with heparin and centrifuged to separate blood form serum. Erythrocytes were than dissolved in physiological solution and deposited on a poly-L-lysine coated petri dish. After 1 h incubation, the poly-L-lysine coated petri dish was gently washed in physiological solution to remove unattached red blood cells.

Subjects affected by iron overload and hyperferritinemia were selected because increased serum ferritin concentration is associated with inflammation processes, which lead to a number of changes in the biophysical properties of red blood cells, including aggregation, sedimentation and deformability [2, 26].

Measurements were performed at 37 °C in liquid environment using a JPK Nanowizard II atomic force microscope (JPK instruments, berlin Germany) coupled to an optical microscope (Axio observer, Carl Zeiss, Milan, Italy). MikroMasch silicon cantilevers with a spring constant of approximately 0.05 N/m and a tip radius of about 10 nm were used (CSC38, MikroMasch). The cantilever spring constant was computed for each measurement by thermal calibration. Force curves were acquired by using an indentation force of 0.5 nN.

To evaluate quantitatively the contribution of dissipative forces, we estimated the energy dissipated during the deformation process (or Hysteresis H). H was computed as the difference between the area (A_E) under the extension force curve $F_E(\delta)$ and that (A_R) under the retract curve ($F_R(\delta)$) normalized by A_E:

$$H = \frac{\int_0^{\delta} F_E(\delta)d\delta - \int_0^{\delta} F_R(\delta)d\delta}{\int_0^{\delta} F_E(\delta)d\delta} = \frac{A_E - A_R}{A_E} \tag{2.1}$$

The nanoscale mapping of H was obtained by a homemade Labview software that allows us for the contemporaneous determination of hysteresis and work of adhesion.

2.3 Results

In Fig. 2.1a, b two representative approach-retract cycles acquired on a healthy (Fig. 2.1a) and a pathological RBC are shown (Fig. 2.1b). The coloured area between the approach and the retract curve provides a graphical representation of H.

One can observe that the pathological red blood cell displays a larger H value than the healthy one. However, the typical biconcave shape of red blood cell hints at a spatial inhomogeneous biomechanical response, suggesting that a single point measure could not be representative of the overall behaviour of the cells.

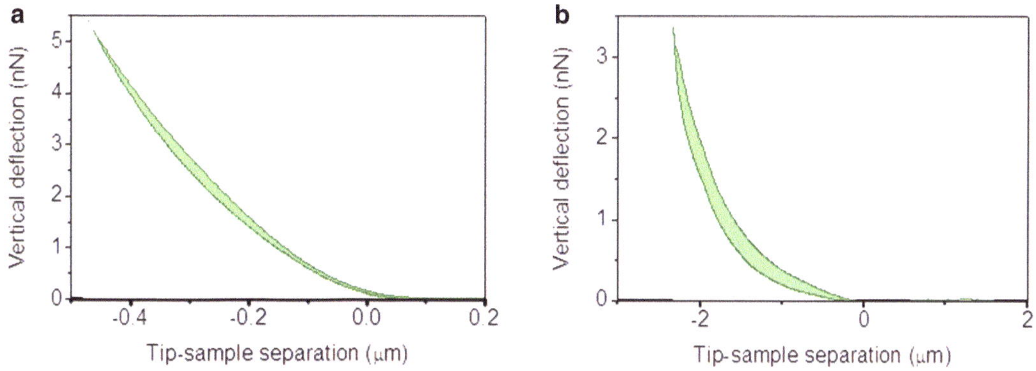

Fig. 2.1 Two representative approach-retract cycles acquired on healthy (**a**) and pathological (**b**) RBCs. The *green* coloured area of the cycle represent the hysteresis

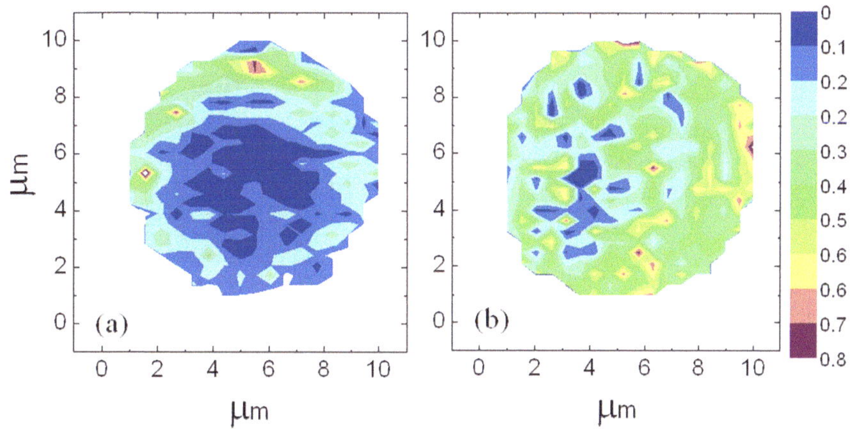

Fig. 2.2 High-resolution hysteresis maps of healthy (**a**) and pathological (**b**) RBCs. Both maps are represented with the same colour scale. These maps display the presence of a cylindrical distribution of H for the healthy RBC, correlated with the biconcave shape of the cell, and the lack of this spatial symmetry for the pathological one. Indeed normal RBC is characterized by value of H in the range 0–0.1 in the centre, and an increase of H in the periphery, while the pathological erythrocyte shows higher value of H, uniformly distributed

As far as RBC stiffness is concerned, this hypothesis was recently confirmed in Ref. [6]. The high resolution nanoscale mapping of E values acquired in physiological solution unveiled that healthy erythrocytes are stiffer in their centre and softer at the cell periphery [6].

Therefore we probed the local response of RBCs by acquiring force-distance curves at different positions over the cell surfaces and evaluating the perceptual of energy dissipated during the indentation process, trough the estimation of H (Eq. (2.1)).

In Fig. 2.2a, b we reported two hysteresis maps acquired on an healthy (a) and pathological red blood cell (b). H values range between 0 and 0.8 for both cells, indicating that a perceptual energy ranging from 0 to 80 % is dissipated during the indentation process.

As far as the healthy RBC is concerned, the nanoscale mapping of H shows a cylindrical symmetry. The cell centre behaves approximately as a pure elastic body, showing H values ranging from 0 to 0.1. An increase in H values can be observed at the cell periphery, where H can be as high as 0.8.

The cylindrical distribution of H values appears to be strongly correlated with the typical biconcave shape of healthy red blood cells. Moreover such distribution well correlates with the Young's modulus distribution detected in healthy red blood cells [6].

A significantly different result is obtained in the pathological case. In this case, the nanoscale map appears to be homogeneously brighter than that of the healthy RBCs indicating that a larger amount of energy is dissipated during the indentation process. Moreover, the cylindrical distribution of H values on healthy RBCs is not observed. Conversely, an almost uniform distribution can be detected.

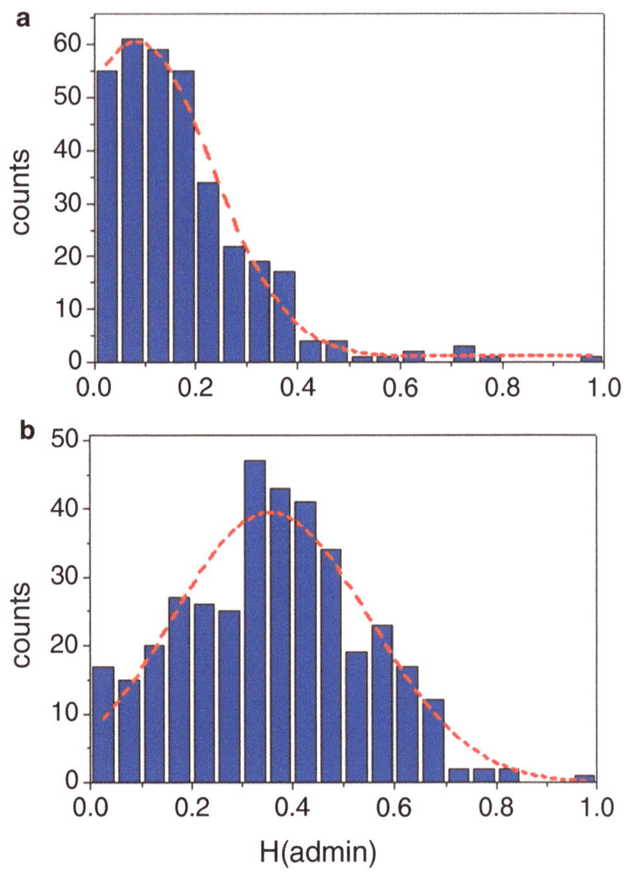

Fig. 2.3 Histograms of H value for the healthy (**a**) and pathological (**b**) RBCs. Data are fitted with a Gaussian curve (*red dashed line*). The distribution of the healthy cell is peaked at 0.08, while distribution's peak of the pathological cell is shifted to 0.35

A frequency histogram of H values is reported in Fig. 2.3a, b for the healthy and pathological RBC, respectively A Gaussian curve was fitted to data (red dashed line).

It can be clearly noted that frequency distribution for the healthy RBC has a peak at approximately 0.08. An average value of 0.17 ± 0.14 can be measured (data are express in term of mean ± standard deviation of the mean). A clear shift of the frequency distribution at larger H values can be observed in the pathological case where a peak at approximately 0.35 can be detected. An average value of 0.35 ± 0.19 can be estimated (data are express in term of mean ± standard deviation of the mean).

2.4 Discussion

The Atomic Force Microscopy is a widely used surface microscopy technique capable to reconstruct the topography of materials and biological samples at the nanometer resolution under virtually any environmental conditions [21, 27–31]. This unique characteristic makes the AFM a key tool to perform high resolution images of cells and tissues in their natural state. Moreover AFM can be used to probe the biomechanical response of biological specimens at both the microscale and the nanoscale level [5, 6, 14–24, 31].

One of the major properties to look quantitatively at the mechanics of biological systems is Young's modulus E, as measured by the analysis of AFM force-distance curves. Mapping E values at the nanoscale level, indeed, has proven to be a valuable parameter with potential application in the clinical practice, in particular in diagnostics [4–6, 31, 32].

Beside Young's modulus, AFM can provide other biomechanical parameters such as adhesion, relaxation time and hysteresis. The latter is rarely considered in AFM experiments on biological samples. Nevertheless it can provide valuable insights in to the molecular modifications occurring during the genesis and the development of many pathologies [6, 33].

In this work we developed a novel scanning probe based methodology that, through mapping, permits to investigate the role of dissipative forces on the whole cell mechanics. The method was tested by using RBCs extracted from healthy donors and patients with hyperferritinemia and iron overload.

Our high resolution hysteresis maps acquired in physiological solution unveil the local changes in the biomechanical response of red blood cells, demonstrating that healthy erythrocytes behaves approximately like an elastic body as far as the cell centre is concerned. Conversely, the cell periphery appears to be more dominated by dissipative forces. This result might be particularly relevant for the theoretical modeling of red blood cells motion under a flowing condition, as it suggests to model the cell centre by using mainly network of elastic springs and the cell periphery by using viscoelastic responders.

The measured spatial distribution of H values on the surface of healthy RBCs shows a cylindrical symmetry that appears to be strongly correlated with the typical RBC biconcave shape. These results hint at the presence of a close relation between the biconcave shape of healthy red blood cells and their mechanical properties. Moreover the spatial inhomogeneity of H values further stresses that a single point measure (as well as some random-landing measures) often used in AFM experiment cannot be representative of the whole RBCs mechanical response.

Healthy red blood cells display an extremely low average H value (0.17 ± 0.14), hinting at a nearly elastic behavior. This finding deserves a more in-depth study as it might have a functional role. In physiological conditions indeed, RBCs undergo extreme deformations when passing trough small vessels and capillaries. Subsequently they easily recover their original shape, suggesting the absence of plastic deformation that might reflect in an increased H value.

Such behaviour appears to be significantly modified in the case of pathological red blood cells, where larger H values were measured (0.35 ± 0.19). This impaired deformability is highly interesting towards a better comprehension of the molecular mechanism underlying the RBC shape, roughness and mechanical modifications occurring at the onset of many pathologies. Our nanoscale mapping highlighted that not only the average H value is changed, but also its spatial distribution. The cylindrical distribution of H values is indeed not observed in the pathological case, hinting at the occurrence of deep structural modifications in the RBC ultrastructure.

The observed changes in H values and their spatial distribution might be due either to changes in viscoelastic properties of the cell membrane or to a modified viscosity of the inner cytosol. Iron overload conditions can indeed be related to an increased haemoglobin (Hb) concentration that might lead to an augmented inner medium viscosity. At the same time the presence of unbound iron causes lipid oxidative stress and consequently impairs the cell membrane. Both phenomena can be linked to the observed modification of the hysteresis map. However, as the major changes occur within the centre of the cell, we suppose that H variation might be mainly related to membrane properties, rather than to a homogeneous change in the inner medium viscosity. A possible method to distinguish between membrane and inner medium contribution would involve the study of the dissipative response of RBC ghosts (RBC membrane lipid bilayer with cytoskeleton), that would help us to avoid the contribution due to the inner medium.

Taken together, our results highlight the importance of mapping the local distribution of hysteresis in the biomechanical characterization of cells. The nanoscale mapping of H indeed permits to clearly distinguish between healthy and pathological red blood cells paving the way to the development of novel diagnostic tools based on the measurement of hysteresis as probed by AFM indentation cycles.

References

1. Tomaiuolo, G., Guido, S.: Start-up shape dynamics of red blood cells in microcapillary flow. Microvasc. Res. **82**(1), 35–41 (2011)
2. Pretorius, E., Kell, D.B.: Diagnostic morphology: biophysical indicators for iron-driven inflammatory diseases. Integr. Biol. **6**(5), 486–510 (2014)
3. Prá, D., Franke, S.I.R., Henriques, J.A.P., Fenech, M.: Iron and genome stability: an update. Mutat. Res. **733**(1), 92–99 (2012)
4. Musielak, M.: Red blood cell-deformability measurement: review of techniques. Clin. Hemorheol. Microcirc. **42**(1), 47–64 (2009)
5. Fornal, M., Lekka, M., Pyka-Fośćiak, G., Lebed, K., Grodzicki, T., Wizner, B., Styczeń, J.: Erythrocyte stiffness in diabetes mellitus studied with atomic force microscope. Clin. Hemorheol. Microcirc. **35**(1–2), 273–276 (2006)
6. Ciasca, G., Papi, M., Di Claudio, S., Chiarpotto, M., Palmieri, V., Maulucci, G., Nocca, G., Rossi, C., De Spirito, M.: Mapping viscoelastic properties of healthy and pathological red blood cells at the nanoscale level. Nanoscale **7**(40), 17030–17037 (2015)
7. Lim, C.T., Li, A.: Mechanopathology of red blood cell diseases—why mechanics matters. Theor. Appl. Mech. Lett. **1**(1), 014000 (2011)
8. Tomaiuolo, G.: Biomechanical properties of red blood cells in health and disease towards microfluidics. Biomicrofluidics **8**(5), 051501 (2014)
9. Cicco, G., Cicco, S.: Hemorheological aspects in the microvasculature of several pathologies. In: Maguire, D., Bruley, D., Harrison, D. (eds.) Oxygen Transport to Tissue XXVIII, pp. 7–15. Springer, Berlin (2008)
10. Wiewiora, M., Sosada, K., Wylezol, M., Slowinska, L., Zurawinski, W.: Red blood cell aggregation and deformability among patients qualified for bariatric surgery. Obes. Surg. **17**(3), 365–371 (2007)
11. Delaunay, J.: The molecular basis of hereditary red cell membrane disorders. Blood Rev. **21**(1), 1–20 (2007)
12. Mokken, F.C., Kedaria, M., Henny, C.P., Hardeman, M., Gelb, A.: The clinical importance of erythrocyte deformability, a hemorheological parameter. Ann. Hematol. **64**(3), 113–122 (1992)
13. Stuart, J., Nash, G.: Red cell deformability and haematological disorders. Blood Rev. **4**(3), 141–147 (1990)

14. Papi, M., Brunelli, R., Familiari, G., Frassanito, M.C., Lamberti, L., Maulucci, G., Monaci, M., Pappalettere, C., Parasassi, T., Relucenti, M.: Whole-depth change in bovine zona pellucida biomechanics after fertilization: how relevant in hindering polyspermy? PLoS One **7**(9), e45696 (2012)
15. Boccaccio, A., Lamberti, L., Papi, M., De Spirito, M., Douet, C., Goudet, G., Pappalettere, C.: A hybrid characterization framework to determine the viscohyperelastic properties of a porcine zona pellucida. Interface Focus **4**(2), 20130066 (2014)
16. Boccaccio, A., Papi, M., De Spirito, M., Lamberti, L., Pappalettere, C.: Effect of the residual stress on soft sample nanoindentation. Appl. Phys. Lett. **102**(13), 133704 (2013)
17. Brunelli, R., Papi, M., Arcovito, G., Bompiani, A., Castagnola, M., Parasassi, T., Sampaolese, B., Vincenzoni, F., De Spirito, M.: Globular structure of human ovulatory cervical mucus. FASEB J. **21**(14), 3872–3876 (2007)
18. De Spirito, M.M., Brunelli, R., Mei, G., Bertani, F.R., Ciasca, G., Greco, G., Papi, M., Arcovito, G., Ursini, F., Parasassi, T.: Low density lipoprotein aged in plasma forms clusters resembling subendothelial droplets: aggregation via surface sites. Biophys. J. **90**(11), 4239–4247 (2006)
19. Maiorana, A., Bugli, F., Papi, M., Torelli, R., Ciasca, G., Maulucci, G., Palmieri, V., Cacaci, M., Paroni Sterbini, F., Posteraro, B.: Effect of alginate lyase on biofilm-grown Helicobacter pylori probed by atomic force microscopy. Int. J. Polym. Sci. **2015**, 989516 (2015)
20. Palmieri, V., Lucchetti, D., Maiorana, A., Papi, M., Maulucci, G., Calapà, F., Ciasca, G., Giordano, R., Sgambato, A., De Spirito, M.: Mechanical and structural comparison between primary tumor and lymph node metastasis cells in colorectal cancer. Soft Matter **11**(28), 5719–5726 (2015)
21. Palmieri, V., Lucchetti, D., Maiorana, A., Papi, M., Maulucci, G., Ciasca, G., Svelto, M., De Spirito, M., Sgambato, A.: Biomechanical investigation of colorectal cancer cells. Appl. Phys. Lett. **105**(12), 123701 (2014)
22. Papi, M., Arcovito, G., Vassalli, M., Tiribilli, B.: Fluid viscosity determination by means of uncalibrated atomic force microscopy cantilevers. Appl. Phys. Lett. **88**(19), 194102 (2006)
23. Papi, M., Lauriola, M., Palmieri, V., Ciasca, G., Maulucci, G., De Spirito, M.: Plasma protein corona reduces the haemolytic activity of graphene oxide nano and micro flakes. RSC Adv. **5**(99), 81638–81641 (2015)
24. Papi, M., Paoletti, P., Geraghty, B., Akhtar, R.: Nanoscale characterization of the biomechanical properties of collagen fibrils in the sclera. Appl. Phys. Lett. **104**(10), 103703 (2014)
25. Vassalli, M., Sbrana, F., Laurita, A., Papi, M., Bloise, N., Visai, L., Bochicchio, B.: Biological and structural characterization of a naturally inspired material engineered from elastin as a candidate for tissue engineering applications. Langmuir **29**(51), 15898–15906 (2013)
26. Ciasca, G., Papi, M., Chiarpotto, M., Rodio, M., Campi, G., Rossi, C., De Sole, P., Bianconi, A.: Transient state kinetic investigation of ferritin iron release. Appl. Phys. Lett. **100**(7), 073703 (2012)
27. Capellini, G., Ciasca, G., De Seta, M., Notargiacomo, A., Evangelisti, F., Nardone, M.: Agglomeration process in thin silicon-, strained silicon-, and silicon germanium-on- insulator substrates. J. Appl. Phys. **105**(9), 093525 (2009)
28. Ciasca, G., Businaro, L., Papi, M., Notargiacomo, A., Chiarpotto, M., De Ninno, A., Palmieri, V., Carta, S., Giovine, E., Gerardino, A., Ciasca, G., Businaro, L., Papi, M., Notargiacomo, A., Chiarpotto, M., De Ninno, A., Palmieri, V., Carta, S., Giovine, E., Gerardino, A.: Self-assembling of large ordered DNA arrays using superhydrophobic patterned surfaces. Nanotechnology **24**(49), 495302 (2013)
29. Chiarpotto, M., Ciasca, G., Vassalli, M., Rossi, C., Campi, G., Ricci, A., Bocca, B., Pino, A., Alimonti, A., De Sole, P.: Mechanism of aluminium biomineralization in the apoferritin cavity. Appl. Phys. Lett. **103**(8), 083701 (2013)
30. Papi, M., Maulucci, G., De Spirito, M., Missori, M., Arcovito, G., Lancellotti, S., Di Stasio, E., De Cristofaro, R., Arcovito, A.: Ristocetin-induced selfaggregation of von Willebrand factor. Eur. Biophys. J. **39**(12), 1597–1603 (2010)
31. Lekka, M., Laidler, P.: Applicability of AFM in cancer detection. Nat. Nanotechnol. **4**(2), 72 (2009)
32. Bester, J., Buys, A.V., Lipinski, B., Kell, D.B., Pretorius, E.: High ferritin levels have major effects on the morphology of erythrocytes in Alzheimer's disease. Front. Aging Neurosci. **5**, 88 (2013)
33. Papi, M., Maiorana, A., Douet, C., Maulucci, G., Parasassi, T., Brunelli, R., Goudet, G., Papi, M., Maiorana, A., Douet, C., Maulucci, G., Parasassi, T., Brunelli, R., Goudet, G., De Spirito, M.: Viscous forces are predominant in the zona pellucida mechanical resistance. Appl. Phys. Lett. **102**(4), 043703 (2013)

Chapter 3
Scale Dependence of the Mechanical Properties of Interfaces in Crustaceans Thin Films

Devendra Verma, Tao Qu, and Vikas Tomar

Abstract Crustacean exoskeletons in the form of thin films have been investigated by several researchers in order to understand the role played by the exoskeletal structure in affecting functioning of species such as shrimps, crabs and lobsters. These species exhibit similar design in their exoskeleton microstructure. Bouligand pattern (twisted plywood structure), layers of different thicknesses across cross section, changes in mineral content through the layers etc. are common feature changes. Different parts of crustacean exoskeletons exhibit a significant variation in mechanical properties based on the variation in the above mentioned features. Mechanical properties have been analyzed by authors using imaging techniques such as SEM (Scanning Electron Microscopy), EDX (Energy Dispersive X-ray) and using mechanical characterization based on nanoindentation. Analyses show that the confinement effect arising from interfaces sandwiched in crustacean microstructure layers along with the strain rates of deformation plays a major role in the deformation of such layered systems. A new constitutive model is proposed that couples the effect of strain-rate and confinement to predict interface deformation behavior. The model predictions are validated based on experiments in glass/epoxy interfaces.

Keywords Shrimp exoskeleton • Biomimetic • Interfaces • Strain rate • Confinement effect

The hierarchical assembly in the structure of natural materials provides them exceptional strength, toughness, and stiffness compared to their counterparts [1–4]. Most recent studies on the natural materials include shrimp exoskeleton [2, 3], crab exoskeletons [5, 6], and sea shells such as nacre, mollusk [7, 8]. These studies have revealed interesting features in the design of such biocomposites that make them much stronger than their constituent materials. The exoskeleton of crustaceans such as shrimps, lobsters, and crabs is in the form of multiple thin films stacked in the Bouligand pattern with a well-defined hierarchical structure. It consists of chitin based fibrils coated in proteins at nanometer level which bind together to form fibers. These fibers are then woven together to form chitin-protein thin film layers. These layers are stacked in a twisted plywood structure known as the Bouligand pattern [9, 10]. The spacing between such woven layers is filled with proteins and biominerals. Raabe et al. [1] have studied structural and mechanical properties of lobster and crab exoskeletons. They reported hardness and reduced stiffness with depth of examination. The exoskeleton features a graded design with different stacking density of chitin based Bouligand structure. Boßelmann et al. [9] have shown a direct correlation between increase in the mineral content and hardness of lobster claw. Our previous experimental work on shrimp exoskeletons have focused on understanding the temperature and size effects [11, 12].

A lot of mechanistic studies have been performed to study the influence of microstructure on biomaterials strength. Feng et al. [8] found that the crack deflection, fiber pull-out, and organic matrix bridging are the three main toughening mechanisms affecting the mechanical strength of a biomaterial such as nacre. In the present paper, basic design principles of the crustaceans and deformation mechanisms responsible for higher strength, stiffness, toughness are highlighted at different hierarchical levels based on information obtained using imaging techniques such SEM (Scanning Electron Microscopy), EDX (Energy Dispersive X-ray), and nanoindentation in the case of two shrimp species *Rimicaris exoculata* and *Pandalus platyceros*.

Nanoindentation is a convenient experimental technique to measure elastic modulus and hardness at nano- and micro-scale. The experimental procedure involves indenting the surface of the material being tested by increasing force in small steps until peak load (P_{max}) or peak depth (h_{max}) is achieved. Figure 3.1 shows a schematic of indentation set up, a standard indentation curve, and parts of curve used for material property calculations. A Berkovich indenter was used in the present work. The unloading part was used for predicting the material properties using a framework based on contact mechanics

D. Verma • T. Qu • V. Tomar (✉)
School of Aeronautics and Astronautics, Purdue University, 701 W. Stadium Avenue, West Lafayette, IN 47907, USA
e-mail: tomar@purdue.edu

C.S. Korach et al. (eds.), *Mechanics of Biological Systems and Materials, Volume 6*,
Conference Proceedings of the Society for Experimental Mechanics Series, DOI 10.1007/978-3-319-41351-8_3

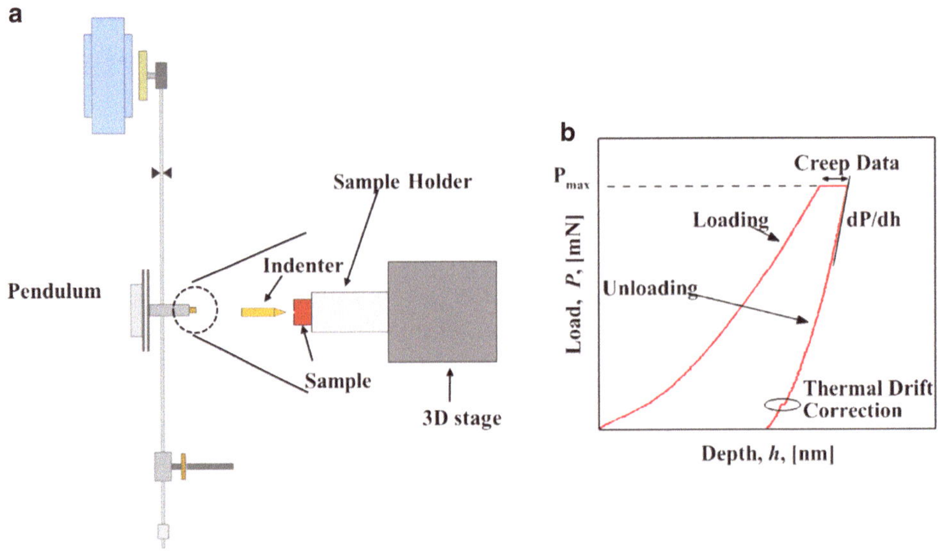

Fig. 3.1 Figure showing (**a**) indentation set up schematic and (**b**) parts of the indentation curve used for stiffness, creep and thermal drift calculations

[13]. During experiments, maximum indentation load $P_{\max}$ and maximum area of indentation A is measured. The hardness, H is given by

$$H = \frac{P_{\max}}{A}. \tag{3.1}$$

The parameter A for an ideal Berkovich indenter used in the present work as a function of contact depth h_c is given as

$$A = 3\sqrt{3}h_c^2 \tan^2 65.3^o \approx 24.5h_c^2. \tag{3.2}$$

The reduced Young's modulus, E_r is related to the slope of the upper part of the unloading curve, given by

$$S = \frac{dP}{dh} = 1.17E_r\sqrt{A}. \tag{3.3}$$

Here, S is the stiffness measured experimentally from the slope of unloading curve. Based on known S and A values, E_r can be calculated. E_r can be used to find the true modulus of the material by the relation

$$\frac{1}{E_r} = \frac{(1-\nu^2)}{E} + \frac{(1-\nu_i^2)}{E_i}. \tag{3.4}$$

Here, E and ν are Young's Modulus and Poisson's ratio of the specimen under test. E_i and ν_i are Young's modulus and Poisson's ratio of the indenter. Experiments were performed at room temperature in a multi module mechanical tester (NanoTest, Micro Materials Ltd., platform 2) [3]. The instrument consists of a vertical pendulum pivoted on a frictionless spring. An indenter is attached to the pendulum that indents sample horizontally. The force on the pendulum is applied through magnetic coils located at the top of the pendulum. The depth of the indents during experiments is measured by capacitor plates located behind the indenter. Depth calibration, load calibration, and frame compliance experiments were performed before conducting actual experiments. The indentations were performed on selected samples at the chosen surface as shown in Fig. 3.2. It was ensured that the sample surface is free of defects before mounting samples. Experiments were performed on ten different samples to capture statistical property differences possibly due to biological changes in samples. The experiments were performed at room temperature of 25 °C. In the present experiments, all measurements were performed using the same indenter and the same mounting technique. SEM images were obtained by FEI Nova nanoSEM. EDX analyses were performed using FEI Quanta 3D FEG Dual-beam SEM.

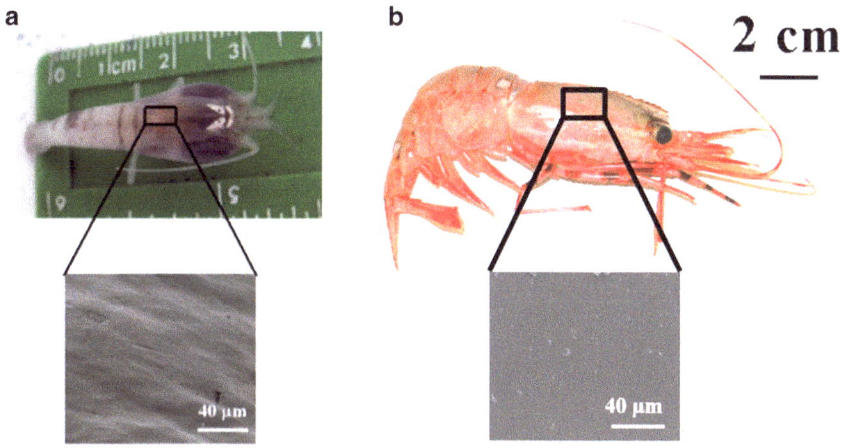

Fig. 3.2 Analyzed samples of (**a**) *R. exoculata* (**b**) *P. platyceros*

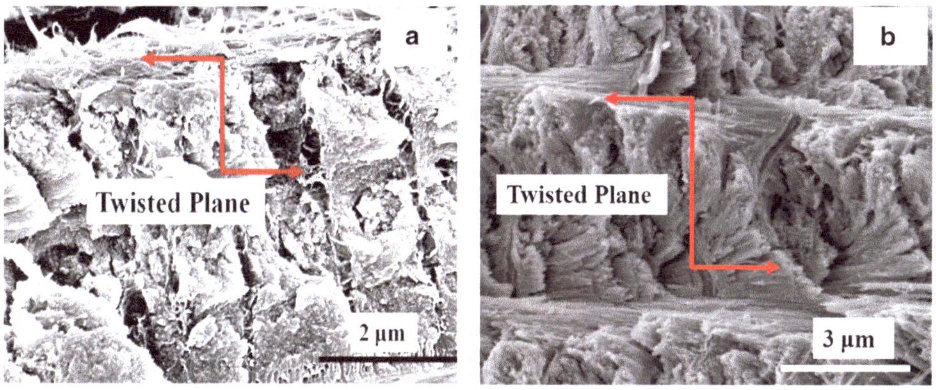

Fig. 3.3 Bouligand structure of shrimp exoskeletons (**a**) *R. exoculata* (**b**) *P. platyceros*

Shrimp samples of *R. exoculata* were provided by Dr. Juliette Ravaux from the mission Momardream in 2007 obtained at a depth of 2300 m. The shrimps were maintained for 10 h at 10 °C with 1 h-heat shock at 30 °C in pressurized aquaria on board. These shrimp samples were further stored in liquid nitrogen before experiments. *Pandalus platyceros* shrimp samples were obtained in fresh unfrozen condition. Shrimps were immediately stored in frozen condition after procurement. Samples for mechanical testing were prepared from such shrimps specimens. Figure 3.2 shows whole shrimps and prepared samples for *R. exoculata* and *P. platyceros*.

Two different shrimp species *R. exoculata* and *P. platyceros* were analyzed during the current study to compare the indentation creep response and its dependence on the structure. As shown in Fig. 3.3, there is a clear non-uniformity in stacking density and thickness of layers across cross section of exoskeletons of both shrimps. Both shrimp exoskeletons have similar features such as Bouligand structure but with difference in the thickness of the layers. The maximum thickness of a layer in the *P. platyceros* is 4 μm as opposed to the maximum thickness being 2 μm in *R. exoculata* exoskeleton as shown in Fig. 3.3. *Rimicaris exoculata* shows non uniform change in the thickness of the layers across the exoskeleton thickness as compared to the case of *P. platyceros*. The minimum thickness is in the range of 20–30 nm in the case of *R. exoculata* while it is in the range of 50–100 nm in the case of *P. platyceros*.

EDX spectrum were also compared to give information on differences in composition of the exoskeletons. Figure 3.4 shows the EDX spectrum of examined *R. exoculata* and *P. platyceros* exoskeleton samples. Both spectrum were collected using same QUANTA 3D SEM at Purdue University. Figure 3.4a shows collected spectrum from *R. exoculata* and Fig. 3.4b shows spectrum collected from *P. platyceros*. It shows that both exoskeletons have peaks of Oxygen and Carbon which are main constituents of the proteins. Figure 3.4a also shows peaks corresponding to Calcium and Phosphorus. The key difference is the presence of one extra peak showing presence of Phosphorus in *R. exoculata*. The similarity in the individual layers of both shrimps can be exhibited from the Bouligand pattern as shown in Fig. 3.3. In addition to the differences in the structure, these species also show differences in the chemical composition of the exoskeletons. Chemical composition at the top layers at the sites of indention was compared by taking EDX spectrums. The elemental composition calculated from the spectrums is given in Table 3.1.

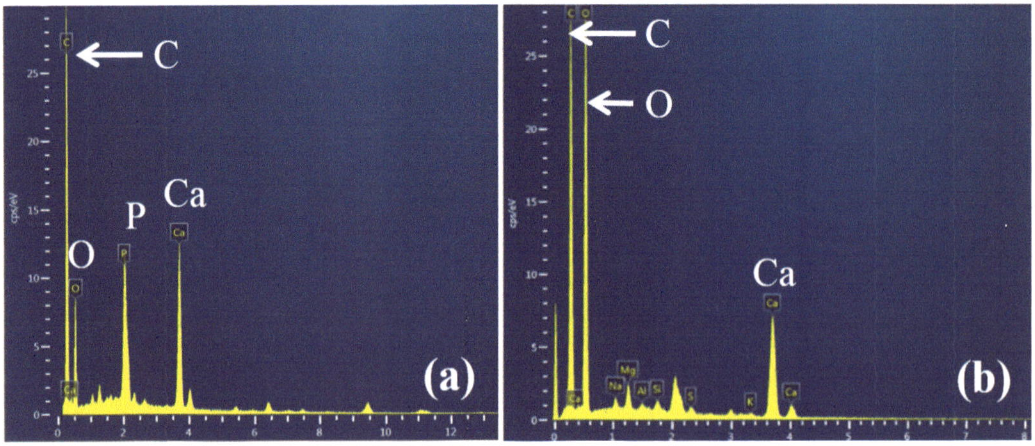

Fig. 3.4 EDX analysis of cross section of shrimp exoskeletons showing elemental spectrums (**a**) *R. exoculata* (**b**) *P. platyceros*

Table 3.1 Comparison of the quantitative data by EDX analysis on cross section of shrimp exoskeletons for *R. exoculata* and *P. platyceros*

Element	*R. exoculata* (atomic %)	*P. platyceros* (atomic %)
C	42.89 ± 0.97	46.20 ± 1.19
O	27.76 ± 1.89	42.41 ± 1.79
P	3.14 ± 1.12	–
Ca	26.2 ± 1.73	9.86 ± 2.20

Table 3.2 Creep compliance fitting parameters from Eq. (3.5) on experimental dwell data

Species	C_0 (Pa^{-1})	C_1 (Pa^{-1})	τ_1 (s)
R. exoculata	1.45e–6 ± 8.21e–8	4.60e–7 ± 4.97e–9	86.33 ± 1.24
P. platyceros	1.08e–6 ± 8.16e–8	1.38e–7 ± 4.80e–9	68.00 ± 1.63

The given composition in Table 3.1 is comparable to the chemical composition obtained by several other researchers on coastal shrimps [14, 15]. *Rimicaris exoculata* samples show some variations in composition with noticeable phosphorus. Table 3.1 gives relative quantitative analysis of individual elements present in exoskeleton of both shrimps. The presence of Ca supports the fact that *R. exoculata* survives at very high pressures and needs more structural strength to survive. Phosphorus is one of the materials from volcanic vents in the nearby habitat of *R. exoculata*. It also became part of shrimp exoskeleton with evolution of shrimps near deep sea vents possibly via food sources. Observations from Fig. 3.3 and Table 3.1 reveal that Rimicaris exoskeleton has higher mineral content and reduced protein layer thickness in comparison to *P. platyceros*.

Further analysis was performed on the creep data obtained from the nanoindentation experiments to highlight the structural differences in mechanical properties. There are several formulations (e.g. Oyen [16] and Lu et al. [17]) to fit the nanoindentation creep data obtained in this work. The model chosen in the current paper was developed by Oyen [16]. It uses the dwell data over the time of holding period to calculate creep compliance parameters. In order to model the load-deformation behavior exhibited by examined shrimp exoskeletons, the creep data is fitted with the creep function given for viscoelastic materials as

$$h^2 = \frac{\gamma^2}{\pi \tan\psi}\left\{C_o k t_R - \sum C_i k \tau_i \exp\left(\frac{-t}{\tau_i}\right)\left[\exp\left(\frac{t_R}{\tau_i}\right) - 1\right]\right\}. \tag{3.5}$$

Here, h is the indentation depth during dwell period, ψ is indenter included angle, γ is a constant relating contact displacement to total displacement, k is loading rate, t_R is start time of dwell period, C_i and τ_i are creep compliance coefficients, and t is time period of loading. One term approximation of Eq. (3.5) is fitted to the experimental creep data to find coefficients C_i and τ_i. Respective coefficients for both *R. exoculata* and *P. platyceros* are tabulated in Table 3.2.

The creep compliance parameters obtained from Eq. (3.5) were then used to plot the creep compliance given in Fig. 3.5 using Eq. (3.6),

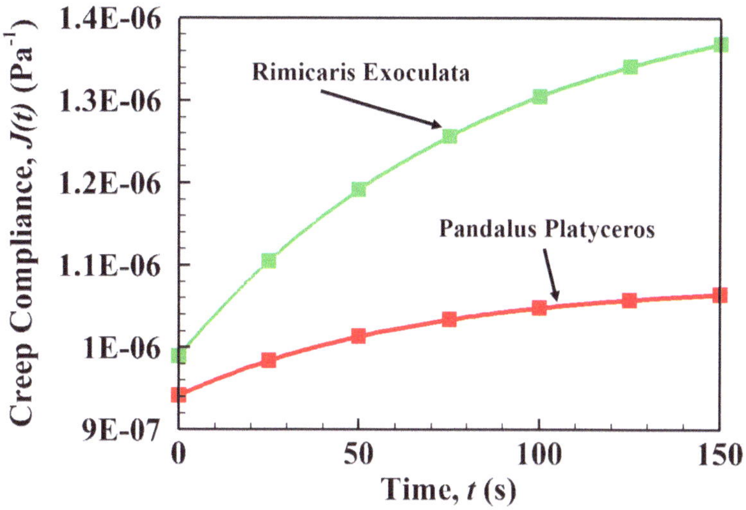

Fig. 3.5 Creep compliance functions $J(t)$ for *R. exoculata* and *P. platyceros* from Eq. (3.6)

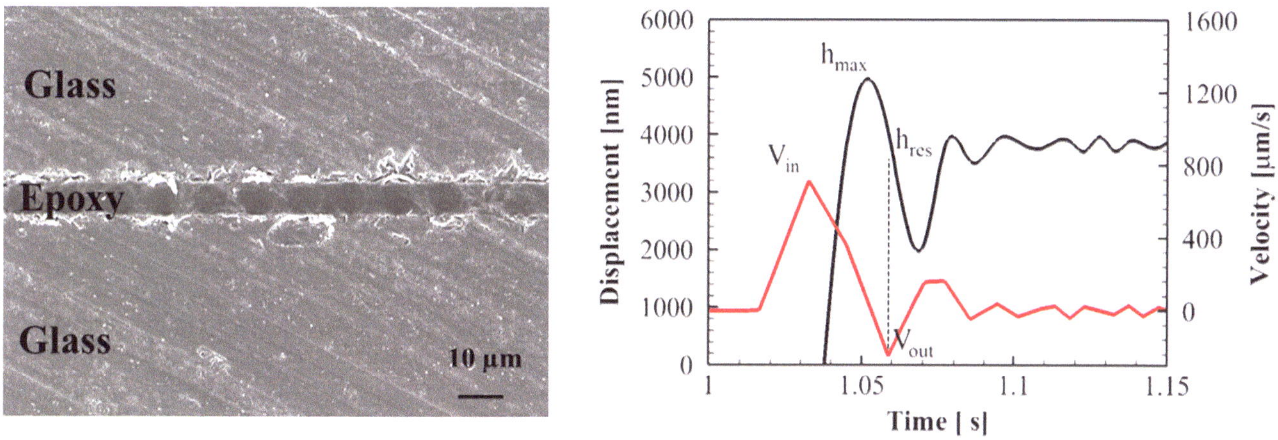

Fig. 3.6 (**a**) Sample showing the epoxy interface and (**b**) raw data of displacement and impact velocity during experiments

$$J(t) = C_0 - \sum_{i=1}^{j} C_i \exp\left(\frac{-t}{\tau_i}\right). \tag{3.6}$$

Here, C_i and τ_i are creep compliance coefficients, and t is time period of loading.

As shown in Fig. 3.5, the creep compliance for *R. exoculata* is higher compared to *P. platyceros* because of lower hardness of *R. exoculata* exoskeleton. Basic building blocks of the *R. exoculata* and the *P. platyceros* exoskeleton are polymer chains made up of chitin proteins and calcium based biominerals. Matrix material is made up of chitin-protein fibers woven together in a planar structure. Spacing between woven fibers is filled with biominerals, mostly of $CaCO_3$. These layers are then arranged to form a twisted plywood structure known as Bouligand pattern [1]. Such hierarchy in their structure makes the exoskeleton similar to polymer composites.

The creep compliance function fits nicely to the creep data of the shrimp exoskeletons. The difference in the creep properties between the two shrimp species exoskeletons could also be attributed to the presence of interfaces between individual layers in the Bouligand structure. *Rimicaris exoculata* has very thin layers in its structure with a large number of interfaces compared to *P. platyceros*. The interfaces in these structures are not of zero thickness as traditionally assumed in fracture mechanics. The interfaces here have their own properties that are dependent on the constituent materials and length scale. The contribution of interface mechanical attributes must be taken into account to fully understand the structure-property relations in these materials.

In order to understand the interface mechanical properties, an idealized system with epoxy interface between two glass slides was examined. The sample surface is shown in Fig. 3.6a and the data output from experiment is shown in Fig. 3.6b.

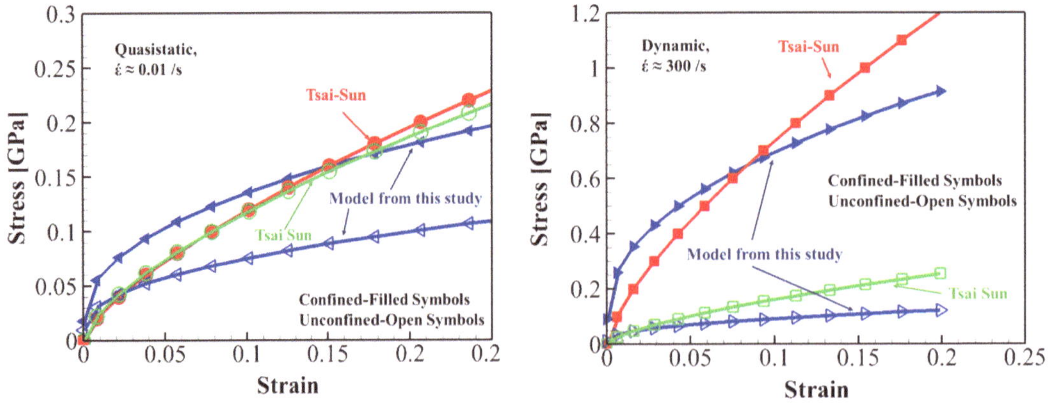

Fig. 3.7 Fit of the current model with stress strain for (**a**) quasistatic loading and (**b**) dynamic loading

The impact experiments were performed on the epoxy interface to extract the stress-strain curve of the interface at different strain rates [18]. It was found that interfaces are affected by both strain rate and confinement effects during deformation. A new constitutive model is developed that couples the effect of both strain rate and lateral stresses given as

$$\sigma = (A + B\varepsilon^n)(1 + C\ \ln\dot{\varepsilon}^*)(1 + k\sigma_1^*). \tag{3.7}$$

Here, σ is the equivalent stress, ε is the equivalent plastic strain, A is the yield stress, B is the strain hardening constant, n is the strain hardening coefficient, C is the strain rate strengthening coefficient and k is the confinement factor. $\dot{\varepsilon}^* = \frac{\dot{\varepsilon}}{\dot{\varepsilon}_{ref}}$ is the dimensionless strain rate normalized with reference strain rate, $\sigma_1^* = \frac{\sigma_1}{\sigma_{compressive\ strength}}$ is the dimensionless lateral stress normalized with the compressive strength of the material. The reference strain rate is taken as 1 s $^{-1}$ and $\sigma_{compressive\ strength}$ is taken as 100 MPa. In the current experiments, the temperature was constant so the temperature effects were neglected.

The stress strain response for the epoxy interface was analyzed at high and low strain rates as shown in Fig. 3.7 with the confined interface behavior labeled with closed symbols and unconfined epoxy behavior labeled with open symbols. At lower strain rates, the difference in stresses is mostly because of the confinement effect while in the dynamic case the strain rate effect also plays a major role. The difference in the stresses is higher in the dynamic case as evident in Fig. 3.7b compared to Fig. 3.7a. It is also compared to the stress strain behavior fitted by Tsai-Sun model on the same plot. Thus, the current model is better suited to model the behavior of materials under confined spaces such as interfaces in the composite materials, metals, ceramics etc. A conventional way is to consider interface as a zero thickness and to not consider interface effect on the material deformation. The model in Eq. (3.7) takes into account the effect of interface mechanical properties on the mechanical deformation and should be considered for cases that have interface dominant geometries such as the biological materials analyzed in this work. The size effect of the interfaces is being analyzed by the authors and will be published in the future papers.

In summary, a series of experiments were performed at room temperature on the exoskeletons of *R. exoculata* and *P. platyceros* in order to understand microstructural dependence of mechanical strength in such exoskeletons. The microstructures were studied extensively using SEM, and EDX analyses. A comparison is drawn between the properties of the examined exoskeletons based on their microstructure and chemical composition. The difference in the mechanical properties for both shrimp species is attributed to their habitat and subsequent changes in the microstructural design such as change in the individual layer thickness, pattern of layer thickness, chemical composition, and interfaces between the layers. The results were fitted with viscoelastic creep compliance functions that predicted Rimicaris exoskeleton to have higher creep rate as compared to the *P. platyceros* exoskeleton due to its lower hardness. The interfaces are believed to play an important role in the observed behavior. The role of interfaces is identified by performing experiments on an idealized system of glass-epoxy interfaces. The confinement effect on the interfaces along with the effect of strain-rate was found to play a major role in the deformation of the examined interfaces. A new model capturing both strain rate and confinement effects is developed for strain rates up to $10^3\,s^{-1}$ in this paper to account for confinement effect and strain rate effect coupling.

Acknowledgments The authors express their sincere thanks to Dr. Juliette Ravaux, Université Pierre et Marie Curie for providing samples of *Rimicaris exoculata* and are grateful to excellent technical assistance of Dr. Christopher J. Gilpin, Chia-Ping Huang and Laurie Mueller with Scanning Electron Microscopy and Energy Dispersive X-ray at Purdue University. Authors would also like to thank their colleagues Dr. Ming Gan, Dr. You Sung Han, Dr. Hongsuk Lee, Dr. Tao Qu, Yang Zhang, Chandra Prakash, Sudipta Biswas, Debapriya Mohanty, Bing Li for helpful discussions and support from grants NSF CMMI 1131112 and DE-FC0011796.

References

1. Raabe, D., Sachs, C., Romano, P.: The crustacean exoskeleton as an example of a structurally and mechanically graded biological nanocomposite material. Acta Mater. **53**(15), 4281–4292 (2005)
2. Verma, D., Tomar, V.: Structural-nanomechanical property correlation of shallow water shrimp (*Pandalus platyceros*) exoskeleton at elevated temperature. J. Bionic. Eng. **11**(3), 360–370 (2014)
3. Verma, D., Tomar, V.: An investigation into environment dependent nanomechanical properties of shallow water shrimp (*Pandalus platyceros*) exoskeleton. Mater. Sci. Eng. C **44**, 371–379 (2014)
4. Bouville, F., et al.: Strong, tough and stiff bioinspired ceramics from brittle constituents. Nat. Mater. **13**(5), 508–514 (2014)
5. Lian, J., Wang, J.: Microstructure and mechanical properties of dungeness crab exoskeletons. In: Proulx, T. (ed.) Mechanics of Biological Systems and Materials, vol. 2, pp. 93–99. Springer, New York (2011)
6. Giraud-Guille, M.M.: Fine structure of the chitin-protein system in the crab cuticle. Tissue Cell **16**(1), 75–92 (1984)
7. Ji, B., Gao, H.: Mechanical properties of nanostructure of biological materials. J. Mech. Phys. Solids **52**(9), 1963–1990 (2004)
8. Feng, Q.L., et al.: Crystal orientation, toughening mechanisms and a mimic of nacre. Mater. Sci. Eng. C **11**(1), 19–25 (2000)
9. Boßelmann, F., et al.: The composition of the exoskeleton of two crustacea: the American lobster *Homarus americanus* and the edible crab *Cancer pagurus*. Thermochim. Acta **463**(1–2), 65–68 (2007)
10. Bouligand, Y.: Twisted fibrous arrangements in biological materials and cholesteric mesophases. Tissue Cell **4**(2), 189–217 (1972)
11. Verma, D., Tomar, V.: A comparison of nanoindentation creep deformation characteristics of hydrothermal vent shrimp (*Rimicaris exoculata*) and shallow water shrimp (*Pandalus platyceros*) exoskeletons. J. Mater. Res. **30**(08), 1110–1120 (2015)
12. Verma, D., Tomar, V.: An investigation into mechanical strength of exoskeleton of hydrothermal vent shrimp (*Rimicaris exoculata*) and shallow water shrimp (*Pandalus platyceros*) at elevated temperatures. Mater. Sci. Eng. C **49**, 243–250 (2015)
13. Oliver, W.C., Pharr, G.M.: Improved technique for determining hardness and elastic modulus using load and displacement sensing indentation experiments. J. Mater. Res. **7**(6), 1564–1583 (1992)
14. Ravichandran, S., Rameshkumar, G., Prince, A.R.: Biochemical composition of shell and flesh of the Indian white shrimp *Penaeus indicus* (H. milne Edwards 1837). Am. Eurasian J. Sci. Res. **4**(3), 191–194 (2009)
15. Emmanuel, I.A., Adubiaro, H.O., Awodola, O.J.: Comparability of chemical composition and functional properties of shell and flesh of *Penaeus notabilis*. Pak. J. Nutr. **7**(6), 741–747 (2008)
16. Oyen, M.: Analytical techniques for indentation of viscoelastic materials. Philos. Mag. **86**(33-35), 5625–5641 (2006)
17. Lu, H., et al.: Measurement of creep compliance of solid polymers by nanoindentation. Mech. Time Depend. Mater. **7**(3–4), 189–207 (2003)
18. Verma, D., et al.: Evaluation of incoherent interface strength of solid-state-bonded Ti64/stainless steel under dynamic impact loading. JOM **67** (8), 1694–1703 (2015)

Chapter 4
Dynamic Analysis of Human Knee

S. Yoshida, U. Tiwari, A. Saladino, M. Nguyen, D. Hollander, B. Boudreaux, and B. Hadley

Abstract Knee joint weakness and disorders are extremely common; with development of a new methodology for early detection of knee disorders in mind, various dynamics tests have been conducted. Subjects were requested to perform exercises of certain patterns including squats and marching at a constant rate for 70 (s). A force plate, acoustic sensor, and set of electromyographic (EMG) sensors were used to measure the ground reaction force and torque to the foot, the vibration from the knee cavity, and the electrical activity of muscles. The acquired signals were analyzed in the frequency domain to associate with certain biomechanical parameters. Preliminary studies indicated clear correlations between several features of the Fourier spectra and biomechanical parameters. The force and torque signals from a knee with a torn meniscus showed considerable broadening in the Fourier spectra around the peak of the exercise frequency (e.g. the 1 Hz peak of the torque associated with the knee extension from 1 Hz marching) and higher transfer function to other degrees of freedom, indicating instability of the injured knee. The acoustic sensor signal from an arthritic knee indicated significant attenuation due to the fluid in the knee cavity, leading to the extremely poorly defined spectral feature. The femoral efficiency, defined by the transfer function from the EMG signal to the force/torque of the dominant degree-of-freedom, showed clear correlation to fatigue in the amplitude and frequency.

Keywords Knee dynamics • Knee disease detection • Electromyography

4.1 Introduction

Knee disorders are very common. It is the second leading cause of physical disability behind ischemic heart disease in the US for persons over 50 [1]. The most prevailing diagnostic methods are X-ray and MRI (Magnetic Resonance Imaging). Although these assessments are well-developed techniques there are two issues. First, they visualize abnormality in the knee joint or surrounding tissues, not diagnose the functionality. Often the disorder is too advanced when the abnormality appears. During routine assessments, MRI images can appear normal but the patient feels intense pain or has a serious functional problem. Obviously there is a latent issue that does not show up in the MRI. Second, the diagnostic is static. The knee joint is one of the most complex joints in the human body. It is possible that an issue in other part of the knee kinetic chain could be contributing to the kinematic pathologies that lead to the pain pathologies. It is rather common that a lower back problem tightens pelvis, creating quadriceps femoris tightening and causes kinematic pathologies at the knee leading to patellar and joint related pain. Some empirical function tests are used for diagnosis but they are qualitative. Dynamic and quantitative diagnostic methods are beneficial not only for an early detection of knee injuries but also for more accurate and overall assessment of the issue.

Considering the above situation, we started a research project [2]. In this project, our focus is dynamic analysis of knee motion based on physical interpretations. We use as many types of data as possible so that we can characterize the dynamics from various viewpoints. With practical applications in mind, we also make the measurement as simple as possible. Recently, we obtained several data sets that we believe verify the usefulness of the proposed methodology. The aim of this paper is to present the experiment and discuss the results of the analysis on the recent experiment.

S. Yoshida (✉) • U. Tiwari
Department of Chemistry and Physics, Southeastern Louisiana University, Hammond, LA 70402, USA
e-mail: syoshida@selu.edu

A. Saladino
Department of Computer Science and Industrial Technology, Southeastern Louisiana University, Hammond, LA 70847, USA

M. Nguyen • D. Hollander • B. Boudreaux • B. Hadley
Department of Kinesiology and Health Studies, Southeastern Louisiana University, Hammond, LA 70845, USA

C.S. Korach et al. (eds.), *Mechanics of Biological Systems and Materials, Volume 6*,
Conference Proceedings of the Society for Experimental Mechanics Series, DOI 10.1007/978-3-319-41351-8_4

4.2 Measurement and Analytical Methods

4.2.1 Test Procedure and Sensor Arrangement

Figure 4.1 illustrates the experimental arrangement along with the coordinate system. A set of data consisted of ground force and torque measured with a force plate, knee vibration measured with a vibration sensor attached to the inner side of the knee, and EMG (Electromyography) signal measured with a set of EMG sensors. The *xyz*-coordinate axes are affixed to the force plate as shown in Fig. 4.1.

The test subjects performed two types of activities; marching and squatting, with one foot on the force plate at a time. Prior to any testing procedures, the test subjects were requested to fill out an Informed Consent and Medical History form, and a questionnaire regarding daylily activities they find difficulties with [3]. Although the testing procedures were stated in the inform consent form, each test subject was briefly informed of the procedures again. On completion of all these steps, the sensors were attached to the subject. The force plate and vibration sensor signals were acquired by a Biopac™ package [4], and the EMG signals were acquired with a Delsys™ package [5].

4.2.1.1 Measurement with Biopac™ System

The acoustic sensor was attached to the subject's knee between the medial condyle of the tibial bone and the medial condyle of the femur. An electrode gel was used on the surface of the sensor. Once the sensor was in place, it was secured with a medical tape wrapped around the knee. The vibration sensor was attached to the leg that was placed on the force plate. This completed the attachment of the vibration sensor. After testing the first knee, the vibration sensor was switched over to the other leg that was placed on the force plate for the second round of the test. The force plate measured the *x*-, *y*- and *z*-components of the force and torque vectors.

4.2.1.2 Measurement with Delsys™ System

Ten EMG sensors (five per leg) and two goniometers (one per leg) were attached to the subject. The goniometer was used to assure that the subject bend knee properly. The five EMG sensors were attached to the belly of the vastus lateralis, vastus medialis, rectus femoris, biceps fermoris, and the anterior tibialis muscle. Once all EMG sensors were in place, the goniometer was attached being lined up with the greater trochanter of the femur bone and the lateral malleolus of the fibula

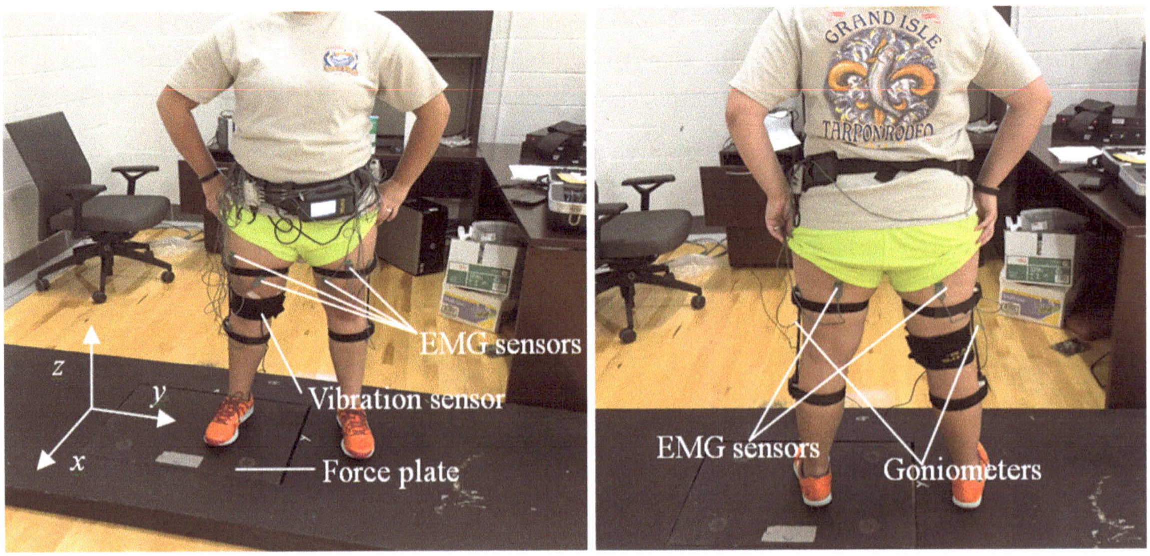

Fig. 4.1 Force plate and other sensor arrangement

bone. There was a white line in the center of the goniometer that was centered with the lateral epicondyle of the tibia bone. Once the goniometer was aligned and in place, it was secured with Velcro wrap.

There were two exercises during the test, marching and squatting. Each exercise was performed at a certain frequency for 70 s. The subject was guided by the sound of a metronome that was set in accordance with the required frequency. Here frequency was defined as the completion of exercise, e.g., 2 Hz marching was defined as two cycles of marching that caused the foot on the force plate land on it twice in 1 s. In other words, the subject touched the ground with the two feet four times in total every second, or 240 times in 1 min. The subject was guided by 240 beats per minute on the metronome for 2 Hz marching. Similarly, 1 Hz squatting meant one crouching and standing motion in 1 s. The subject was guided by 120 beats on the metronome to crouch and stand.

4.2.2 Frequency Domain Analysis

The signal from each sensor was sent to a channel of the Biopack or Delsys system. The two systems were synchronized manually. All the signals were analyzed in the frequency domain via Fourier transform of the time series collected by the Biopack and Delsys systems at the sampling rate of 1 kHz.

4.2.3 Subjects

Subjects were selected from a variety of age groups. In this paper, four particular subjects are discussed. Subject 1 was an 18-year-old male with no specific knee issue. Subject 2 was a 69-year-old male with severe arthritis in both knees. Subject 3 was a 25-year-old female with a torn meniscus on her right knee. Subject 4 was a 68-year-old male complained chronic pain in both knees (under exercises) but the MRI or other medical examination did not indicate any issue.

4.3 Results and Discussion

4.3.1 1 Hz March (Subject 1 and Subject 2)

Figure 4.2 presents vibration sensor (called "acoustic" in the legend) and torque around the three coordinate axes obtained from Subject 1 and 2 under 1 Hz marching. The following observations can be made.

Observation 1

For both subjects, among the three degrees of freedom in torque, R_z (the torque around the z-axis) is the strongest. This indicates that when their foot lands on the force plate, the foot adjusts internal-external rotation by exerting the torque R_z.

Observation 2

Both subject show a peak at 0.5 Hz, which makes sense because the foot lands on the plate every 2 s. Subject 1's vibration signal has a peak at 1 Hz (the double frequency of the peak on the torque spectra) whereas Subject 2's vibration signal barely shows a peak at 1 Hz. Unlike the force plate, the vibration sensor records the signal whichever foot lands the force plate. The 1 Hz peak indicates that the vibration sensor detects a signal when the foot is raised from the force plate and the foot is lowered to the force plate. The fact that Subject 2 does not show the 1 Hz peak indicates that the vibration signal is weaker either when the foot is raised or lowered.

Observation 3

Torque around the y-axis (R_y) is the dominant degree of freedom for the flexion-stretch motion in the marching activity. Subject 2 shows much broader peak spectral shape in R_y at the frequency of activity. This indicates instability of the subject's flexion-and-extension motion.

Observation 4

The vibration sensor signal in Subject 2 is much weaker (approximately in two-orders of magnitude at the peak value) than Subject 1. This can be interpreted as the fluid in the subject's knee cavity attenuated the vibration signal.

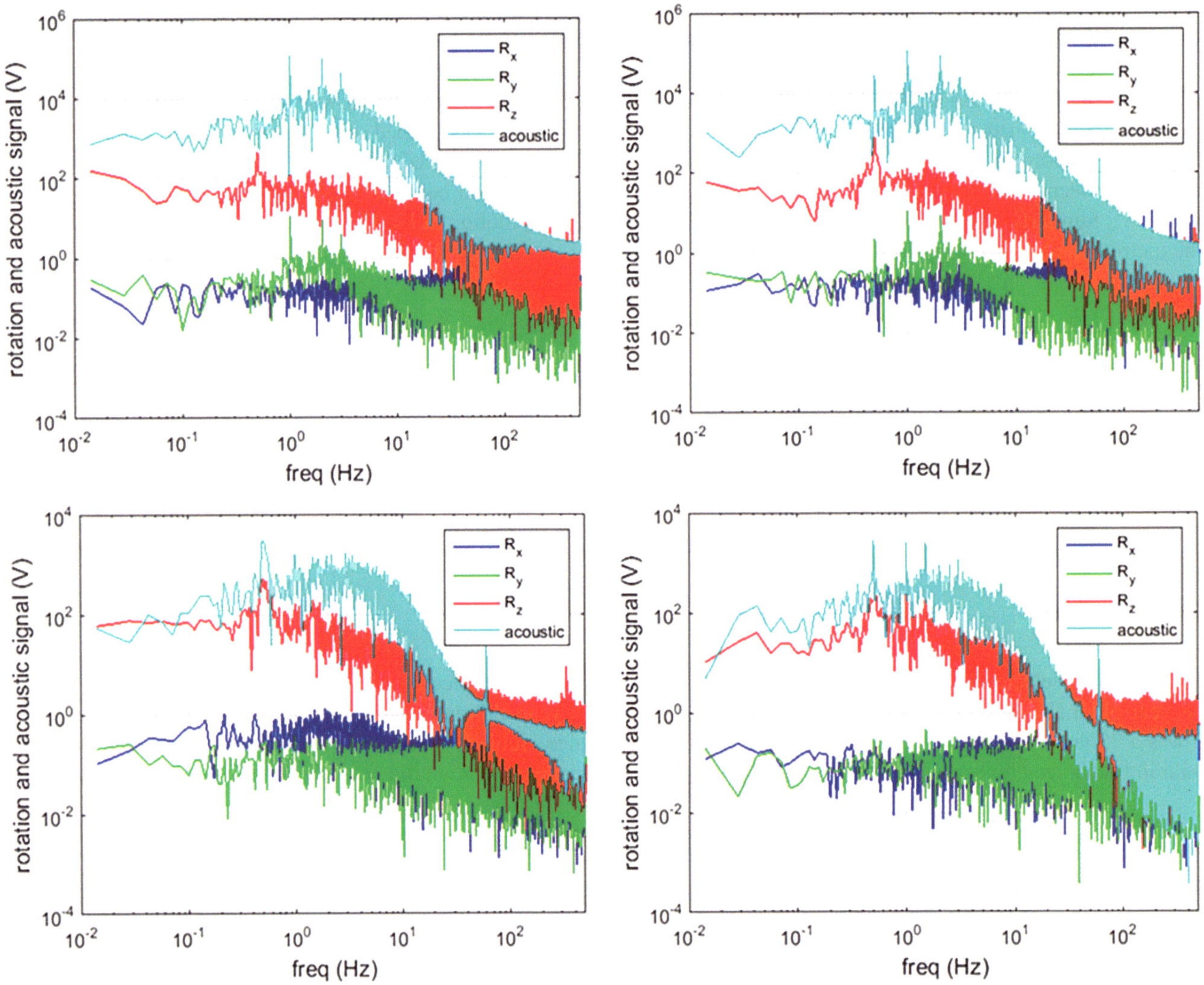

Fig. 4.2 Marching Fourier spectra from subject 1 (*upper*) and subject 2 (*lower*). The *left graphs* are signals from the left knees and the *right graphs* are signals from the right knees

Observation 5

R_y in Subject 2 is an order of magnitude lower than Subject 1 whereas R_z is at the same level as Subject 1. The combination of the lower R_y and higher R_z indicate the subject's instability in the flexion-and-extension motion compensated by the internal-external rotational motion.

Observation 6

Subjects 1 and 2 show the same level of R_x. This is understandable because the rotational motion around the x-axis is the most irrelevant degree of freedom to the marching activity. The fact that the signal level in R_x is the same signal level indicates that the foot plate was well calibrated and justifies the above-discussed comparison in other torques between the subjects.

4.3.2 1 Hz Squat (Subject 3 and Subject 4)

Figure 4.3 shows the Fourier spectra (the vibration, R_y and R_z signals) of the left and right knees when subject 3 underwent 1 Hz squat activity. It is clearly seen that in both cases, a peak appears at the activity frequency of 1 Hz. The right knee shows much broader peak, clearly indicating instability due to the torn meniscus.

Figure 4.4 plots the EMG signals from the femoris and hamstring of Subject 3. The EMG measurements were made for 70 s, and Fourier transform was computed for 5 of 15 s of the total duration. This set of measurement was repeated twice. The

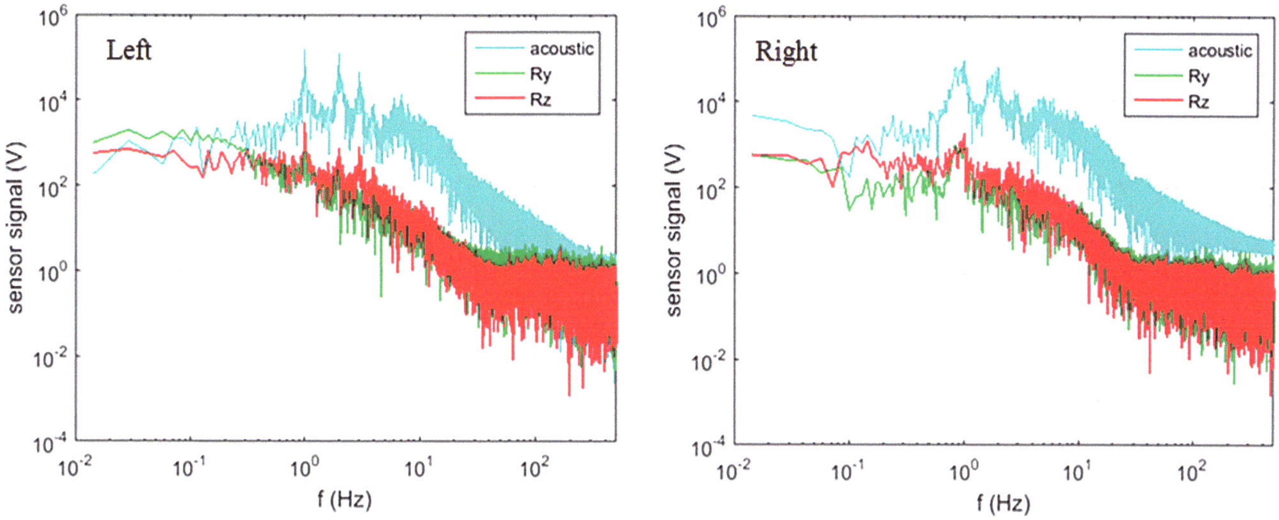

Fig. 4.3 Squat Fourier spectra of subject 1 with meniscus disorder on right knee

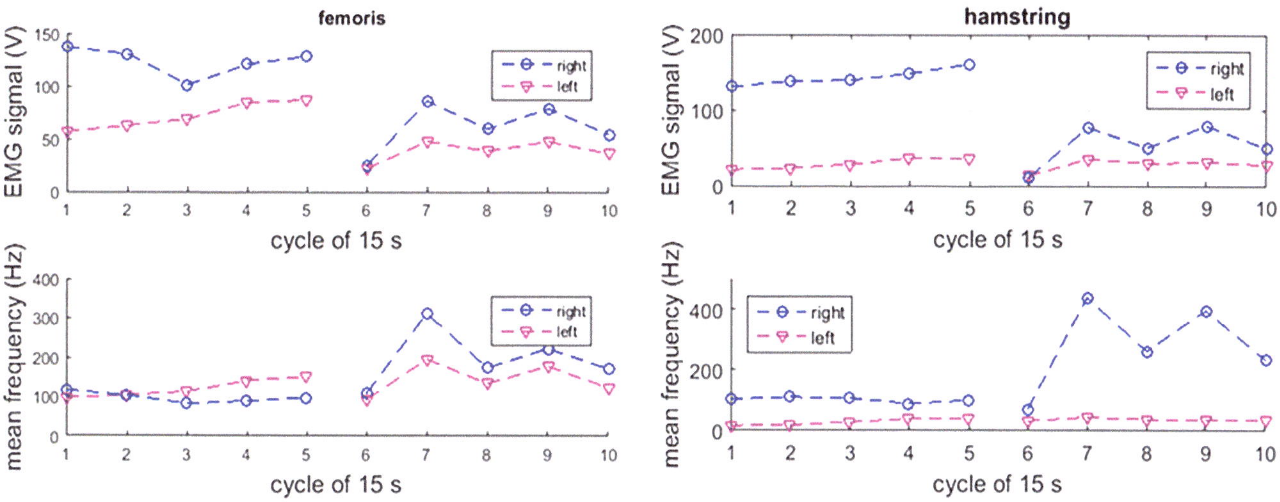

Fig. 4.4 Squat fatigue characteristics of subject 3 with meniscus disorder on right knee

horizontal axis of Fig. 4.4 indicates the cycle of the 15 s. There was approximately 5 min between the first and second sets. The top graph of Fig. 4.4 is the magnitude of the EMG signal's Fourier spectrum and the bottom is the mean frequency for each 15 s segment. The left graph is for the femoris and the right graph is for the hamstring.

The EMG signal indicates the muscle's activity. It is known that when a muscle is fatigued, the EMG-signal increases in the amplitude and decreases in the mean frequency [6]. Figure 4.4 does not show clear tendency of fatigue in either leg or muscle. In fact, the mean frequency appears higher in the second set than the first. It is possible that the frequency is lower in the first set because the muscles are not warmed up yet.

It is clear that the right femoris and hamstring exhibit higher chemical potential than the left. This indicates that the right muscles try to work harder because of the meniscus issue. Since the EMG signal is an indicator of muscle's energy and the torque around the *y*-axis (the torque for the flexion-and-extension) is important for the squatting activity, the ratio of the torque to the EMG (R_y/EMG magnitude) can be used to evaluate the efficiency of the muscles for the activity. Thus, Fig. 4.5 plots the ratio (called the EMG to R_y transfer function) for the femoris (left graph) and hamstring (right graph). Contrary to our prediction, the right femoris shows higher efficiency than the left femosir. On the other hand, the left hamstring shows higher efficiency than the right hamstring. We do not have a clear explanation for this observation at this time.

Figures 4.6, 4.7, and 4.8 are the same type of plots as Figs. 4.3, 4.4, and 4.5 for Subject 4 (68-year-old male complaining knee pain under exercise with no medical evidence of disorder).

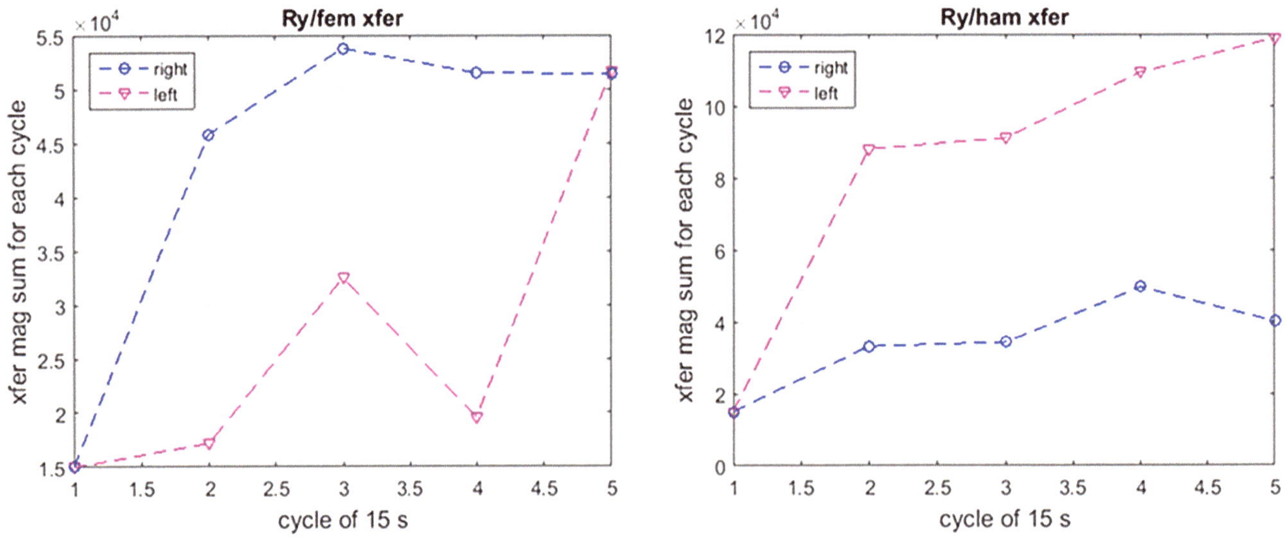

Fig. 4.5 Femoris EMG to Ry transfer function of subject 3 with meniscus disorder on right knee

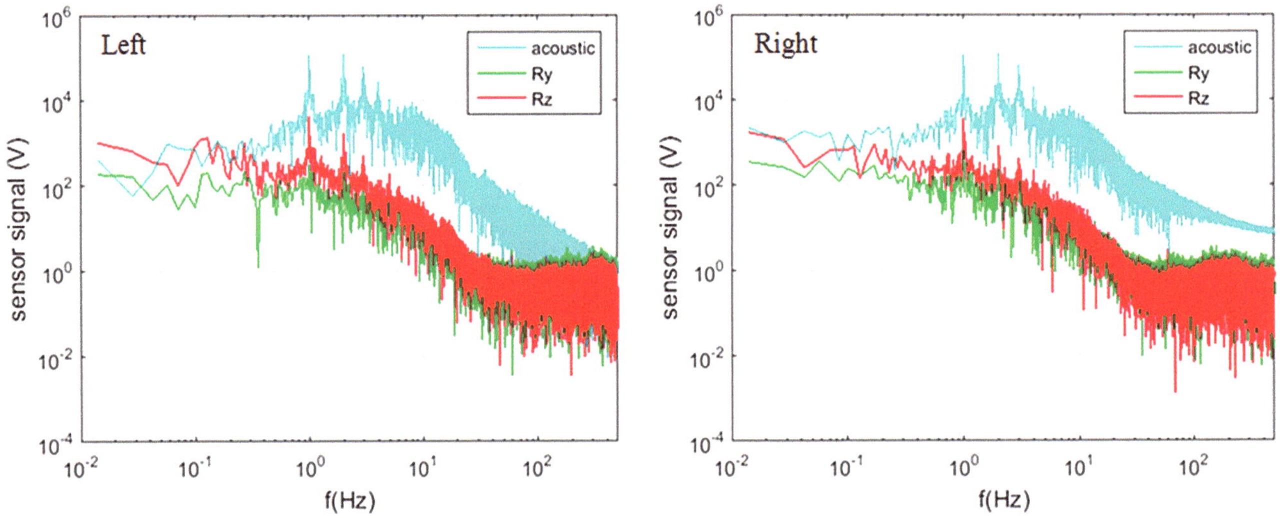

Fig. 4.6 Squat Fourier spectra of subject 4 with no identified knee disorder

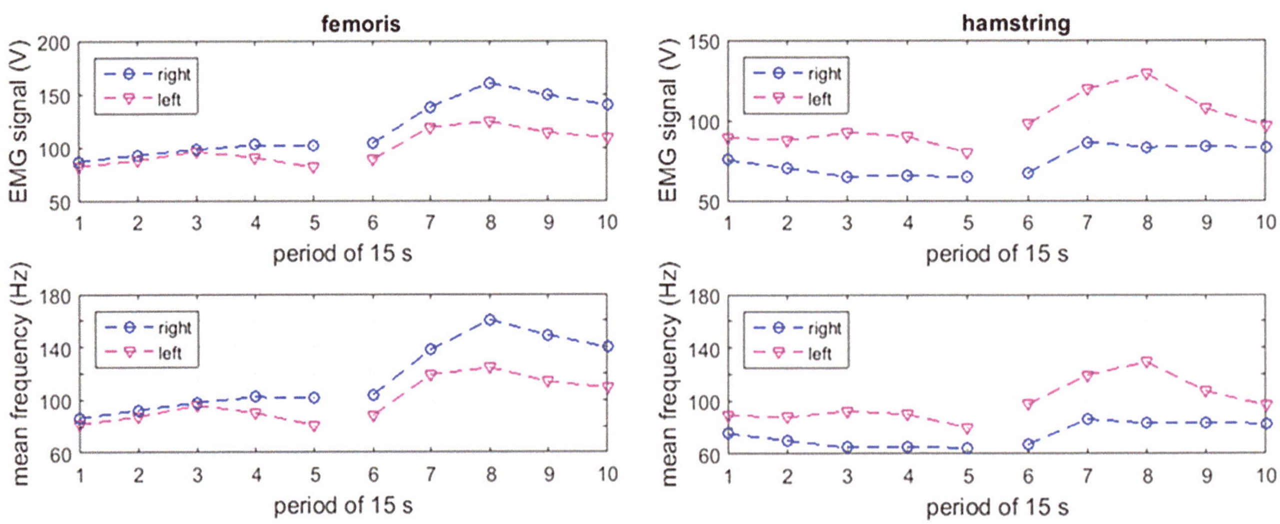

Fig. 4.7 Squat fatigue characteristics of subject 4 with no identified knee disorder

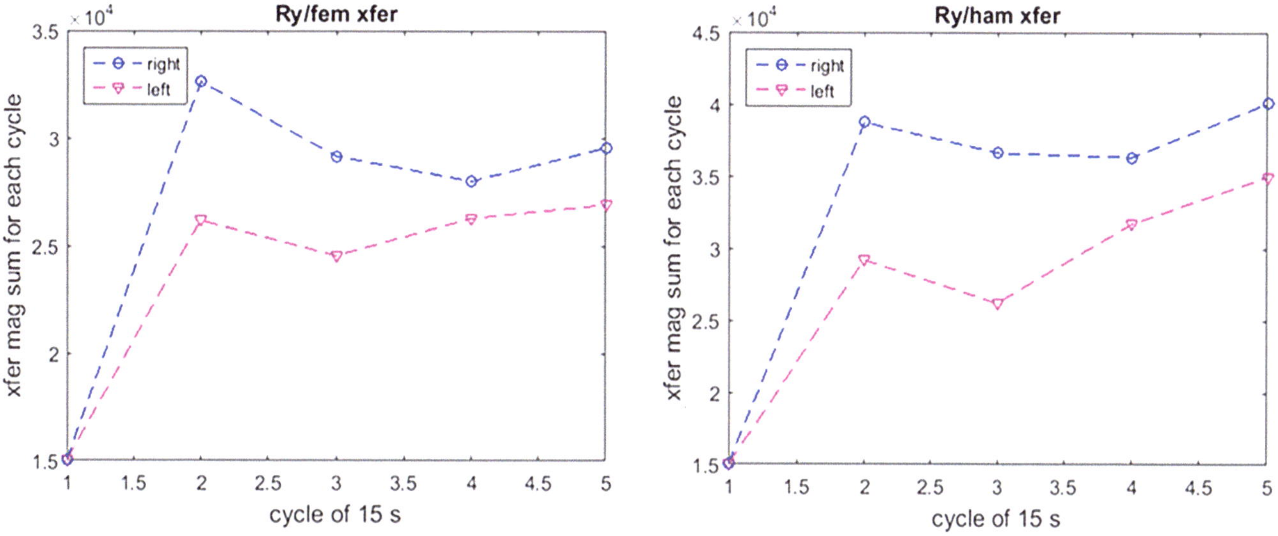

Fig. 4.8 Femoris EMG to Ry transfer function of subject 4 with no identified knee disorder

The Fourier spectra of the vibration and torque signals (Fig. 4.6) are very similar between the left and right. The peaks at the frequency of exercise are in fact very sharp. Unlike the case of Subject 3, the EMG signals' magnitude and frequency appear similar between the left and right (Fig. 4.7). The efficiency also appears the same between the left and right.

4.4 Summary

Human knee functionality has been dynamically tested with torque exerted by the ground, vibration signal from the knee and EMG from thigh muscles. The signals have been analyzed in the frequency domain. For each test subject, the set of data have been discussed in association with the knee condition specific to the subject. Overall, reasonable correlations have been found between various features observed in the experimental data and the known knee condition/disorder.

Acknowledgement The authors are grateful to Jesse Hatchet for his support in hardware conditioning and software installation.

References

1. Arden, N., Nevitt, M.C.: Osteoarthritis epidemiology. Clin. Rheumetol. **20**, 3–35 (2006)
2. Tiwari, U., Yoshida, S., Saladino, A., Hadley,B., Nguyen, M., Hatchett, J., Hollander, D.: Study of knee joints with spectral analysis. In: 82nd Annual Meeting of the APS Southeastern Section, **60**(18), Mobile, 18–21 November 2015
3. Saladino, A.: Framework for analysis of knee joint sound signals and muscle strength for non-invasive early detection of knee joint conditions. Master's Thesis, Integrated Science and Technology Program, Southeastern Louisiana University (2015)
4. BIOPAC, Systems, Inc.: https://www.biopac.com/
5. DELSYS, Inc.: http://www.delsys.com/
6. Viitasalo, J.H.T., Komi, P.V.: Signal characteristics of EMG during fatigue. Eur. J. Appl. Physiol. Occup. Physiol. **37**(2), 111–121 (1977)

Chapter 5
Viscohyperelastic Calibration in Mechanical Characterization of Soft Matter

E. Ficarella, L. Lamberti, M. Papi, M. De Spirito, and C. Pappalettere

Abstract The usual assumption made in mechanical characterization of soft tissues with Atomic Force Microscopy (AFM) is that the specimen behaves as a purely elastic material. Recent investigations have shown that in the case of immature porcine zona pellucida (ZP) samples viscous effects become more significant for sharp tips. Interestingly, a linear relationship between the limit indentation rate and the geometry of the AFM probe was derived, which is similar to the relationship existing between hardness and radius of curvature of the indenter typically observed for elasto-plastic materials. This study will analyze soft biotissues with different elasticity and viscous properties showing different response to nanoindentation. The relationship between limit indentation rate and tip geometry will be derived for each material to confirm previous experimental observations.

Keywords Soft matter • Mechanical characterization • Atomic force microscopy • Probe geometry • Indentation rate • Viscous effects

5.1 Introduction

Atomic Force Microscopy (AFM) is very suitable for mechanical characterization of soft matter because the nanometric dimensions of the probe allow to analyze the local behavior of the sample while other techniques causing microscopic deformations of the sample can only monitor equivalent properties of a large amount of material. In the last 10 years, AFM was used for characterizing mechanical behavior of cells and biopolymer networks [1–8]. Aspects related to calibration of nanoindentation measurements [9], sensitivity of finite indentation response to probe geometry [10] and residual stresses [11] also were studied.

The typical assumption made in atomic force spectroscopy is that materials are purely elastic: different constitutive laws ranging from linear elasticity to hyperelasticity were hypothesized. Viscoelastic effects can be neglected if quasi-static tests are performed [12]. However, recent investigations [13] proved that viscous forces may govern the nanoindentation response of biopolymer networks even at rather low indentation rate.

Experimental determination and modeling of viscoelastic behavior of soft materials is well documented in literature [13–22]. The present authors developed a visco-hyperelastic model to describe the nanoindentation response of immature porcine zona pellucida (ZP) membranes [23]. The hyperelastic behavior was described by the Arruda-Boyce constitutive law while the viscous response was defined by a N-terms Prony series expansion of the dimensionless relaxation modulus. By comparing finite element computations and AFM data at different indentation rates, it was possible to separate hyperelastic behavior and viscous effects. In [24], a systematic study attempted to determine for each geometric configuration of the AFM tip, the limit indentation rate above which viscous effects must be taken into account to correctly determine mechanical properties. It was shown that viscous effects become more significant for sharp tips. In [25], a linear relationship between the limit indentation rate and the geometry of the AFM probe was derived, which is similar to the relationship existing between hardness and radius of curvature of the indenter typically observed for elasto-plastic materials.

E. Ficarella • L. Lamberti (✉) • C. Pappalettere
Dipartimento di Meccanica, Matematica e Management, Politecnico di Bari, 70126 Bari, Italy
e-mail: luciano.lamberti@poliba.it

M. Papi • M. De Spirito
Istituto di Fisica, Università Cattolica del Sacro Cuore, 00168 Rome, Italy

C.S. Korach et al. (eds.), *Mechanics of Biological Systems and Materials, Volume 6*,
Conference Proceedings of the Society for Experimental Mechanics Series, DOI 10.1007/978-3-319-41351-8_5

This study will analyze two soft cell membranes, immature porcine ZP and immature bovine ZP, which have a considerably different level of elasticity. In [13], based on a classical linear elastic analysis of AFM data, the equine ZP was found to be about ten times weaker than porcine ZP. The apparent Young modulus increased by about 400 % for both materials when indentation rate was increased to from 0.5 to 10 μm/s. Here, we will attempt to derive more detailed information on the relationship between viscoelastic response and tip geometry for each material.

5.2 Methodology

A parametric axisymmetric finite element model was developed in order to simulate the nanoindentation process. The general purpose finite element software ABAQUS® Version 6.12 (Dassault Systèmes, France) was utilized for this purpose. The axis of symmetry of the FE model coincides with the direction of indentation. The indenter is modeled as a rigid blunt cone while the sample is modeled as an incompressible visco-hyperelastic slab (Fig. 5.1).

The bottom edge nodes of the model are clamped while the axis of symmetry cannot displace in the horizontal direction. For equine ZP, a vertical displacement of $\delta = 300$ nm is imposed to the rigid body part of the FE model simulating the AFM tip. The indenter displacement is limited to $\delta = 100$ nm for the porcine ZP. The contact interaction between the AFM tip and the soft material sample was hypothesized to be frictionless and the hard contact option available in ABAQUS was selected.

The AFM tip radius of curvature R is selected as 10, 30 and 50 nm while the blunt cone angle of aperture α is set equal to 40°. Tip dimensions are consistent with commercially available products and accounted for changes of probe geometry that might occur in AFM measurements.

The samples were first modeled as purely hyperelastic materials following the Arruda-Boyce constitutive law [26], originally developed to simulate the mechanical response of polymeric chains. The same sample was then modeled as a visco-hyperelastic material where viscous behavior is described by a N-terms Prony series expansion of the dimensionless relaxation modulus. The effective relaxation modulus $\mu_{R8chain}(t)$ is the product between the instantaneous shear modulus μ_{8chain} and the dimensionless relaxation function.

For the purely hyperelastic model and each AFM tip geometric configuration, one time-independent finite element analysis was performed by giving in input to ABAQUS the $\mu_{8chain} = 1.688$ kPa, $\lambda_L = 2.958$ hyperelastic parameters determined in [23] for the porcine ZP. In the case of equine ZP, material properties were determined as $\mu_{8chain} = 1.6$ kPa and $\lambda_L = 1.05$ via the optimization based inverse process described in Sect. 5.2.1.

In the visco-hyperelastic model, for each AFM tip configuration, time-dependent FE analyses were performed for the following indentation rates v_i: 100, 500, 1000, 2000, 4000, 6000, 8000 and 10,000 nm/s. For each indentation rate, the displacement given to the AFM indenter was ramped over the time $t_{indent\_i} = \delta/v_i$. In the case of porcine ZP, visco-hyperelastic parameters given in input to ABAQUS were $\mu_{8chain} = 1.688$ kPa, $\lambda_L = 2.958$, $g_I = 0.9681$ and $\tau_I = 0.00133$ s [23]. Inverse analysis was carried out for the equine ZP leading to determine the following values for material properties: $\mu_{8chain} = 1.234$ kPa, $\lambda_L = 1.256$, $g_I = 0.0490$ and $\tau_I = 0.00594$ s.

For each indentation depth and each AFM tip configuration, the reaction forces $F^{visco-hyperelastic}_{\delta=100/300\,nm}$ and $F^{hyperelastic}_{\delta=100/300\,nm}$ (acting on the tip) computed by ABAQUS by assuming, respectively, the visco-hyperelastic behavior and the purely hyperelastic behavior were compared. The percent difference ε between these values of indentation force was computed as:

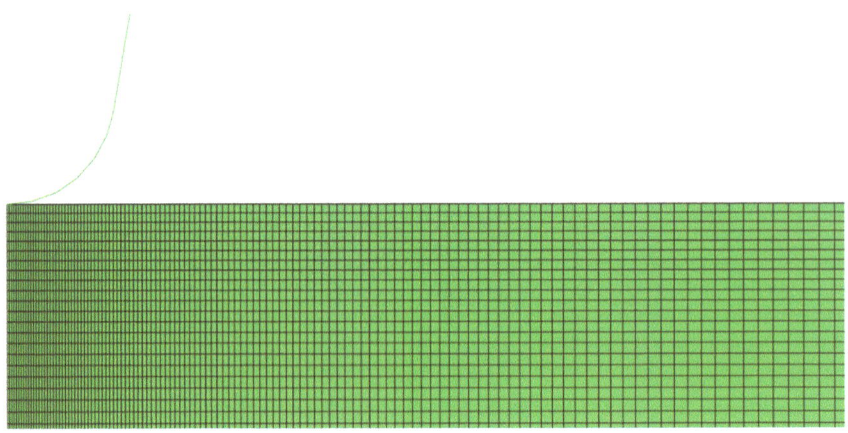

Fig. 5.1 Finite element model to simulate the nanoindentation process

$$\varepsilon = \left| \frac{F^{visco-hyperelastic}_{\delta=100/300\,nm} - F^{hyperelastic}_{\delta=100/300\,nm}}{F^{hyperelastic}_{\delta=100/300\,nm}} \right| \times 100 \tag{5.1}$$

for each indentation rate and each AFM tip geometry considered in this study. The purely hyperelastic model was judged in agreement with the visco-hyperelastic model if $\varepsilon < 25\ \%$. Indentation rates yielding $\varepsilon > 25\ \%$ were rated not acceptable, that means it is necessary to utilize a visco-hyperelastic model to correctly determine values of mechanical properties for the material under investigation.

5.2.1 Determination of ZP Visco-Hyperelastic Properties

The visco-hyperelastic properties of the ZP membranes were determined via an optimization-based identification algorithm. A similar approach was utilized in other mechanical characterization studies of nonlinear materials at the micro and nanoscale [27–29]. The identification algorithm, coded in the MATLAB® (The Mathworks Inc., Austin, TX, USA) software environment, minimizes the difference between nanoindentation data and finite element analyses via nonlinear optimization. Initial values are assigned to the four unknown material parameters $\mu_{8chain}{}^{o}$, $\lambda_{L,}{}^{o}$ $g_1{}^{o}$ and $\tau_I{}^{o}$ included as optimization variables. The computed force-indentation curve is compared with the target curve measured experimentally in order to evaluate the difference between numerical results and experimental data. This leads to formulate an optimization problem including the unknown material properties as design variables where the objective is to minimize the error functional Ω:

$$\begin{cases} \text{Min} \left[\Omega(\mu_{8\text{chain}}, \lambda_L, g_1, \tau_1) = \sqrt{\dfrac{1}{N_{CNT}} \sum_{j=1}^{N_{CNT}} \left(\dfrac{F_{FEM}{}^{j} - \overline{F^{j}}}{\overline{F^{j}}} \right)^2} \right] \\ 0.1\ \text{kPa} \le \mu_{8\text{chain}} \le 100\ \text{kPa} \\ 1 \le \lambda_L \le 10 \\ 0.001 \le g_1 \le 1 \\ 10^{-5} \le \tau_1 \le 10^{-1} \end{cases} \tag{5.2}$$

As mentioned above, there are four unknown material properties: two hyperelastic constants μ_{8chain} and λ_L, and two viscous parameters g_1 and τ_1. The bounds of the shear modulus μ_{8chain} were chosen so as to cover the range of variation of the apparent Young modulus of porcine/equine oocytes indicated in literature for the same indentation rate range and to account for the fact that values of elastic moduli determined under the assumption of hyperelastic behavior may be up to one or two orders of magnitude smaller than those determined with the classical Hertzian model which assumes instead linear elasticity, "infinite space", perfect spherical indenter, and small contact area. In general, material property bounds should always be set large enough to increase design freedom and rapidly converge to global optimum.

In Eq. (5.2), $F_{FEM}{}^{j}$ and $\overline{F^{j}}$, respectively, are the indentation force values for the j-th load step computed by ABAQUS and those measured experimentally by AFM. The number of control points N_{CNT} corresponds to the number of load steps executed for completing nonlinear FE analysis.

The computed error functional Ω is compared with a pre-defined convergence limit ε_{CONV} set as 0.0001 (i.e. 0.1 %). If it occurs $\Omega < \varepsilon_{CONV}$, the identification process terminates and material properties are listed in output. Conversely, if it occurs $\Omega > \varepsilon_{CONV}$, material parameters μ_{8chain}, λ_L, g_1 and τ_1 are perturbed in the subsequent design cycles until convergence.

The above described identification process was carried out to determine the corresponding material properties μ_{8chain}, λ_L, g_1 and τ_1 for each indentation rate. The inverse problem stated in Eq. (5.2) was solved with the Sequential Quadratic Programming (SQP) method, a globally convergent gradient-based optimization algorithm, coded in the MATLAB® Version 7.0 commercial software. For each indentation rate, optimization runs were started from five different sets of material properties randomly generated. This multi-start optimization strategy together with the large range of variability chosen for material parameters allowed to cover the whole search space and increased the probability of finding the global optimum. Serial optimization runs were carried out for each test case to avoid premature convergence. This process ended as soon as relative variations of error functional $\left|(\Omega_K - \Omega_{k-1})/\Omega_{\bar{K}}\, 1\right|$ and design vector $\|\mathbf{X}_K - \mathbf{X}_{K-1}\|/\|\mathbf{X}_{K-1}\|$ between the last two serial runs became smaller than 0.0001 (i.e. 0.1 %).

5.3 Results and Discussion

Although viscoelastic behavior of human and animal cells/tissues was extensively investigated in recent years [13–23], systematic studies on the relationships between AFM tip geometry, indentation rate and visco-hyperelastic response of soft materials were carried out only recently, aimed to determine the limit indentation rate $v_{threshold}$ below which viscous effects become negligible and a purely hyperelastic model may reliably describe the material behavior thus simplifying considerably the mechanical characterization process.

The present study compared two hyperelastic materials with a different level of elasticity. The equine ZP is nominally 10 times weaker than porcine ZP [13]. This suggested us to select considerably different indentation depths: 300 nm for the equine ZP vs. only 100 nm for the porcine ZP. The results of the characterization process seem to be consistent with the different elasticity of the materials. In particular, the porcine ZP was found to have almost the same shear modulus as the equine ZP (1.688 kPa vs. 1.6 kPa) and considerably higher distensibility than equine ZP (2.958 vs. 1.05). However, the difference is well below the one order of magnitude difference reported in [13].

The viscous properties were found to be in the ratio 1 to 2: in particular, the parameters $g_I = 0.0490$ and $\tau_I = 0.0028$ found for the equine ZP are about half or twice as their counterpart found for porcine ZP. This suggests that the equine ZP is quite less sensitive to viscous effects than porcine ZP over the long period. Such a conclusion is again in agreement with the findings of [23] where the increase in apparent elastic modulus caused by indenting specimens at different rate was larger for the porcine ZP, 433 % vs. only 400 % of the equine ZP.

The analysis of data gathered for tips of different geometry with radius of curvature 10, 30 or 50 nm provided the following results. The shape of the contact surface between the tested sample and the indenter strongly depends on the geometry of the AFM tip. In particular, deformations of indented samples were highly localized near the probe in the case of sharper tips. The percent difference ε between indentation forces computed by ABAQUS assuming a visco-hyperelastic behavior or a purely hyperelastic behavior increased considerably with indentation rate for the porcine ZP but much less for equine ZP. For example, even for the very high indentation rate of 10 μm/s, the error on maximum force was only 150 % for the equine ZP while it was about 800 % for the porcine ZP. This occurred in spite of having considered a threefold indentation depth in the case of the equine ZP.

The average error made in the case of equine ZP was about 150 % and decreased linearly with the tip radius. The same behavior was observed either at the final indentation depth of 300 nm and at the intermediate indentation rate of 100 nm, the same as that selected for the porcine ZP. This behavior is somehow consistent with that observed for the porcine ZP which shows a linear variation of the limit indentation rate with the radius of curvature. This suggests that the two materials may follow a similar viscohyperelastic behavior although viscous effects are much more pronounced for the porcine ZP.

Ultrastructural investigations will have to relate the inherent structure of the two ZPs with their mechanical response. Viscoelastic properties depend on the length and density of polymeric network included in the ZP membrane. As a matter of fact, the porcine ZP is much more distensible than the equine ZP and this implies that it can undergo a much higher stretch before that polymeric network is locked.

5.4 Conclusions

This study analyzed the viscohyperelastic behavior of two cell membranes characterized by a considerably different stiffness (in the ratio 1 to 10). The limit indentation rate and the error on the maximum indentation force made with respect to experimental data when viscous effects are neglected in the characterization process of soft biological tissues/cells were found to change linearly with the radius of curvature of the AFM tip.

The methodology described in the article is absolutely general and can be applied in principle to any kind of soft material to understand when viscoelastic effects become not significant in the mechanical characterization process. Furthermore, sensitivity analysis can serve to "correct" values of material properties determined by assuming that probe geometry does not change during the indentation test. However, target force-indentations curve should always be referred to the actual tip geometry measured after having completed the experiments.

References

1. Suresh, S.: Biomechanics and biophysics of cancer cells. Acta Biomater. **3**(4), 413–438 (2007)
2. Tang, Y., Chen, X., Yoo, J., Yethiraj, A., Cui, Q.: Numerical simulation of nanoindentation and patch clamp experiments on mechanosensitive channels of large conductance in *Escherichia coli*. Exp. Mech. **49**(1), 35–46 (2009)
3. Ikai, A.: A review on: atomic force microscopy applied to nano-mechanics of the cell. Adv. Biochem. Eng. Biotechnol. **119**, 47–61 (2010)
4. Christ, A.F., Franze, K., Gautier, H., Moshayedi, P., Fawcett, J., Franklin, R.J.M., Karadottir, R.T., Guck, J.: Mechanical difference between white and gray matter in the rat cerebellum measured by scanning force microscopy. J. Biomech. **43**(15), 2986–2992 (2010)
5. Jacot, J.G., Martin, J., Hunt, D.L.: Mechanobiology of cardiomyocyte development. J. Biomech. **43**(1), 93–98 (2010)
6. Bernick, K.B., Prevost, T.P., Suresh, S., Socrate, S.: Biomechanics of single cortical neurons. Acta Biomater. **7**(3), 1210–1219 (2011)
7. Papi, M., Brunelli, R., Familiari, G., Frassanito, M.C., Lamberti, L., Maulucci, G., Monaci, M., Pappalettere, C., Parasassi, T., Relucenti, M., Sylla, L., Ursini, F., De Spirito, M.: Whole-depth change in bovine zona pellucida biomechanics after fertilization: how relevant in hindering polyspermy? PLoS One **7**, e45696 1–7 (2012)
8. Boccaccio, A., Frassanito, M.C., Lamberti, L., Brunelli, R., Maulucci, G., Monaci, M., Papi, M., Pappalettere, C., Parasassi, T., Sylla, L., Ursini, F., De Spirito, M.: Nanoscale characterization of the biomechanical hardening of bovine zona pellucida. J. R. Soc. Interface **9**(76), 2871–2882 (2012)
9. Barone, A.C., Salerno, M., Patra, N., Gastaldi, D., Bertarelli, E., Carnelli, D., Vena, P.: Calibration issues for nanoindentation experiments: direct atomic force microscopy measurements and indirect methods. Microsc. Res. Tech. **73**(10), 996–1004 (2010)
10. Costa, K.D., Yin, F.C.P.: Analysis of indentation: implications for measuring mechanical properties with atomic force microscopy. J. Biomech. Eng. **121**(5), 462–471 (1999)
11. Boccaccio, A., Papi, M., De Spirito, M., Lamberti, L., Pappalettere, C.: Effect of the residual stress on soft sample nanoindentation. Appl. Phys. Lett. **102**, 133704,1–4 (2013)
12. Radmacher, M.: Measuring the elastic properties of living cells by the atomic force microscope. In: Jena, B.J., Horber, J.K.H. (eds.) Atomic Force Microscopy in Cell Biology, vol. 68, pp. 67–90. Elsevier, San Diego (2002)
13. Papi, M., Maiorana, A., Douet, C., Maulucci, G., Parasassi, T., Brunelli, R., Goudet, G., De Spirito, M.: Viscous forces are predominant in the zona pellucida mechanical resistance. Appl. Phys. Lett. **102**, 043703,1–4 (2013)
14. Guilak, F., Tedrow, J.R., Burgkart, R.: Viscoelastic properties of the cell nucleus. Biochem. Biophys. Res. Commun. **269**, 781–786 (2000)
15. Mathur, A.B., Collinsworth, A.M., Reichert, W.M., Kraus, W.E., Truskey, G.A.: Endothelial, cardiac muscle and skeletal muscle exhibit different viscous and elastic properties as determined by atomic force microscopy. J. Biomech. **34**(12), 1545–1553 (2001)
16. Trepat, X., Grabulosa, M., Puig, F., Maksym, G.N., Navajas, D., Farré, R.: Viscoelasticity of human alveolar epithelial cells subjected to stretch. Am. J. Physiol. Lung Cell. Mol. Physiol. **287**(5), L1025–L1034 (2004)
17. Puig-de-Morales-Marinkovic, M., Turner, K.T., Butler, J.P., Fredberg, J.J., Suresh, S.: Viscoelasticity of the human red blood cell. Am. J. Physiol. Cell Physiol. **293**(2), C597–C605 (2007)
18. Darling, E.M., Topel, M., Zauscher, S., Vail, T.P., Guilak, F.: Viscoelastic properties of human mesenchymally-derived stem cells and primary osteoblasts, chondrocytes, and adipocytes. J. Biomech. **41**(2), 454–464 (2008)
19. Lubarda, V.A., Marzani, A.: Viscoelastic response of thin membranes with application to red blood cells. Acta Mech. **202**(1–4), 1–16 (2009)
20. Lee, Y.J., Patel, D., Park, S.: Local rheology of human neutrophils investigated using atomic force microscopy. Int. J. Biol. Sci. **7**(1), 102–111 (2011)
21. Pravincumar, P., Bader, D.L., Knight, M.M.: Viscoelastic cell mechanics and actin remodelling are dependent on the rate of applied pressure. PLoS One **7**, e43938,1–10 (2012)
22. Jérusalem, A., Dao, M.: Continuum modeling of a neuronal cell under blast loading. Acta Biomater. **8**(9), 3360–3371 (2012)
23. Boccaccio, A., Lamberti, L., Papi, M., De Spirito, M., Douet, C., Goudet, G., Pappalettere, C.: A hybrid characterization framework to determine the visco-hyperelastic properties of a porcine zona pellucida. Interface Focus **4**(2), 20130066,1–10 (2014)
24. Boccaccio, A., Lamberti, L., Papi, M., De Spirito, M., Pappalettere, C.: Effect of AFM probe geometry on visco-hyperelastic characterization of soft materials. Nanotechnology **26**(32), 325701 (2015)
25. Boccaccio, A., Lamberti, L., Papi, M., De Spirito, M., Pappalettere, C.: A deeper look into immature porcine zona pellucida viscohyperelasticity. Proceedings of the 2015 SEM Annual Conference & Exhibition on Experimental and Applied Mechanics, Costa Mesa (CA), June 2015. In: Tekalur, S.A., Zavattieri, P., Korach, C.S. (eds.) Conference Proceedings of the Society for Experimental Mechanics Series; Mechanics of Biological Systems and Materials, Chapter 9, vol. 6, pp. 85–89. Springer, New York (2016)
26. Arruda, E.M., Boyce, M.C.: A three-dimensional constitutive model for the large stretch behavior of rubber elastic materials. J. Mech. Phys. Solids **41**(2), 389–412 (1993)
27. Genovese, K., Lamberti, L., Pappalettere, C.: A new hybrid technique for in-plane mechanical characterization of orthotropic materials. Exp. Mech. **44**(6), 584–592 (2004)
28. Cosola, E., Genovese, K., Lamberti, L., Pappalettere, C.: Mechanical characterization of biological membranes with moiré techniques and multi-point simulated annealing. Exp. Mech. **48**, 465–478 (2008)
29. Cosola, E., Genovese, K., Lamberti, L., Pappalettere, C.: A general framework for identification of hyper-elastic membranes with moiré techniques and multi-point simulated annealing. Int. J. Sol. Struct. **45**, 6074–6099 (2008)

Chapter 6
Contact Zone Evaluation of Dental Implants Using Digital Photoelasticity

M.P. Hariprasad and K. Ramesh

Abstract Use of dental implants in case of missing natural dentition is now common in clinical dentistry. The tilted implant configurations, which are inevitable in many clinical situations, are prone to higher stresses. The orientation of implants has a great influence on its structural integrity as high stresses around the implants adversely affect the osseointegration process and eventually fail due to the bone resorption. Hence, complete understanding of such complex systems demand better understanding of the bio-mechanics involved. Photoelasticity, a whole-field optical technique, is the right choice for analysing such implant configurations. Recent advancements in digital photoelasticity make it possible to evaluate isochromatics as well as isoclinic parameters over the entire model domain with considerable accuracy. However, these techniques have not been exploited in the field of implant dentistry and photoelasticity has been mostly used as a visualisation tool. This paper explores the use of modern photoelastic techniques for analysing the stress distribution due to an implant with an inclination of 15°. Further, this work also studies the high stress zones in the coronal region of this angled implant. The isochromatic data is post processed using least squares method for determining the contact parameters such as effective contact length and frictional coefficient.

Keywords Digital photoelasticity • Stress analysis • Biomechanics • Implant dentistry • All-On-Four

6.1 Introduction

Implantology is one of the promising areas in dentistry, which poses challenging problems for both clinical practitioners as well as for design engineers. Extensive research done on dental implants has resulted in newer designs and concepts like All-On Four procedure, which has revolutionised the concepts of total rehabilitation in the case of complete edentulism, where the entire teeth is replaced by prosthetics [1]. Complete understanding of such complex systems demand better understanding of the bio-mechanics involved and photoelastic experiments are the right choice for analysing the stress distribution in such implant designs.

Photoelasticity can provide whole field information regarding the principal stress difference (isochromatics) and the orientation of principal stresses (isoclinics). With the advancement in image processing techniques and computational efficiency, modern digital photoelasticity [2, 3] is capable of providing rich whole field data for problems with complex specimen shapes. In concepts like All-On Four® procedures, placement of implants in the posterior region of the mandibular arches is often complex owing to the anatomical structure and critical nervous system present in those areas [4, 5]. To circumvent this situation, the distal implants are often tilted posteriorly. However, the inclination with respect to the implant axis is patient specific. In such situations, the position and orientation of implants have a great influence on the structural integrity between the bone and the implants. Higher stresses around the implants can adversely affect the osseointegration process, by which the bone density decreases around the implants which leads to the failure of the implant.

Studies had been carried out to understand the influence of different implant-abutment connection systems such as internal/external hexagonal, platform switching etc. on the stress distribution of the bone using photoelasticity [6, 7]. However, the previous studies in this area were confined only to extract the fringe order information at discrete locations either using a simple colour matching methodology or by mere visualization of the isochromatics. This paper, initially discuss

M.P. Hariprasad (✉) • K. Ramesh
Department of Applied Mechanics, Indian Institute of Technology Madras, Chennai 600036, India
e-mail: hprasadmp@gmail.com

C.S. Korach et al. (eds.), *Mechanics of Biological Systems and Materials, Volume 6*,
Conference Proceedings of the Society for Experimental Mechanics Series, DOI 10.1007/978-3-319-41351-8_6

about an effective in-house developed method for obtaining a proper photoelastic model for such implant related studies. Further, the fringe order information extracted is post-processed using digital photoelastic algorithms to understand the biomechanics of the high fringe order zones. An effective methodology for the evaluation of contact zone near the implant bone interface is established and the influence of the geometric attributes of the implant-abutment connection in stress generated is investigated.

6.2 Experimental Methodology

6.2.1 Identification of Proper Photoelastic Model for Implant Dentistry

In Two Dimensional (2-D) photoelastic approaches, model analysis is easier as standard photoelastic sheets are available commercially. However a 2-D block does not capture the essential anatomical structure of the mandible/maxilla. This demands a Three Dimensional (3-D) photoelastic model, which has the essential anatomical feature of the real human mandible. Even though photoelastically sensitive models with high accuracy could be made using rapid prototyping methods, due to the layered manufacturing, the fringe definitions are not good and further the model has residual stress [8]. This demanded a proper method in which stress free photoelastic models can be casted, which has the essential features of the human mandible. Different methods are explored and an effective methodology is established for producing stress free photoelastic models for such studies [9]. A human mandibular bone is reproduced with wax pattern and implant system are placed in the wax model as in All-On-Four concept, at predetermined positions. This is then duplicated using liquid silicon such that the mould has uniform minimum thickness around the alveolar ridge. Photoelastic model used in this study is made by a mixture of araldite resin (HUNTSMAN Araldite CY230) and hardener in 10:1 ratio. Uniform heat dissipation is facilitated by providing an ice bath around the mould during pouring and curing stage.

Once the stress free photoelastic model with 15° implant inclination is obtained, bar attachments are used to connect the implants along the alveolar bridge. The bars are investment cast using Ni–Cr alloy. In situations involving complex models, even though stress freezing and slicing could be used to view any segment of the 3-D model, if certain key areas of the model could be viewed directly under live load conditions and analyse them directly, it would be then preferable in many situations. The model obtained using the in-house casting procedure are well suited for such analysis [9]. Implants placed in the molar region can be evaluated by adopting an appropriate angled view approach as shown in Fig. 6.1a. Dark field isochromatics observed for a load of 273 N is shown in Fig. 6.1b.

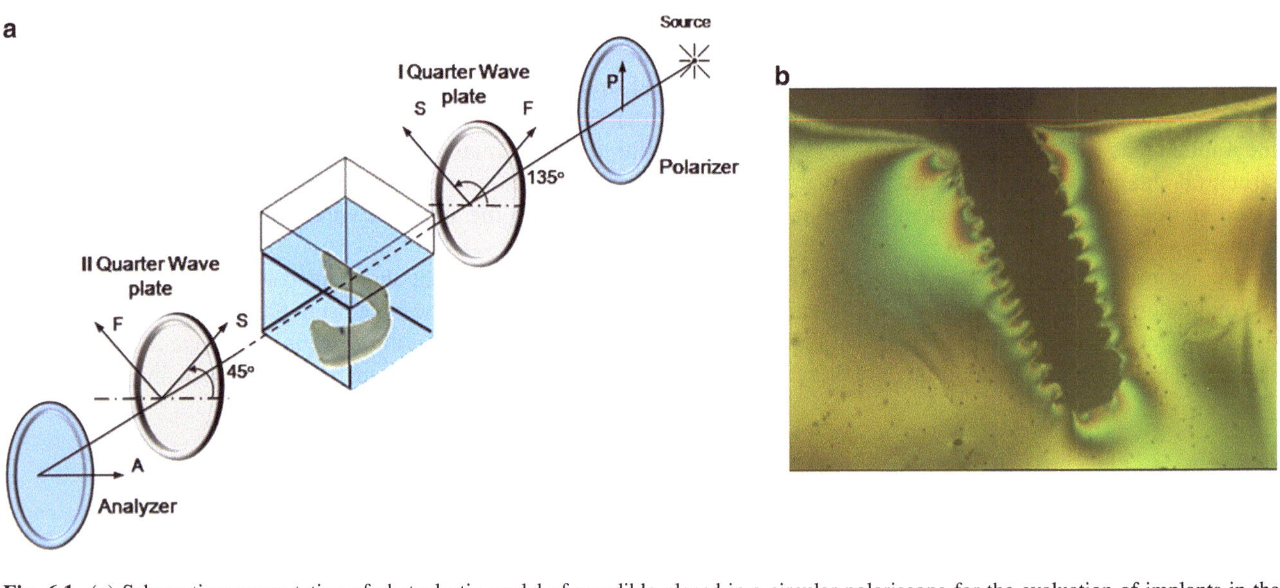

Fig. 6.1 (**a**) Schematic representation of photoelastic model of mandible placed in a circular polariscope for the evaluation of implants in the molar area. (**b**) Dark field isochromatic observed for a load of 273 N when viewed according to 6.1 (**a**)

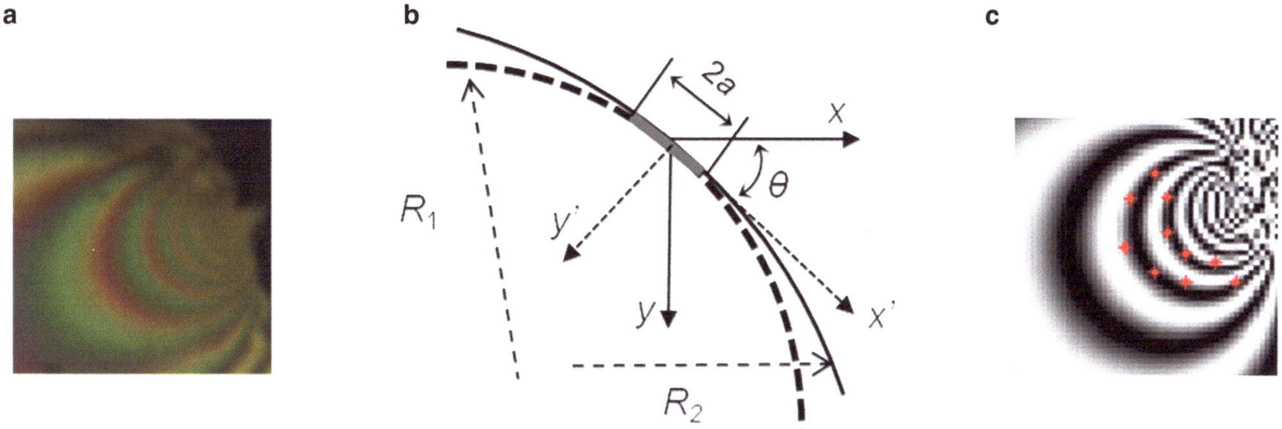

Fig. 6.2 (**a**) Magnified view of dark field isochromatic for the coronal region shown in 6.1 (**b**). (**b**) Schematic representation of two bodies in contact. (**c**) Reconstructed dark field isochromatics in monochrome from the results of least squares analysis with data points echoed back for the contact zone shown in 6.2 (**a**)

6.2.2 Isochromatic Evaluation Using Three Fringe Photoelsaticity

Three-fringe photoelasticity [10] uses a single isochromatic color image for finding the total fringe order at each point in the model domain. This is done by comparing RGB intensity values in the dark field isochromatic image of the application specimen with that of a calibration table. To address the issue of fringe order continuity in complex specimen shapes, Sohan and Ramesh [11] has proposed a new scanning method with mask based algorithm known as Advancing Front Scanning for TFP (AFSTFP). Modified window search method is used for refinement which is based on the fringe order increments and is given as

$$e_i = \sqrt{(R - R_i)^2 + (G - G_i)^2 + (B - B_i)^2}; \quad N \in [N_p - \Delta N, N_p + \Delta N] \tag{6.1}$$

where R, G, B corresponds to the intensities of the pixel under consideration and R_i, G_i, B_i is the corresponding intensities in the calibration table. The average value of the fringe orders of all the neighboring resolved pixels is used as Np in Eq. (6.1) and a value of 0.4 for ΔN is found to be suitable for the current study.

The isochromatic fringe orders are evaluated using TFP for a quantitative appreciation of the isochromatic data. Maximum fringe order is observed in the coronal region of the model. Figure 6.2a shows the magnified experimental dark field isochromatic for the coronal region shown in Fig. 6.1b. It is interesting to note that the fringe features shown in Fig. 6.2a are similar to that of an isochromatics corresponding to a typical contact problems. This needs further investigation on coronal region to understand the mechanics involved better. The next section of this paper deals with the evaluation of the high stressed areas in the coronal region with a focus on analyzing the contact zone using photoelasticity.

6.3 Contact Zone Evaluation of the Coronal Region

Structural integrity between the bone and the implants has great importance, as higher stresses around the implants can lead to bone resorption. It is also stated that the total area of contact between the implant and bone is significant in deciding the osseointegration strength of implant-bone interface [12]. The higher stresses produced in the coronal region is very crucial as it can lead to peri-implant bone lose. Generally, the stresses near the contact zones are characterized by the contact length and frictional coefficient. The potential of digital photoelasticity can be used to determine these contact parameters from the isochromatic data in the coronal area of the model.

6.3.1 Evaluation of Contact Stress Parameters Using Fringe Order Information

The idea of using isochromatic data in conjunction with least squares method for the evaluation of contact stress parameters like contact area and coefficient of friction between two bodies had shown its usefulness in the stress analysis of the engineering components [2, 13]. The stress fields near the contacting zones of two bodies can be expressed in terms of its contact length ($2a$) and frictional coefficient (μ) as [2, 14]

$$\begin{Bmatrix} \sigma_x \\ \sigma_x \\ \tau_{xz} \end{Bmatrix} = \begin{bmatrix} -\frac{a}{\pi\zeta}\left[\left(a^2 - 2x^2 + 2z^2\right)\frac{z\psi_1}{a} - \frac{2\pi z}{a} - 3xz\psi_2 + \mu\left\{\left(2x^2 - 2a^2 - 3z^2\right) + 2\pi x + 2\left(a^2 - x^2 - z^2\right)\frac{x\psi_1}{a}\right\}\right] \\ -\frac{a}{\pi\zeta}\left(a\psi_1 - x\psi_2 + \mu z\psi_2\right) \\ -\frac{a}{\pi\zeta}\left[z^2\psi_2 + \mu\left\{\left(a^2 + 2x^2 + 2x^2\right)\frac{z\psi_1}{a} - \frac{2\pi z}{a} - 3xz\psi_2\right\}\right] \end{bmatrix} \tag{6.2}$$

In the above sets of equations, a represents the semi-contact length and ξ is a function of the geometrical and elastic properties of the bodies in contact (Fig. 6.2b) and is given as

$$\zeta = \frac{1}{\frac{1}{2}\left(\frac{1}{R_1} - \frac{1}{R_2}\right)}\left(\frac{1 - {\nu_1}^2}{E_1} + \frac{1 - {\nu_2}^2}{E_2}\right) \tag{6.3}$$

The parameters $\psi_1$and ψ_2 are defined as

$$\psi_1 = \frac{\pi(r_1 + r_2)}{r_1 r_2\sqrt{2r_1 r_2 + 2x^2 + 2z^2 - 2a^2}} \quad \psi_1 = \frac{\pi(r_1 - r_2)}{r_1 r_2\sqrt{2r_1 r_2 + 2x^2 + 2z^2 - 2a^2}} \tag{6.4}$$

where $r_1 = \sqrt{(a + x)^2 + z^2}$, $\quad r_2 = \sqrt{(a - x)^2 + z^2}$

The stress-optic law relates the difference in principal stress ($\sigma_1 - \sigma_2$), and the fringe order (N) as

$$\sigma_1 - \sigma_2 = \frac{NF_\sigma}{h} \tag{6.5}$$

where F_σ is the material fringe value and h is the thickness of the model. Using Eq. (6.2), the expression for $\sigma_1 - \sigma_2$ can be obtained and it can be seen that it is non-linear in terms of the semi contact length a and the coefficient of friction μ. The concept of nonlinear least squares analysis for getting the contact parameters, utilizing the fringe order information from the experimental result is implemented in Matlab® as described in [2]. The coronal region of the molar implant shown in Fig. 6.2a is analyzed for evaluating the contact parameters using the non-linear least squares method.

The implant abutment connection system used in this study had a profile curvature of 2.32 mm which represents R_2 in Fig. 6.2b. The photoelastic model area under the influence of the implant abutment connection region is idealized with slightly increased radius of curvature of 3 mm. The input parameters used for the current study are given in Table 6.1.

The fringe orders and the corresponding positional coordinates can be collected either from whole field solutions obtained by TFP, Phase shifting technique (PST) or from the fringe skeleton image. Once the data is collected near the contact zone, it is then fed in to the least squares algorithm. Convergence error is calculated at each iteration of the solution procedure using fringe order minimization criteria. The iteration procedure is stopped when the convergence error reaches a minimum value. Once the parameters (a, μ) are obtained, the algorithm calculates the fringe orders at every point in the data field using

Table 6.1 Input parameters used for the analysis

Specifications	Units	Photoelastic model (epoxy)	Implant (Ti alloy)
Young's modulus (E)	GPa	3.3	110
Poisson's ratio, (γ)	–	0.33	0.37
Material stress fringe value (F_σ)	N/mm/fringe	12.25	–
Radius of curvature (R)	mm	2.23	3

Eq. (6.2), for visual comparison with the experimental fringe field. Using the method described, an effective contact length of 0.17 mm and a frictional coefficient of 0.28, is obtained for the 15° implant tilted model. The reconstructed dark field image for a monochromatic illumination with data point echoed back for the evaluated contact parameter is shown in Fig. 6.2c.

The comparison of the fringe features between Fig. 6.2a, c clearly shows that, the essential features of the fringe field due to the contact between the implant around the coronel region are captured well by the predicted contact stress parameters. Hence it is clear from this analysis that the higher stresses produced in the coronal region of the implants are not purely a function of the implant tilts but also depend up on other factors like the contact parameters and the profile radius of the implant-abutment connection system.

6.4 Conclusion

This paper has shown the usefulness of digital photoelastic techniques in understanding the biomechanics related to implant dentistry better. The methodologies involved in getting a proper photoelastic model for implant related studies are established. Further photoelastic model having an implant inclinations of 15° is quantitatively analyzed using Three fringe Photoelasticity. A detailed study on the zones of high stressed area in the coronal region using nonlinear least squares method in conjunction with the isochromatic data has been carried out to get the effective contact length and coefficient of friction in the bone-implant interface. This has reviled that, even though the implant inclination affects the stress distribution in the other areas of the model, the higher stresses produced in the coronal area is also dictated by the geometric aspects of the implant abutment connection used. Better understanding obtained in this study will be of great interest for medical practitioners as well as for the engineers who is into the design aspects of such implants.

Acknowledgment The authors would like to acknowledge Dr. Bhuvaneswari. S, from Vinayaka Mission's Sankaracharyar Dental College-Salem, India, for helping in the model preparation for this study.

References

1. Maló, P., Rangert, B., Nobre, M.: All-on-Four® immediate-function concept with Brånemark System® implants for completely edentulous mandibles: a retrospective clinical study. Clin. Implant. Dent. Relat. Res. **5**(1), 2–9 (2003)
2. Ramesh, K.: Digital Photoelasticity: Advanced Techniques and Applications. Springer, Berlin (2000)
3. Ramesh, K.: Digital photoelasticity. In: Rastogi, P. (ed.) Digital Optical Measurement Techniques and Applications, pp. 289–344. Artech House, London (2015)
4. Ten Bruggemkate, C.M., Sutter, F., Oosterbeek, H.S., Schroeder, A.: Indications of angled implants. J. Prosthet. Dent. **67**, 85–93 (1992)
5. Begg, T., Geerts, G.A., Gryzagoridis, J.: Stress patterns around distal angled implants in the All-on-Four concept configuration. Int. J. Oral Maxillofac. Implants **67**, 663–693 (2009)
6. Asvanund, P., Morgano, S.M.: Photoelastic stress analysis of external versus internal implant-abutment connections. J. Prosthet. Dent. **106**(4), 266–271 (2011)
7. Rossi, F., Zavanelli, A.C., Zavanelli, R.A.: Photoelastic comparison of single tooth implant-abutment- bone of platform switching vs conventional implant designs. J. Contemp. Dent. Pract. **12**(2), 124–130 (2011)
8. Ashokan, K., Prasath, R.G.R., Ramesh, K.: Noise-free determination of isochromatic parameter of stereolithography-built models. Exp. Tech. **36**(1), 70–75 (2012)
9. Hariprasad, M.P., Bhuvaneswari, S., Ramesh, K., Simon, P., Mohan, J.: Evolution of suitable photoelastic model in implant dentistry. In: Proceedings of 9th International Symposium on Advanced Science and Technology in Experimental Mechanics (9th ISEM 14), New Delhi, 1–6 November 2014
10. Ramesh, K., Deshmukh, S.S.: Three fringe photoelasticity—use of colour image processing hardware to automate ordering of isochromatics. Strain **32**(3), 79–86 (1996)
11. Kale, S., Ramesh, K.: Advancing front scanning approach for three-fringe photoelasticity. Opt. Lasers Eng. **51**, 592–599 (2013)
12. Caram, S.J., Huynh-Ba, G., Schoolfield, J.D., Jones, A.A., Cochran, D.L., Belser, U.C.: Biologic width around different implant-abutment interface configurations. A radiographic evaluation of the effect of horizontal offset and concave abutment profile in the canine mandible. Int. J. Oral Maxillofac. Implants **29**, 1114–1122 (2014)
13. Sukla, A., Nigam, H.: A numerical-experimental analysis of the contact stress problem. J. Strain Anal. Eng. Des. **20**(4), 241–245 (1985)
14. Smith, J.O., Liu, C.K.: Stresses due to tangential and normal loads on an elastic solid with application to some contact stress problems. J. Appl. Mech. **20**, 157–166 (1953)

Chapter 7
Evolution of the Skin Microstructural Organization During a Mechanical Assay

B. Lynch, S. Bancelin, C. Bonod-Bidaud, F. Ruggiero, M.-C. Schanne-Klein, and J.-M. Allain

Abstract Skin is a complex multi-layered tissue, consisting of three main parts: the epidermis, the dermis and the hypodermis. The dermis is responsible for most of the complex mechanical properties of skin, such as viscoelasticity, non-linearity and anisotropy. At the microscopic level the dermis consists for the greater part of extracellular matrix, compounded mainly of collagen fibers forming an orderless network. The mechanical properties of skin have been studied in the past, but their exact link with the microscopic organization is still an open question. The goal of our study is to measure the evolution of the microstructure during a mechanical assay and to improve existing mechanical models of skin with relevant parameters identified at the microscopic level.

We perform uniaxial tensile test on ex vivo mouse skin. The mechanical tests are performed in situ under a second harmonic generation microscope. This allows us to determine quantitatively and simultaneously the mechanical response and the microstructural reorganization of the tissue. This technique can be used to better understand the link between pathological alterations of collagen synthesis, fibers organization, and alteration of the biomechanical properties of skin, as in the Ehlers-Danlos syndrome (EDS).

Keywords Biomechanics, Multiscale mechanics, Mice skin, Collagen, Multiphoton imaging

7.1 Introduction

Connective tissues such as skin, tendon or cornea are mostly made of extracellular matrix proteins, with only few cells embedded. This matrix is in fact a mixture of different proteins, the most abundant proteins belonging to the collagen family. The ability of the certain types of collagen to self-organize into supramolecular cables—called fibrils—leads to the formation of a fibrillar network inside of the tissue, surrounded by a less organized medium of proteins and water. These fibers play a key role in the behavior of the connective tissues, by providing stiff structures and mechanical support to the residing cells. Alteration in the collagen structure or organization leads to degraded mechanical properties and poor regeneration, by affecting the cell interaction with the surrounding matrix. This has a strong impact on wound repair or ageing [1].

Macroscopic mechanical properties of skin are well known. They show an initial non-linear part, the heel region, followed by a linear part and a saturation, indicating the rupture of the tissue [2–4]. Similar responses are found in most connective tissues as aorta [5, 6], tendon [7–9] and cornea [10, 11]. However, the link with the microstructure has been less studied in disorganized tissues. In highly organized tendons, different methods have been used: polarized-light microscopy [12], OCT [13], multiphoton [9, 14] and confocal [8, 15] microscopy and X-ray diffractions [16, 17]. They all concluded that the heel region is associated with an uncrimping of the collagen fibers—which means that the initial macroscopic undulations of the fibers disappear under tension. In the linear part, the response of a tendon combines stretching of the fibers and sliding between adjacent fibers [16, 17]. Like for tendon, the mechanical response of skin has been interpreted as an initial alignment of the fibers in the heel region, followed by the stretching of the fibers in the linear part [18, 19]. This has led to a large set of theoretical works on the so-called microstructural models [20–23], which took as an entry the local

B. Lynch • J.-M. Allain (✉)
LMS, Ecole Polytechnique, CNRS, Université Paris-Saclay, Palaiseau, France
e-mail: allain@lms.polytechnique.fr

S. Bancelin • M.-C. Schanne-Klein
LOB, Ecole Polytechnique, CNRS, INSERM, Université Paris-Saclay, Palaiseau, France

C. Bonod-Bidaud • F. Ruggiero
Institut de Génomique Fonctionnelle de Lyon, ENS-Lyon, CNRS, Université Lyon 1, Lyon, France

C.S. Korach et al. (eds.), *Mechanics of Biological Systems and Materials, Volume 6*,
Conference Proceedings of the Society for Experimental Mechanics Series, DOI 10.1007/978-3-319-41351-8_7

organization of fibers, and predict the macroscopic response based on the mechanical properties of the fibers and of the surrounding medium. However, only few papers have investigated experimentally this interpretation at different stretch level, on the same sample, in skin [18, 19]: it is only recently that set-ups have been designed allowing such observation in a direct way.

To obtain a continuous monitoring of the tissue microstructure during a mechanical assay, we have combined multiphoton microscopy with traction assay [24]. Multiphoton microscopy is an optical technique which is perfectly adapted to investigate the three-dimensional (3D) architecture of connective tissues as skin [25]. its creates stacks of images, using intrinsic optical sectioning. The images can be obtained through different modes of contrasts, the most interesting being the Two-Photon Excited Fluorescence (2PEF) and the Second Harmonic Generation (SHG). SHG is specific for dense non-centrosymmetric media, such as fibrillar collagen [26, 27]; it provides a way to image the collagen organization of the tissue at the micrometer-scale. This approach has been combined with traction assays first on tendon [9, 14], but is now used on more complex tissues as aorta [28, 29], fetal membrane [30, 31] or cornea [11]. Very recently, application to skin have been reported [24, 32].

In this present study, we investigated the relationship between the microstructure of the dermis and its macroscopic mechanical properties at all stretch levels. Mechanical measurement were performed while imaging optical sections of ex vivo mouse skin using multiphoton microscopy. Analysis of the image allowed to extract the tissue microstructure and to relate it with the stress/stretch relationship. A comparison with the classical affine assumption of the microstructural models is then done.

7.2 Materials and Methods

The experimental set-up is similar to the one presented in [24]. We remind briefly the main steps.

7.2.1 Samples Preparation

We used skin of the back of 1 month old mice. Hairs and epidermis were removed, and the right foreleg was spotted by a black ink dot to identify the head-tail axis. Skin samples were stored in culture medium and used within 5 days for the mechanical experiments.

Before traction, the samples were cut into a dog-bone shape to ensure homogeneous tensile load in the central portion. The sample was cut so that the traction will be in the head-tail direction. Then, the sample was placed in the traction device, with the papillary dermis up (facing the objective lens). A drop of immersion gel ensured optical contact with the objective lens while preventing skin dehydration. If needed, the drop may be replaced during the traction assay.

We use a custom-build traction device, with two motors and two load cells. The symmetry of the traction allows imaging always the same region of interest (ROI) of the sample. When the sample was in place in the traction

7.2.2 Multiphoton Microscopy

Multiphoton imaging was performed using a custom-built laser scanning microscope (see Fig. 7.1). Image stacks were typically about 500 x 500 x 50 μm^3, for a pixel size of 0. 5 μm, and axial steps of 2 μm. Recording a full stack required 5 min, plus few minutes of preparation. No degradation of the skin was observed under these conditions.

7.2.3 Traction Device and Mechanical Assays

We use a custom-build traction device, with two motors and two load cells (see Fig 7.1). The symmetry of the traction allows imaging always the same region of interest (ROI) of the sample. When the sample was in place, it is slowly stretched up to the moment we do not observe any vertical motion of the surface on the microscope: the skin is then horizontal and not bend under its own weight. No measurable force have been recorded at that step. We use this configuration as the reference

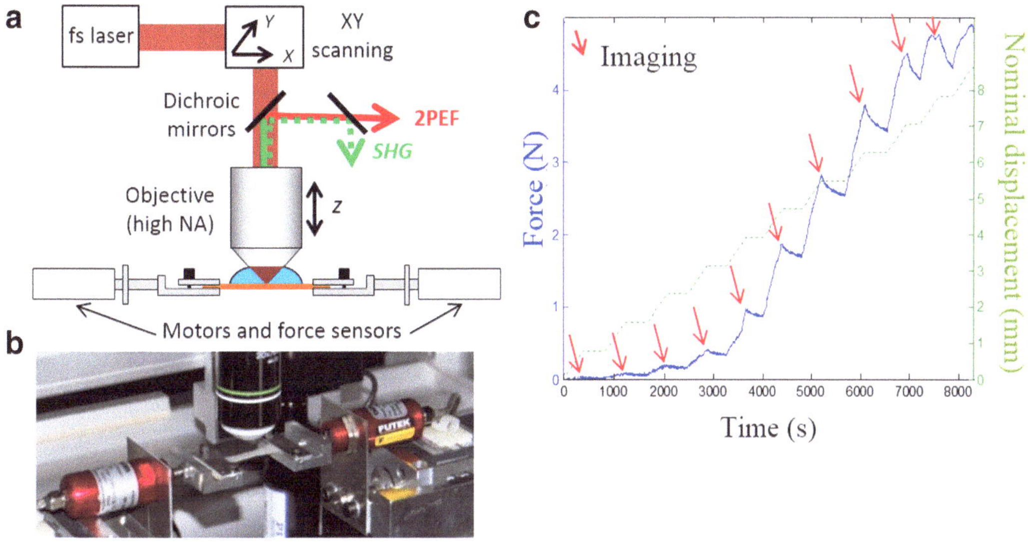

Fig. 7.1 Experimental setup. (**a**) Scheme of the combined MPM and traction device. 2PEF and SHG are detected in the backward direction; (**b**) View of the sample in the traction device under the microscope; (**c**) Experimental timetable showing the increasing stretching, by steps (*green*), with the force measurement (*blue*) and the times for sample imaging (*red arrows*). Image from [24]

state of the sample. The dimensions of the sample were then measured using a digital caliper: we had typically samples of 20 x 8 x 1 mm^3.

The tensile test was performed at a fixed slow strain rate (10^{-4} s^{-1}, around 2 μm s^{-1}): this is needed to observe the same ROI. If needed, small displacements of the ROI are accommodated by moving the whole traction device through micrometer stages. At every 5 % of stretch, we stopped the traction device to perform the imaging without being blurred by the sample motions. The traction is performed until breakage of the sample.

7.2.4 *Mechanical Data Processing*

We define the global stretch ratio as

$$\lambda = \frac{l}{l_0} \tag{7.1}$$

where l is the length between the jaws and l_0 the reference length. The nominal stress is defined as the measured force divided by the initial skin section.

To quantify the mechanical properties of the sample, we use four parameters: the tangent modulus (the slope of the linear part), the length of the heel region (the variation of stretch between the first measurable force and the linear part), the failure stretch ratio and ultimate strength (the rupture).

7.2.5 *SHG Image Processing*

SHG images showed the collagen fibrillar organization (see Fig. 7.2). The round structures with no SHG signal are the hair follicles. Two quantitative informations were then extracted: the local stretch at the ROI size (about 300 x 300 μm^2) and the local organization of the fibrillar collagen. This was done using a custom-written script (Matlab).

The local stretch was determined by tracking the centers of the hair follicles (see Fig. 7.2). The centers of the follicles in the reference configuration define a network of points; the displacement of the follicles are then measured by following the same follicles at each stretching step. Then, the surface strain tensor is determined by a standard derivation on this Delaunay mesh. To avoid errors due to false points, we used the average of the strain of the ROI.

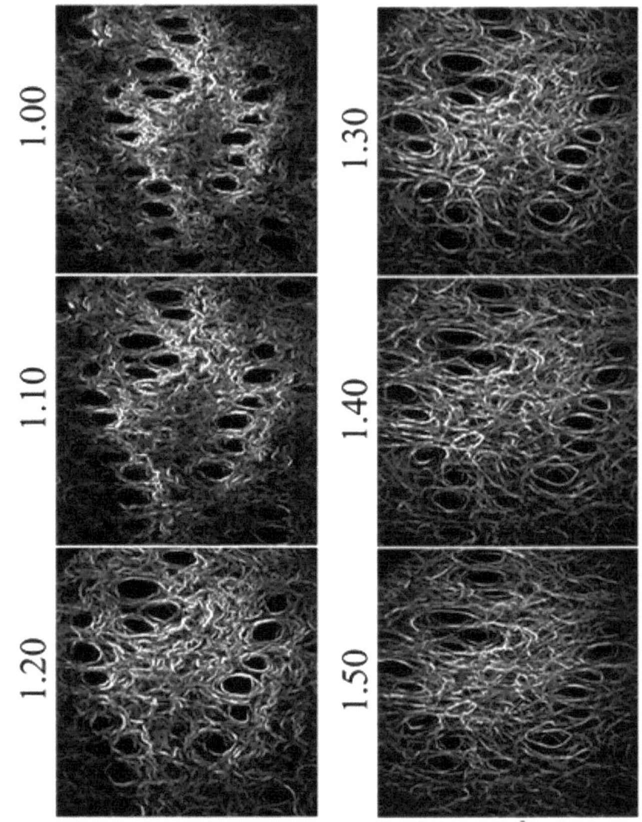

Fig. 7.2 SHG imaging from mice skin, as a function of increased skin global deformation. Traction is done in the horizontal axis. Image size: 480 x 480 μm^2. Image adapted from [24]

The fibers orientations are extracted using morphological filtering by a rotating linear element. We obtain then a map of orientations, at each stretching step. From this map, we extract the histogram of the orientation on this map. For simplicity, we construct two scalar parameters: the main orientation θ_{max} in the ROI and the orientation index (OI) in the main orientation:

$$OI = \left[2 \frac{\int_{-\pi/2}^{\pi/2} I(\theta) cos^2(\theta - \theta_{max}) d\theta}{\int_{-\pi/2}^{\pi/2} I(\theta) d\theta} - 1 \right] .100 \tag{7.2}$$

The OI represents roughly the fraction of fibers aligned in the main direction of orientation.

7.2.6 Theoretical Fibers Reorientation

If we assume that the fibers follow the motion of the volume of tissue in which they are present, it is possible to predict the evolution of the orientation of the fiber distribution based only on the stretch of the surrounding volume. This computation was done for surface strain [33]. In the case where the gradient of the transformation is diagonal (which is the case here), it leads to the following formula:

$$\Lambda_1(\beta) = \Lambda_0(\theta) \left(\frac{\lambda_{yy}}{\lambda_{xx}} sin^2(\theta) + \frac{\lambda_{xx}}{\lambda_{yy}} cos^2(\theta) \right) \tag{7.3}$$

where $\Lambda_0(\theta)$ is the distribution in the reference configuration, while $\Lambda_1(\beta)$ is the one in the current configuration. λ_{xx} and λ_{yy} are respectively the dilatation in the x and in the y direction.

7.3 Results

These experiments were carried out on skin samples from 1 month old WT (n = 25) mice. Similar mechanical assays were also performed without multiphoton imaging in a set of other mice skin samples to improve the statistics of mechanical data (n = 7). No significant difference was observed on the mechanical properties (tangent modulus, length of the heel region, ultimate tensile stress and failure stretch ratio) between experiment under multiphoton microscope (with relaxation during imaging) and outside the microscope (without relaxation), meaning that the pauses during SHG imaging did not strongly affect the mechanical behavior of dermis at this loading rate.

SHG images revealed a meshwork of collagen fibrillar structures. Upon traction, the collagen meshwork increasingly aligned along the traction direction and the follicle dark regions became more elliptical (Fig. 7.2), as expected [18, 32].

We first processed the SHG images to measure the local deformation tensor. Figure 7.3 shows the average value of these tensor components in the SHG image as a function of the global stretch ratio λ applied to the tissue for the same sample as in Fig. 7.2. The tensor component along the traction direction, λ_{xx}, exhibited a linear behavior with a slope $A_{xx} \approx 1$ 1 and a slope intercept $\lambda_{xx} \approx 1$. It proved that there was no sliding of the sample between the jaws and showed that the sample was homogeneous at the ROI scale (around 300 x 300 μm^2). The behavior along the y-direction, transverse to the traction, was more complex. It first slightly increased, which corresponded to a surface increase and would be associated to a negative Poisson ratio in linear materials. This peculiar behavior was likely due to water absorption at the beginning of the stretching [2], although structural effects may also explain it. Then it exhibited a linear contraction, for higher stretch ratios. Finally, there was no observable shear at the scale of the ROI, the average sliding angle remaining small (<5deg) throughout the experiment. These results were highly reproducible for all the analyzed skin samples, which showed A_{xx} values close to 1 with a good fitting quality (R2 parameter), and exhibited a bimodal behavior for λ_{yy}.

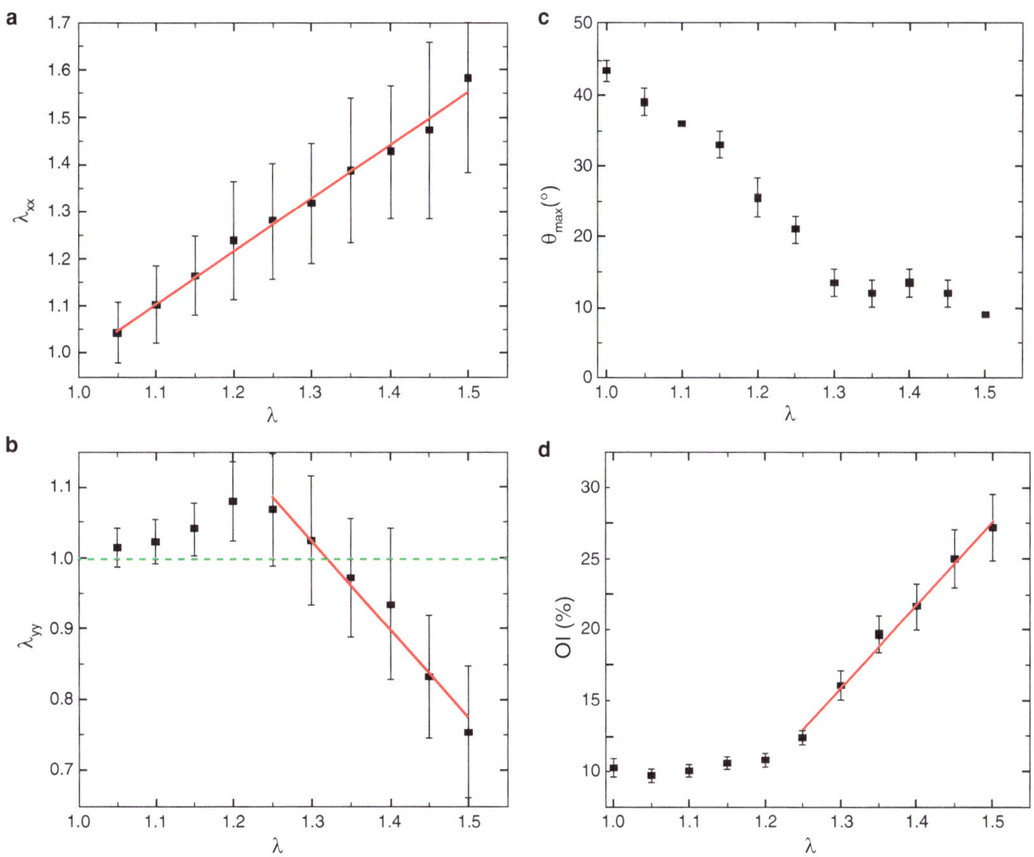

Fig. 7.3 Local deformation and microstructural remodeling of skin sample. (**a**) λ_{xx}, (**b**) λ_{yy} components of the local deformation tensor measured from the centers of the hair follicles motion. x corresponds to the direction of traction. *Red lines* are linear fitting. (**c–d**) Collagen reorganization probed by SHG imaging: (**c**) Main orientation θ_{max} of the collagen fibers, (**d**) Orientation Index (OI) of the collagen fiber histograms. These points are obtained by averaging over six consecutive images (about 10 μm of depth). *Red lines* are linear fitting. Figure adapted from [24]

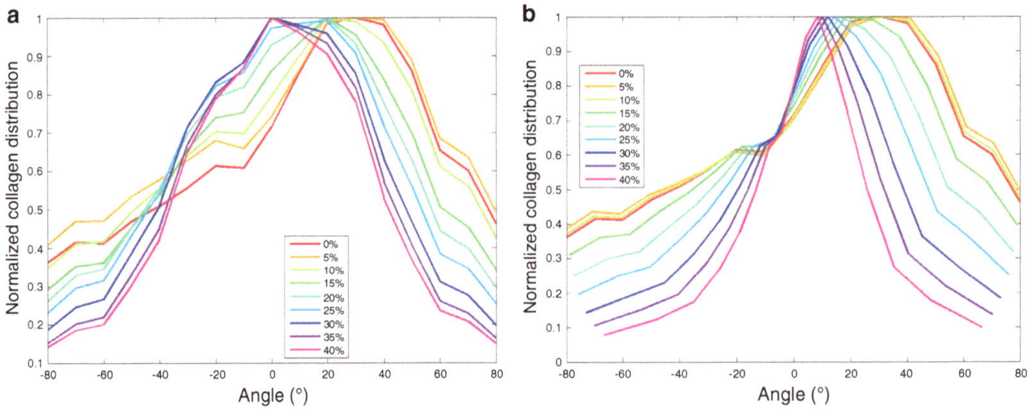

Fig. 7.4 Experimental (**a**) and theoretical (**b**) fibers orientation histogram and theoretical ones. The experimental data come from the analysis of the SHG image at increasing stretch level; the theoretical histograms are obtained from the initial experimental distribution and the strain measured through the motion of the centers of the hair follicles, under the assumption of affine transformation

Collagen reorganization was quantified by extracting collagen orientation from SHG images at every deformation step. The initial histogram shows usually two wide peaks, or one wide peak with a shoulder, which seems to indicate pretension lines in murine skin similar to Langer's lines in humans [34]. These orientation histograms clearly showed that the collagen network rearranged during the traction assay from a more or less disorganized state (Fig. 7.2, stretch ratio 1.0 initial configuration), to a peaked distribution (Fig. 7.2, stretch ratio 1.5) where the majority of fibers are parallel to the traction direction. Figure 7.3 display the maximal angle and the OI of the tissue as a function of the global stretch ratio for the same skin sample as in Fig. 7.2. As expected, the main orientation angle tended to 0. Moreover, the OI increased with the global stretch ratio, which indicated that more and more fibers were aligned in the main direction (Fig. 7.3d). Interestingly, the reorganization of the collagen fibers in the tissue appeared as a bimodal process. First the main orientation went to 0 with only a slight increase of OI. Then θ_{max} was stable and the tissue better organized locally as indicated by the strong, almost linear, increase in OI.

Based on an affine assumption [33], it is possible to predict a theoretical histogram, based on the knowledge of the strain and of the initial histogram. Figure 7.4 show an example of such computation: sub-figure (a) shows the experimental observation, while (b) gives the theoretical one. One notices that, at low and high stretch the distributions are rather similar. However, at intermediate stretches, the experiment and the model differ significantly.

Finally, we plotted the OI variation with stretch ratio in the same graph as the stress/stretch curve (Fig. 7.4). We observed that both parameters exhibited the same behavior, although OI quantifies collagen remodeling at microscopic scale and stress measures the mechanical response at macroscopic scale. This result was consistent for all the samples.

7.4 Discussion

Advanced investigations of skin biomechanics have been impeded so far by technical limitations since it requires simultaneous measurements of the mechanical response at macroscopic scale and of the tissue reorganization at microscopic scale, which is highly challenging in such a complex tissue. To overcome this bottleneck, we developed an experimental set-up combining a traction device with a SHG microscope and we imaged in situ the collagen network in stretched murine skin up to skin rupture. All these measurements were performed in *ex vivo* murine skin.

Our multiscale approach thus enables robust characterization of skin biomechanics. It provides quantitative information on the collagen reorganization at micrometer scale and stress at macroscopic scale as a function of controlled stretch measured at a scale of few hundredth micrometers (the resolution being limited by the distance between hair follicles).

Our measured stress/stretch relationship of the skin is similar to previously reported [2–4]: a heel region followed by a linear part before breaking. Our approach distinctively allows a quantitative challenge of the microstructural origin of the stress/stretch curve, which classically associates the heel region to an alignment of the fibers and the linear one to the stretching of the fibers. Extrapolating this interpretation in term of fibers orientation, we should observe an increase of the OI in the heel region, followed by a plateau (or a reduced increase) in the linear region, and a decrease after breaking. Our OI measurements did not sustain this interpretation. The evolution of the OI with the stretch ratio (Fig. 7.5) indicates that the

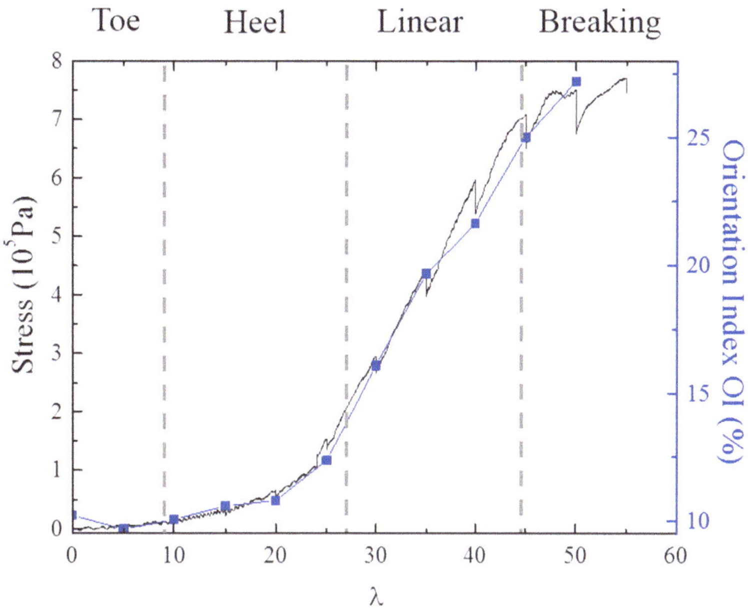

Fig. 7.5 Nominal stress (*black*) and Orientation Index (*blue*) versus the global stretch ratio. Figure adapted from [24]

heel region is associated to only a very partial alignment of the fibers [24, 32] and that the linear region corresponds to the maximal alignment of the collagen fibers, as also observed on the orientation histograms (Fig. 7.4).

The increase of the stress in the linear part might thus be associated to the increase of the fraction of fibers aligned in the direction of traction instead of the stretching of the fibers. In that case, each aligned fiber would provide a given stress to the tissue, whatever the stretch ratio is. Such behavior remains a plastic deformation of the fibers, and could originate from the sliding of fibers as reported in tendon [16]. It might be that sliding occurs above a critical value of load, associated with breaking of cross-bridges and that each stretched fiber contributes only by this critical load to the total stress applied to the sample.

It is also interesting to notice that our observations allow to challenge the classical affine assumption, which is central in most microstructural models. Our results indicate that this assumption is not able to capture properly the reorganization of the collagen fibers, especially for intermediate stretches. This indicates that microstructural interpretations of the collagen reorganization should take into account either structural effects of the collagen network or viscous behavior in the interaction between the fibrils and the surrounding medium.

In conclusion, the approach developed here allows a careful and robust analysis of the changes in microstructural organization of the dermis as a response to mechanical loading, challenging the classical interpretation. It shows a strong correlation between the macroscopic force and the fraction of aligned fibers, which might indicate that each fiber provides only a given force. This work also provides an efficient tool to investigate the biomechanics of collagen-rich tissues in normal and pathological context and to guide tissue engineering with appropriate biomechanical responses

Acknowledgements This work was supported by grants from Ecole Polytechnique (interdisciplinary project), from Agence Nationale de la Recherche ($ANR-13-BS09-0004-02$) and from Association du Syndrome Ehlers-Danlos (AFSED).

References

1. Ricard-Blum, S., Ruggiero, F.: The collagen superfamily: from the extracellular matrix to the cell membrane. Pathol. Biol. **53**(7), 430–442 (2005)
2. Veronda, D.R., Westmann, R.A.: Mechanical characterization of skin finite deformations. J. Biomech. **3**(1), 111–124 (1970)
3. Fung, Y.: Biomechanics. Mechanical Properties of Living Tissues. Springer, New York (1993)
4. Xu, F., Lu, T.J., Seffen, K.A.: Biothermomechanical behavior of skin tissue. Acta Mech. Sinica **24**(1), 1–23 (2008)
5. Sokolis, D.P., Kefaloyannis, E.M., Kouloukoussa, M., Marinos, E., Boudoulas, H., Karayannacos, P.E.: A structural basis for the aortic stress strain relation in uniaxial tension. J. Biomech. **39**(9), 1651–1662 (2006)
6. Zeinali-Davarani, S., Chow, M.-J., Turcotte, R., Zhang, Y.: Characterization of biaxial mechanical behavior of porcine aorta under gradual elastin degradation. Ann. Biomed. Eng. **41**(7), 1528–1538 (2013)
7. Abrahams, M.: Mechanical behaviour of tendon in vitro. Med. Biol. Eng. **5**(5), 433–443 (1967)

8. Screen, H., Toorani, S., Shelton, J.: Microstructural stress relaxation mechanics in functionally different tendons. Med. Eng. Phys. **35**(1), 96–102 (2013)
9. Houssen, Y.G., Gusachenko, I., Schanne-Klein, M.-C., Allain, J.-M.: Monitoring micrometer-scale collagen organization in rat-tail tendon upon mechanical strain using second harmonic microscopy. J. Biomech. **44**(11), 2047–2052 (2011)
10. Boyce, B., Jones R., Nguyen T., Grazier, J.: Stress-controlled viscoelastic tensile response of bovine cornea. J. Biomech. **40**(11), 2367–2376 (2007)
11. Benoit, A, Latour, G., Schanne-Klein, M.-C., Allain, J.-M.: Simultaneous microstructural and mechanical characterization of human corneas at increasing pressure. J. Mech. Behav. Biomed. Mater. **60**, 93–105 (2016)
12. Franchi, M., Fini, M., Quaranta, M., De Pasquale, V., Raspanti, M., Giavaresi, G., Ottani, V., Ruggeri, A.: Crimp morphology in relaxed and stretched rat Achilles tendon. J. Anat. **210**(1), 1–7 (2007)
13. Hansen, K.A., Weiss, J.A., Barton, J.K.: Recruitment of tendon crimp with applied tensile strain. J. Biomech. Eng. **124**(1), 72 (2002)
14. Gusachenko, I., Tran, V., Houssen, Y., Allain, J.-M., Schanne-Klein, M.-C.: Polarization-resolved second-harmonic generation in tendon upon mechanical stretching. Biophys. J. **102**(9), 2220–2229 (2012)
15. Cheng, V.W.T., Screen, H.R.C.: The micro-structural strain response of tendon. J. Mater. Sci. **42**(21), 8957–8965 (2007)
16. Puxkandl, R., Zizak, I., Paris, O., Keckes, J., Tesch, W., Bernstorff, S., Purslow, P., Fratzl, P.: Viscoelastic properties of collagen: synchrotron radiation investigations and structural model. Philos. Trans. R. Soc. B **357**(1418), 191–197 (2002)
17. Gupta, H., Seto, J., Krauss, S., Boesecke, P., Screen, H.: In situ multi-level analysis of viscoelastic deformation mechanisms in tendon collagen. J. Struct. Biol. **169**(2), 183–191 (2010)
18. Brown, I.A.: Scanning electron-microscope study of effects of uniaxial tension on human skin. Br. J. Dermatol. **89**(4), 383–393 (1973)
19. Belkoff, S.M., Haut, R.C.: A structural model used to evaluate the changing microstructure of maturing rat skin. J. Biomech. **24**(8), 711–720 (1991)
20. Lanir, Y., Lokshin, O.: Viscoelasticity and preconditioning of rat skin under uniaxial stretch: microstructural constitutive characterization. J. Biomech. Eng. **131**(3), 031009 (2009)
21. Jor, J.W.Y., Nash, M.P., Nielsen, P.M.F., Hunter, P.J.: Estimating material parameters of a structurally based constitutive relation for skin mechanics. Biomech. Model. Mechanobiol. **10**(5), 767–778 (2010)
22. Groves, R.B., Coulman, S.A., Birchall, J.C., Evans, S.L.: An anisotropic, hyperelastic model for skin: experimental measurements, finite element modelling and identification of parameters for human and murine skin. J. Mech. Behav. Biomed. Mater. **18**, 167–180 (2013)
23. Tonge, T.K., Voo, L.M., Nguyen, T.D.: Full-field bulge test for planar anisotropic tissues: Part II A thin shell method for determining material parameters and comparison of two distributed fiber modeling approaches. Acta Biomater. **9**(4), 5926–5942 (2013)
24. Bancelin, S., Lynch, B., Bonod-Bidaud, C., Ducourthial, G., Psilodimitrakopoulos, S., Dokladal, P., Allain, J.-M., Schanne-Klein, M.-C., Ruggiero, F.: Ex vivo multiscale quantitation of skin biomechanics in wild-type and genetically-modified mice using multiphoton microscopy. Sci. Rep. **5**, 17635 (2015)
25. Pena, A.-M., Strupler, M., Boulesteix, T., Godeau G., Schanne-Klein, M.-C.: Spectroscopic analysis of keratin endogenous signal for skin multiphoton microscopy. Opt. Express **13**(16), 6268–6274; Erratum: **13**(17), 6667 (2005)
26. Zipfel, W.R., Williams, R.M., Webb, W.W.: Nonlinear magic: multiphoton microscopy in the biosciences. Nat. Biotechnol. **21**(11), 1369–1377 (2003)
27. Strupler, M., Pena, A.-M., Hernest, M., Tharaux, P.-L., Martin, J.-L., Beaurepaire, E., Schanne-Klein, M.-C.: Second harmonic imaging and scoring of collagen in fibrotic tissues. Opt. Express **15**(7), 4054–4065 (2007)
28. Keyes, J.T., Haskett, D.G., Utzinger, U., Azhar, M., Geest, J.P.V.: Adaptation of a planar microbiaxial optomechanical device for the tubular biaxial microstructural and macroscopic characterization of small vascular tissues. J. Biomech. Eng. **133**(7), 075001 (2011)
29. Keyes, J.T., Lockwood, D.R., Simon, B.R., Vande Geest, J.P.: Deformationally dependent fluid transport properties of porcine coronary arteries based on location in the coronary vasculature. J. Mech. Behav. Biomed. Mater. **17**, 296–306 (2013)
30. Mauri, A., Perrini, M., Mateos, J., Maake, C., Ochsenbein-Koelble, N., Zimmermann, R., Ehrbar, M., Mazza, E.: Second harmonic generation microscopy of fetal membranes under deformation: normal and altered morphology. Placenta **34**(11), 1020–1026 (2013)
31. Mauri, A., Ehret, A.E., Perrini, M., Maake, C., Ochsenbein-Koelble, N., Ehrbar, M., Oyen, M.L., Mazza E.: Deformation mechanisms of human amnion: quantitative studies based on second harmonic generation microscopy. J. Biomech. **48**, 1606–1613 (2015)
32. Nesbitt, S., Scott, W., Macione, J., Kotha, S.: Collagen fibrils in skin orient in the direction of applied uniaxial load in proportion to stress while exhibiting differential strains around hair follicles. Materials **8**(4), 1841–1857 (2015)
33. Fan, R., Sacks, M.S.: Simulation of planar soft tissues using a structural constitutive model: finite element implementation and validation. J. Biomech. **47**(9), 2043–2054 (2014)
34. Langer, K.: Anatomy and physiology of skin 2. skin tension. Br. J. Plast. Surg. **31**, 93–106 (1978)

Chapter 8
A Numerical Study of a Biaxial Sollicitation to Set-Up the Displacement Field Measurement of Ex Vivo Mouse Skin

J.-S. Affagard, F. Wijanto, R. Rubio Amador, C. Bonod-Bidaud, F. Ruggiero, and J.-M. Allain

Abstract The behavior of the skin is an important issue in many applications, but is always not well understood. Nevertheless, skin is well-known to be mainly composed of collagen and has a very hierarchical microstructure that influences its mechanical behavior at different scales. To correlate and to characterize the macroscopic behavior with the microstructure, a bi-axial tensile test coupled with a macroscopic (digital image correlation) and microscopic (second harmonic generation) measurement has been developed. This study aims at presenting the method used to identified the mechanical properties of mice ex vivo skin. A full-field displacement is firstly acquired using digital image correlation (DIC) and, then, a specific finite element model is used to optimize the mechanical properties. The skin is modelled with a Holzapfel hyperelastic behavior which presents the particularity of being strongly based on the microstructure. The purpose of this article is to prepare the identification process. To this aim, a sensitivity approach is developed to choose the configuration with the highest sensitivity. Finally, the displacement and strain field is measured with DIC.

Keywords Biomechanics • Biaxial traction • Mice skin • Digital image correlation (DIC) • Anisotropy

8.1 Introduction

The behavior of the skin is not always well understood, but is an important issue in many applications where the understanding of soft tissues appears as quite important in order to better understand the organs behaviors in the physiological or pathological context. For instance, the design of devices such as car seats, wheelchairs or razors where stress distribution could be an ergonomic indicator.

The understanding of the phenomena involved can be difficult to analyze through in vivo mechanical tests. For this main reason, ex vivo experiments where the mechanical stresses are controlled, are much more frequent. Ex vivo studies can be separated in two categories: uniaxial traction and multiaxial sollicitation.

Firstly, with a monotonic tension test, basic biomechanical performances such as elastic modulus and ultimate tensile strength were characterized [1–4]. To go further in the skin mechanical behavior description, the viscoelastic properties were characterized through various strain rates [5–7], relaxation or creep test [8]. Finally, the preconditioning and fatigue were assessed thanks to uniaxial cyclic loading [9]. The multiaxial sollicitation are more relevant to characterize the influence of the microstructure [10, 11]. Indeed, in [12, 13] a biaxial tension was compared with a uniaxial traction. The results highlighted the relaxation differences induced by the transverse loading and a structural theory of the relationships in flat collagenous tissues was then proposed [14]. In [15], a multi-axial experiment was designed to characterise the two dimensional elastic properties of human skin, and, to go further in the mechanical behavior characterisation, from uniaxial and multiaxial sollicitation, to better describe the heterogeneities, the displacement field is measured. In [3, 16] the deformation field is measured on skin tissue. Before analysis, a fine grid is drawn to obtain the displacement field. The results show strong non-linearities from the reorganization of the collagen fibers during the test. In [17, 18], the displacement field was quantified from a uniaxial test to identify the Young's modulus anisotropy of rat and human skin. Nevertheless, uniaxial testing provides few information on anisotropy and it is therefore necessary to develop multi-axial test.

To identify a behavior close to the anisotropic one proposed in [19], a multiaxial tests were performed and coupled with a cross correlation technique [20]. In multiaxial tests, and more particularly in biaxial tests, the difficulties associated with

J.-S. Affagard (✉) • F. Wijanto • J.-M. Allain
Laboratoire de Mécanique des Solides (LMS), Ecole Polytechnique, CNRS, Université Paris-Saclay, 91128 Palaiseau, France
e-mail: affagard@lms.polytechnique.fr

R. Rubio Amador • C. Bonod-Bidaud • F. Ruggiero
Institut de Génomique Fonctionnelle de Lyon, ENS-Lyon, CNRS, Université Lyon 1, Lyon, France

C.S. Korach et al. (eds.), *Mechanics of Biological Systems and Materials, Volume 6*,
Conference Proceedings of the Society for Experimental Mechanics Series, DOI 10.1007/978-3-319-41351-8_8

geometry and structure of the specimens were highlighted in the literature. Indeed, in [21, 22] the sensitivity of a biaxial test with suture attachment points (spacing and alignment) which can lead to stress concentration was highlighted in order to optimize the stress distribution. In [23, 24], finite element simulations of a biaxial clamp experiment were performed to quantify the stress-shielding. This information allowed to formulate a correction factor. An illustration of the dependence on sample geometry and material anisotropy was showed. In [25], a FEMU approach was developed on sheet steel metal to numerically compare the identification accuracy between several biaxial tests. In [26], an original approach based on the sensitivity analysis was proposed to design and to optimize the shape of a biaxial clamp test to identify the elastoplastic properties of stainless steel.

In this present study, the challenge is to choose the best configuration to identify the Holzapfel mechanical properties [19]. A sensitivity analysis is first performed to illustrate the identification accuracy on an equibiaxial test and on biaxial test with an alternative loading. Next, the sensitivity of the experimental configuration is observed and a displacement field is measured thanks to DIC (Digital Image Correlation).

8.2 Materials and Methods

8.2.1 Experimental Protocol

The biaxial test (Fig. 8.1a) was adapted from the custom-made setup published in [27]. Ex vivo mouse skin samples were obtained from 1 month old mice with the epidermis and the hypodermis being previously removed. The cutting is performed with a punch having a radius of 14 or 16 mm (Fig. 8.1b) depending the sample size. These dermal samples are then subjected to a biaxial tension. The sample dimensions are measured using digital calipers.

The skin was then covered with graphite powder to create a speckle pattern enabling the measurement of the displacement field with DIC (Digital Image Correlation) (Fig. 8.1c). Each arm of the skin sample was afterward inserted and fixed in the holding jaws. Each jaw was itself screwed onto a force sensor. The force sensor was mounted using a custom made plate (Fig. 8.1c) on a motor.

Displacement fields are measured by DIC [28–30] between image pairs. The images were acquired with an 8-bit CCD camera (resolution: 4384x6576 pixels) every 5 s. A telecentric lens was mounted on the camera (Opto engineering TC16M036) allowing a pixelsize of 5. 5 μm. The displacement field measurement was performed with CMV software (CorrelManuV [28]).

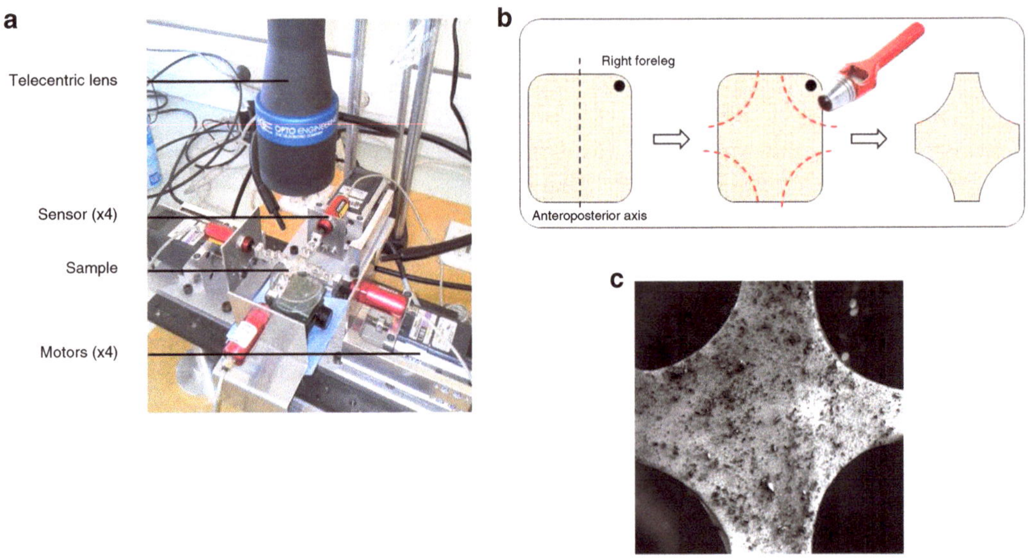

Fig. 8.1 (**a**) Custom-made device. (**b**) Illustration of the sample cutting procedure. (**c**) Sample covered with graphite powder for the measurement of the displacement field

DIC aims at measuring the displacement field between two images by comparing the gray level between subimages (≈ 128 pixels). By considering f and g a reference and a deformed image with $u(x)$ the in-plane displacement at the reference position x. The gray level conservation can be written as:

$$g(x + u(x)) = f(x) + b(x) \tag{8.1}$$

where $b(x)$ is the noise induced by the image acquisition. In practice, an objective functional, Ξ, is minimized as a quadratic difference:

$$\Xi(u(x)) = \int [g(x + u(x)) - f(x)]^2 dx \tag{8.2}$$

In order to validate the displacement field the strain field was represented. Due to noise in the displacement field, the strain field was filtered by diffuse approximation [31] that consists on hang over the local solution with a polynomial approximation. A radius of 3 was chosen to represent the strain fields.

8.2.2 *Mesh Generation*

The computational modeling ran on Abaqus software [32] and was specifically built from the experimental geometry and from the full-field displacement. A custom routine (Matlab) was developed to export the position of the correlated points and to refine the mesh with the reference image previously automatically segmented (Fig. 8.2a).

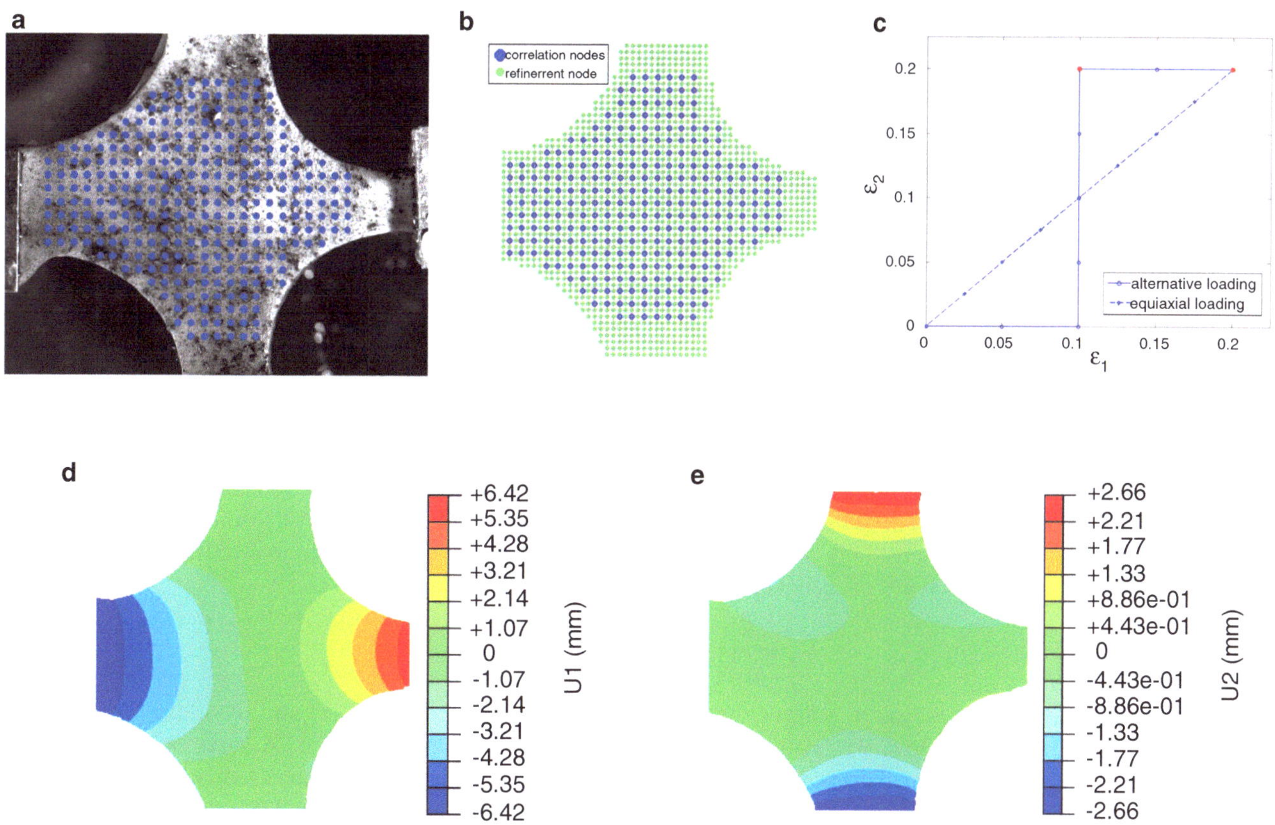

Fig. 8.2 (**a**) Image of the mouse skin with a representation of the measured displacement field position. (**b**) FE model with a refinement of 2 (i.e.: element side length of 86 pixels). (**c**) Investigated loading path. (**d**)–(**e**) FE displacement field respectively in the U_1 and U_2 direction for a imposed x-strain of 20 % and y-strain of 10 % with $C_{10} = 40$ kPa, $k_1 = 0.8$ kPa, $k_2 = 0.5$ kPa, $\kappa = 0.28$, $\alpha = 10°$

A 2D FE model of the biaxial experiment was developed, consider the plane stress assumption. The geometry of the mice skin was obtained with the reference image acquired during the DIC process. Then, the region of interest was segmented with Matlab using an automatic routine. A subset that includes all the nodes of the displacement field was created and then a refinement was performed (Fig. 8.2a, b). Thereafter, the 3-node linear mesh (CPS3) was created with matlab (202, 775 nodes). Figure 8.2d, e presents the simulated displacement respectively in the U_1 and U_2 direction for a imposed x-strain of 20 % and y-strain of 10 % with $C_{10} = 40$ kPa, $k_1 = 0.8$ kPa, $k_2 = 0.5$ kPa, $\kappa = 0.28$, $\alpha = 10°$.

The skin tissue was assumed to be homogeneous. The epidermis and hypodermis were removed before the experimental test and the residuals were included in the dermis. It was assumed that the incorporated tissues contributed to the average, in terms of hyperelastic properties.

8.2.3 *Constitutive Behavior*

The skin was assumed to be anisotropic and hyperelastic, governed by the assumption of large strain theory. A Holzapfel behavior was chosen to describe the mechanical response [19, 33] where the strain energy function was expressed as (8.3):

$$W = C_{10}(\overline{I}_1 - 3) + \frac{k_1}{2k_2}\sum_{\alpha=1}^{N}\left\{e^{k_2 <\overline{E}_\alpha>^2} - 1\right\} \tag{8.3}$$

where C_{10} describe the non-collagenous isotropic ground material behavior and k_1 and k_2 the contributions from the different families of collagen fibers. N is the number of fiber families ($N \leq 3$). In this study, the number of fiber families is chosen equal to 1. $\overline{I}_1$ is the first deviatoric strain invariant defined as: $\overline{I}_1 = J^{-2/3}I_1, J = det\left(\underline{\underline{F}}\right)$ and $I_1 = tr\left(F^T \cdot F\right)\left(\underline{\underline{F}}^T\underline{\underline{F}}\right)$ with $\underline{\underline{F}}$ the deformation gradient tensor. $\overline{E}_\alpha$, written as (8.4), is a strain-like quantity that characterizes the deformation of fibers in the mean direction. The operator $<.>$ is defined as $<x>=\frac{1}{2}(|x| + x)$, (i.e.: $<\overline{E}_\alpha>\geq 0$).

$$\overline{E}_\alpha = \kappa(\overline{I}_1 - 3) + (1 - 3\kappa)(\overline{I}_{4(\alpha\alpha)} - 1) \tag{8.4}$$

where the κ parameter describes the level of fiber dispersion along the mean direction with ($0 \leq \kappa \leq 1/3$). When $\kappa = 0$ the fibers are perfectly aligned and when $\kappa = 1/3$ the fibers are randomly distributed and the material is isotropic. The pseudo-invariant $\overline{I}_{4(\alpha\alpha)}$ represents the square of the stretch ratio in the fiber direction α.

8.2.4 *Sensitivity Analysis*

To simulate the experimental loading, a displacement was imposed at the extremities of the arms. An alternated test was simulated where the loading steps was imposed at 5 % (Fig. 8.2c). The loading path was first a loading along the x-axis of 10 % ($\epsilon_1 = 0.1, \epsilon_2 = 0$), then a loading along the y-axis of 20 % ($\epsilon_1 = 0.1, \epsilon_2 = 0.2$), and finally a loading along the x-axis of 20 % ($\epsilon_1 = 0.2, \epsilon_2 = 0.2$). The sensitivity S of the displacement field and resultant forces, collectively denoted by P, quantifies the effect of a variation $\delta\theta_k$ of each parameter θ_k independently. S_{θ_k} is given by (8.5):

$$S_{\theta_k} = P(x, t, \theta_k + \delta\theta_k) - P(x, t, \theta_k) \tag{8.5}$$

The higher the sensitivity, the more influent the parameter on the affected quantities and therefore the easier the identification will be. The derivative was estimated from a finite difference approximation. A variation ($\delta\theta_k$) of 1 %, corresponding to the level of expected measurement noise, was chosen. The reference parameters were evaluated from [27] where the correlation between the microstructure and the mechanical behavior was deduced and [20] where an identification of an anisotropic hyperelastic behavior, close to Holzapfel one, was performed. The references parameters were: $C_{10} =$ 40 kPa, $k_1 = 0.8$ kPa, $k_2 = 0.5$ kPa, $\kappa = 0.28$, $\alpha = 10°$.

8.3 Results

8.3.1 Sensitivity for a Real Geometry

Conventionally, biaxial tests are carried out with a homogeneous loading on each axis [21–24]. This kind of loading reveled a lower sensitivity of almost parameters which led us to realize a heterogeneous loading. Figure 8.3 shows the displacement sensitivity for a real geometry where the radius and the arms are not identical. The results show a higher sensitivity closer to the center compared to the perfect geometry. Moreover, a higher maximal value is observed for each parameter ($max(k_2) > 10^{-3}$). In addition, the same influence is revealed for the force sensitivity (Fig. 8.4). Therefore, an unsymmetrical geometry revealed higher sensitivity of each Holzapfel parameter which implies a better identification of these parameters.

8.3.2 Full-Field Measurement

The results of the correlation for 3 different loading steps are presented in Fig. 8.5. The results of the measured displacement fields are consistent with an applied loading where a 10 % displacement along X is applied and an incremental displacement along Y is performed (0, 10, 20 %). Along the X axis, the first image (Fig. 8.5a) illustrates the 10 % traction. Then, the Poisson's ratio is highlighted (Fig. 8.5b, c). Along the direction of stretching (Y axis), the measured field is proportional between the loading (Fig. 8.5d, e, f) which validates the DIC measurement.

The strain fields are also shown (Fig. 8.6) and reveal that the larger strains are located close to the arms of the sample. ε_{yy} field is proportional between the loading (Fig. 8.6d, e, f) which corresponds well to the loading increment. However, the maximal value is lower than the 0. 2 as expected. This difference can be explained by the fact that displacement measurement points are far from the clamp and thus do not cover the whole arm of the specimen where the displacement field is more sensitive in the sample arms. ε_{xx} have a similar maximal value for the 3 loadings that illustrates the constant loading in this direction. Nevertheless, the strain field evolution might be due to Poisson's ratio. Finally, ε_{xy} reveals an inhomogeneous field. This shape might be explained by the radius or by the microstructure heterogeneities which can be useful in the parameter identification or the loading heterogeneities.

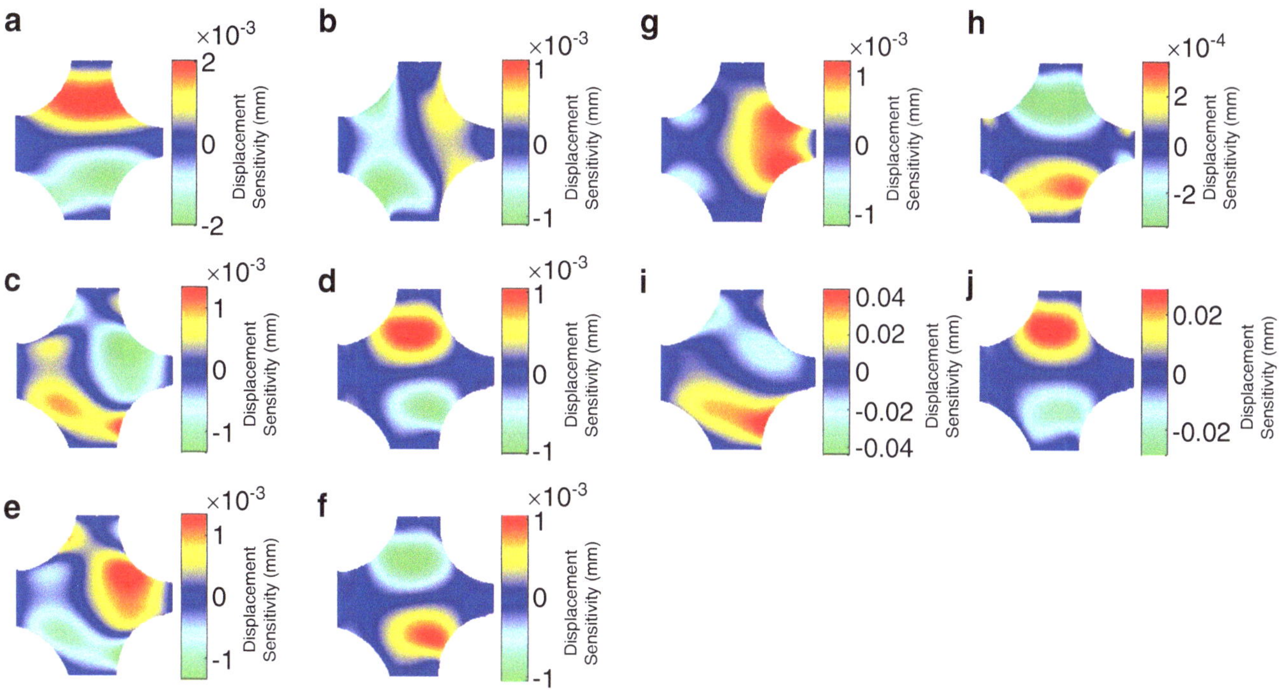

Fig. 8.3 Cartographies of the displacement sensitivity for the alternated loading experiment applied to a real geometry: (**a**) angle for U_1, (**b**) angle for U_2, (**c**) C_{10} for U_1, (**d**) C_{10} for U_2, (**e**) k_1 for U_1, (**f**) k_1 for U_2, (**g**) k_2 for U_1, (**h**) k_2 for U_2, (**i**) κ for U_1, (**j**) κ for U_2

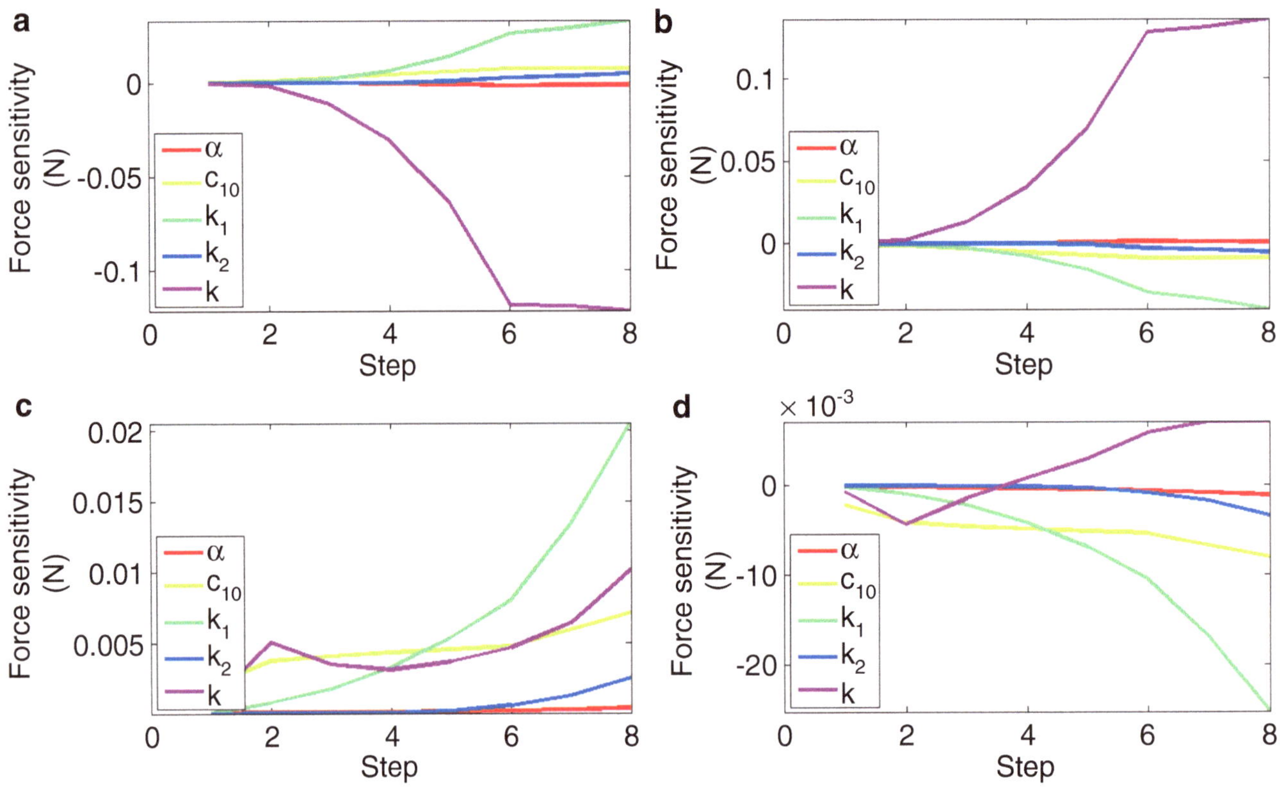

Fig. 8.4 Evolution of the sensitivity for the alternated experiment applied to a real geometry (**a**) Sensitivity along X for the right arm and (**b**) for the left arm. (**c**) Sensitivity along Y for the top arm and (**d**) bottom arm

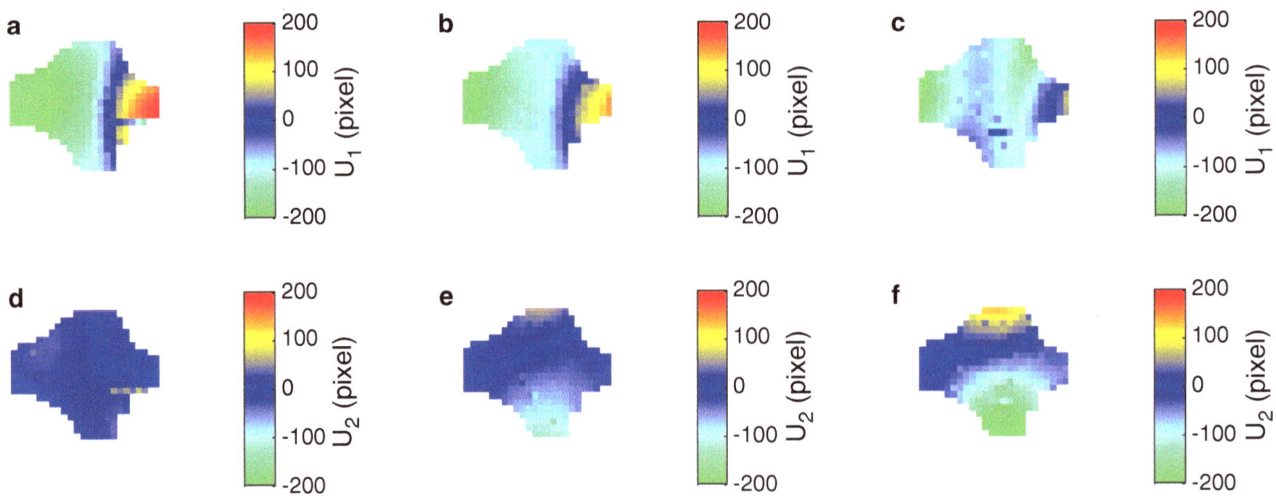

Fig. 8.5 Displacement field measurement in the X direction in the first line and in the Y direction in the second line, for an experimental loading of (**a**, **d**) 10–0 %, (**b**, **e**) 10–10 % and (**c**, **f**) 10–20 %

8.4 Discussion

In this work, a 2*D* simulation with in plane stress was carried out. This hypothesis of an uniformly thin thickness throughout the skin was performed and the out of plane stresses were considered null. A way to improve this 2*D* simulation would be to perform a full 3*D* model.

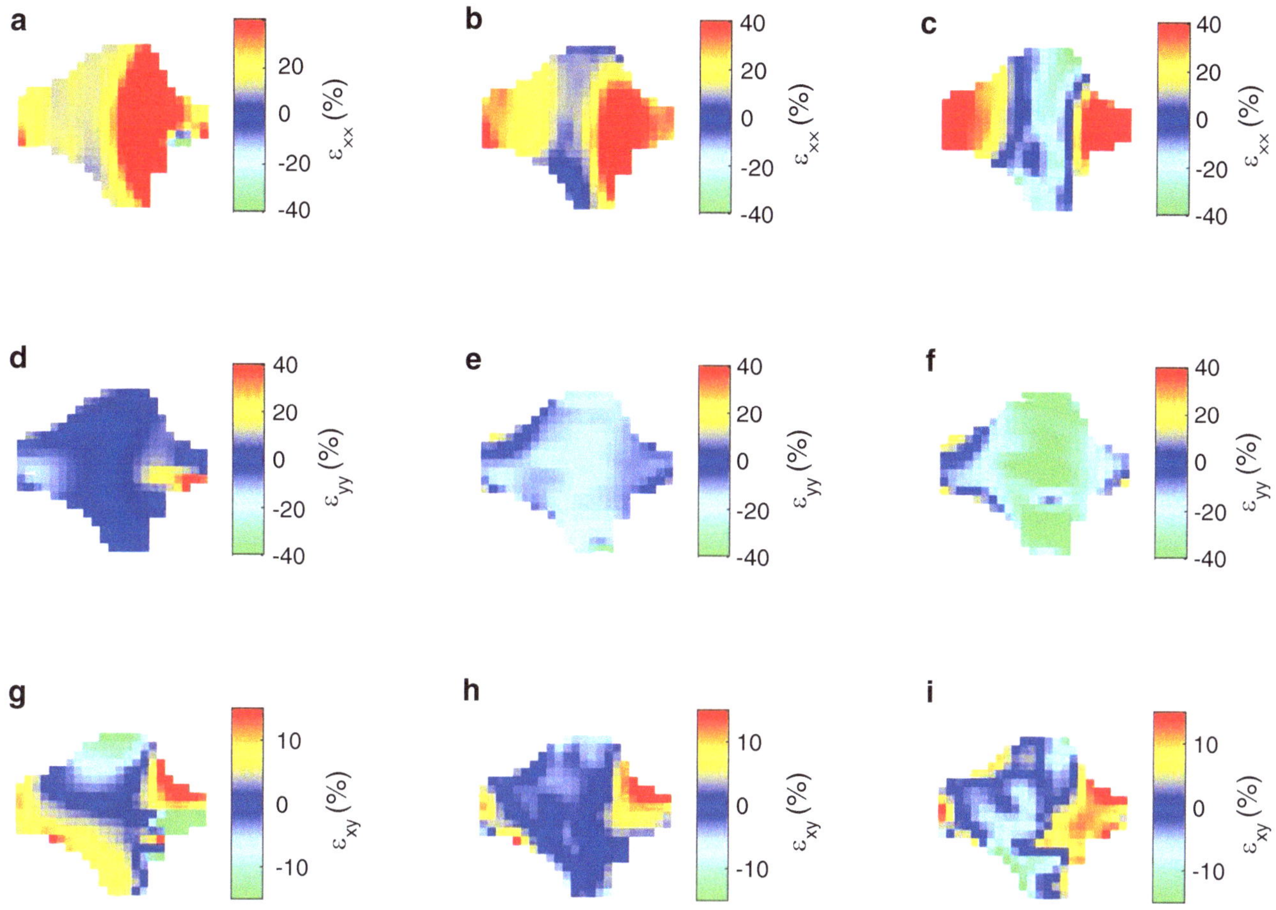

Fig. 8.6 ε_{xx} in the first line, ε_{yy} in the second and ε_{xy} in the third for an experimental loading of 10–0%, 10–10 % and 10–20 %

Skin tissue has often been considered as linear, elastic, and incompressible. Thereby, a Young's modulus or a tangential modulus was often characterized by a Poisson's ratio close to 0. 5 [1–9]. In the present study, a hyperelastic model, based on the microstructure, with consideration of the anisotropy was chosen. The choice of this Holzapfel model was performed in order to compare numerical simulation with microstructural because this model assumes an affine transformation [34].

The result of the sensitivity analysis showed a high dependence on the geometry and loading. An unsymmetric geometry and loading allowed to be very sensitive to all parameters of the Holzapfel model. For this reason, questions may arise concerning the accuracy of the identification of mechanical parameters in several studies. Nevertheless, it is still important to note that the choice of geometry and loading will be dependent on the adopted behavior. Therefore, this sensitivity analysis is essential to design experiments for an optimal parameter identification.

Concerning the displacement field measurement, the subimages are quite large ($\approx$ 128 pixels). This large area could be explained by the fact that the camera has a resolution smaller than the graphite powder size. Moreover, graphite powder is difficult to fix on mouse skin and in addition, at large scale, some graphite powder came off. Nevertheless the choice of graphite powder was performed because, unlike painting, graphite powder does not alter the sample.

8.5 Conclusion

This study proposed a sensitivity analyse as an approach to design the experiment (i.e. loading and shape). This methodology allows us to determine an alternative loading which is more efficient to characterize the Holzapfel properties. The next step will consist of identifying the Holzapfel parameters. Thus, the orientation index of fibers could be deduced and compared with SHG measurements.

Acknowledgements This work was supported by grants from Agence Nationale de la Recherche ($ANR - 13 - BS09 - 0004 - 02$) and from Association Française du Syndrome Ehlers-Danlos (AFSED).

References

1. Wilkes, G., Brown, I., Wildnauer, R.: The biomechanical properties of skin. CRC Crit. Rev. Bioeng. **1**(4), 453 (1973)
2. Jansen, L., Rottier, P.: Elasticity of human skin related to age. Dermatology **115**(2), 106–111 (1957)
3. Daly, C.H.: Biomechanical properties of dermis. J. Investig. Dermatol. **79**, 17–20 (1982)
4. Veronda, D., Westmann, R.: Mechanical characterization of skin finite deformations. J. Biomech. **3**(1), 111–124 (1970)
5. Iatridis, J.C., Wu, J., Yandow, J.A., Langevin, H.M.: Subcutaneous tissue mechanical behavior is linear and viscoelastic under uniaxial tension. Connect. Tissue Res. **44**(5), 208–217 (2003)
6. Silver, F.H., Freeman, J.W., De Vore, D.: Viscoelastic properties of human skin and processed dermis. Skin Res. Technol. **7**(1), 18–23 (2001)
7. Pan, L., Zan, L., Foster, F.S.: Ultrasonic and viscoelastic properties of skin under transverse mechanical stress in vitro. Ultrasound Med. Biol. **24**(7), 995–1007 (1998)
8. Ridge, M., Wright, V.: A bio-engineering study of the mechanical properties of human skin in relation to its structure. Br. J. Dermatol. **77**(12), 639–649 (1965)
9. Kang, G., Wu, X.: Ratchetting of porcine skin under uniaxial cyclic loading. J. Mech. Behav. Biomed. Mater. **4**(3), 498–506 (2011)
10. Waldman, S.D., Lee, J.M.: Effect of sample geometry on the apparent biaxial mechanical behaviour of planar connective tissues. Biomaterials **26**(35), 7504–7513 (2005)
11. Waldman, S.D., Lee, J.M.: Boundary conditions during biaxial testing of planar connective tissues. Part 1: dynamic behavior. J. Mater. Sci. Mater. Med. **13**(10), 933–938 (2002)
12. Lanir, Y., Fung, Y.: Two-dimensional mechanical properties of rabbit skin - i. experimental system. J. Biomech. **7**(1), 29–34 (1974)
13. Lanir, Y., Fung, Y.: Two-dimensional mechanical properties of rabbit skin - ii. experimental results. J. Biomech. **7**(2), 171–182 (1974)
14. Lanir, Y.: A structural theory for the homogeneous biaxial stress-strain relationships in flat collagenous tissues. J. Biomech. **12**(6), 423–436 (1979)
15. Reihsner, R., Balogh, B., Menzel, E.: Two-dimensional elastic properties of human skin in terms of an incremental model at the in vivo configuration. Med. Eng. Phys. **17**(4), 304–313 (1995)
16. Schneider, D.C., Davidson, T.M., Nahum, A.M.: In vitro biaxial stress-strain response of human skin. Arch. Otolaryngol. **110**(5), 329–333 (1984)
17. Guan, E., Smilow, S., Rafailovich, M., Sokolov, J.: Determining the mechanical properties of rat skin with digital image speckle correlation. Dermatology **208**(2), 112–119 (2004)
18. Annaidh, A.N., Bruyère, K., Destrade, M., Gilchrist, M.D., Otténio, M.: Characterization of the anisotropic mechanical properties of excised human skin. J. Mech. Behav. Biomed. Mater. **5**(1), 139–148 (2012)
19. Holzapfel, G.A., Gasser, T.C., Ogden, R.W.: A new constitutive framework for arterial wall mechanics and a comparative study of material models. J. Elast. Phys. Sci. Solids **61**(1–3), 1–48 (2000)
20. Jor, J.W., Nash, M.P., Nielsen, P.M., Hunter, P.J.: Estimating material parameters of a structurally based constitutive relation for skin mechanics. Biomech. Model. Mechanobiol. **10**(5), 767–778 (2011)
21. Eilaghi, A., Flanagan J.G., Brodland, G.W., Ethier, C.R.: Strain uniformity in biaxial specimens is highly sensitive to attachment details. J. Biomech. Eng. **131**(9), 091003 (2009)
22. Sacks, M.S.: Biaxial mechanical evaluation of planar biological materials. J. Elast. Phys. Sci. Solids **61**(1–3), 199–246 (2000)
23. Jacobs, N.T., Cortes, D.H., Vresilovic, E.J., Elliott, D.M.: Biaxial tension of fibrous tissue: using finite element methods to address experimental challenges arising from boundary conditions and anisotropy. J. Biomech. Eng. **135**(2), 021004 (2013)
24. Nolan, D., McGarry, J.: On the correct interpretation of measured force and calculation of material stress in biaxial tests. J. Mech. Behav. Biomed. Mater. **53**, 187–199 (2016)
25. Schmaltz, S., Willner, K.: Comparison of different biaxial tests for the inverse identification of sheet steel material parameters. Strain **50**(5), 389–403 (2014)
26. Bertin, M., Hild, F., Roux, S., Mathieu, F., Leclerc, H., Aimedieu, P.: Integrated digital image correlation applied to elasto-plastic identification in a biaxial experiment. J. Strain Anal. **51**(2), 118–131 (2016)
27. Lynch, B.: Étude du lien entre propriétés mécaniques macroscopiques et organisation microstructurale de la peau. 21ème Congrès Français de Mécanique, 26 au 30 août 2013, Bordeaux, France (FR), 2013
28. Bornert, M., Hild, F., Orteu, J.-J., Roux, S.: Digital image correlation (chapter 6). In: Grédiac M, Hild F (eds) Full-field measurements and identification in solid mechanics. Wiley-ISTE, pp 157–190 (2012)
29. Bornert, M., Orteu, J.-J., Roux, S.: Corrélation d'images. Mesures de champs et identification, 2011
30. Bornert, M., Brémand, F., Doumalin, P., Dupré, J.-C., Fazzini, M., Grédiac, M., Hild, F., Mistou, S., Molimard, J., Orteu, J.-J., et al.: Assessment of digital image correlation measurement errors: methodology and results. Exp. Mech. **49**(3), 353–370 (2009)
31. Feissel, P., Schneider, J., Aboura, Z., Villon, P.: Use of diffuse approximation on DIC for early damage detection in 3d carbon/epoxy composites. Compos. Sci. Technol. **88**, 16–25 (2013)
32. Systèmes, D.: Abaqus 6.14 online documentation. 2014, 2015
33. Gasser, T.C., Ogden, R.W., Holzapfel, G.A.: Hyperelastic modelling of arterial layers with distributed collagen fibre orientations. J. R. Soc. Interface **3**(6), 15–35 (2006)
34. Sacks, M.S.: Incorporation of experimentally-derived fiber orientation into a structural constitutive model for planar collagenous tissues. J. Biomech. Eng. **125**(2), 280–287 (2003)

Chapter 9
Dynamic Polarization Microscopy for In-Situ Measurements of Collagen Fiber Realignment During Impact

Xianyu Wu, Hsiao-Ying Shadow Huang, Mark Pankow, and Kara Peters

Abstract The long term goal of this work is to better understand the tendon-to-bone insertion injury due to medium strain rate impact (e.g. sports activity). Specifically, we imaged collagen fiber realignment during impact, to investigate the ability of the tendon-to-bone insertion to these survive harsh dynamic events. A polarized light microscopy (PLM) setup was built in the lab and used to monitor the birefringence property changes of a known material under changing stress conditions. Initially polycarbonate dogbone specimens were tested quasi-statically to validate the setup and analysis algorithm. Polarized light retardation and alignment direction images are generated to quantitatively analyze the birefringence property change under different stress and compared to theoretical predictions. To perform dynamic experiments a drop weight tower was modified for medium strain rate testing (10–100 %/s) and the PLM setup is being incorporated for imaging. Several dynamic experiments have been conducted using this modified drop tower on porcine tendon specimens. A high-speed camera is used to record their dynamic response and deformation.

Keywords Polarized light microscopy • Impact • Tendon-to-bone insertion • Drop tower • Mechanical testing

9.1 Introduction

Tendon tissue is flexible but strong fibrous connective tissue which attaches muscle to bone. The transition from tendon to bone is a complex region that transfers the load effectively to prevent injury and provide proper joint function [1]. The structure and composition of the tendon-to-bone insertion varies along its longitudinal direction gradually, as seen in Fig. 9.1. In the tendon-to-bone insertion, as in many soft connective biological tissues (i.e., tendon, ligaments, etc.), the mechanical properties are influenced by the collagen fiber alignment direction. The zone near the tendon side consists of collagen fibers with small amounts of the proteoglycan decorin and fibrocartilage [2]. The zone near the bone side contains mineralized fibrocartilage and predominant collagen [2]. Such tissues are highly orthotropic due to the fiber alignment directions. The insertion region contains collagen fibers that align with the loading direction and the architecture of collagen fibers affects the mechanical properties of the tissue.

Tendon-to-bone insertions are subjected to physiologic loading on a daily basis; however, these highly specialized tissues are often damaged from onetime events such as car accidents or sports activities. Based on the studies of human Achilles tendon and patellar tendon, it has been reported that the modulus and maximum stress for tendon can range from 500 to 1000 MPa and 60 to 100 MPa, respectively, depending on tissue organization, anatomical location, and species [3–8]. Critically, the failure stresses and strains are also highly strain-rate dependent [6, 8–10].

Surgical repair operations of tendon-to-bone insertion tears have a high failure rate. The most common example of tendon-to-bone insertion tears occur at the rotator cuff which is a large tendon connecting muscles around the upper end of the arm (Teres minor, Infraspinatus, supraspinatus, subscapularis) and the humerus. The failure rates in rotator cuff repair surgeries range between 11 and 95 % [11]. For massive rotator cuff tears, the failure rates for rotator cuff repair are as high as 94 % [12–14]. This high rate of failure is due to the fact that the tendon-to-bone insertion is not regenerated during the healing process [11]. Since the mechanical properties of the tendon and tendon-to-bone insertions are loading-rate sensitive and most injuries occur during impacting events. Therefore, understanding the high loading rate behavior and failure mechanisms of the insertion region will lead to better clinical repair techniques.

X. Wu (✉) • H.-Y.S. Huang • M. Pankow • K. Peters
North Carolina State University, Raleigh, NC, USA
e-mail: xwu13@ncsu.edu

C.S. Korach et al. (eds.), *Mechanics of Biological Systems and Materials, Volume 6*,
Conference Proceedings of the Society for Experimental Mechanics Series, DOI 10.1007/978-3-319-41351-8_9

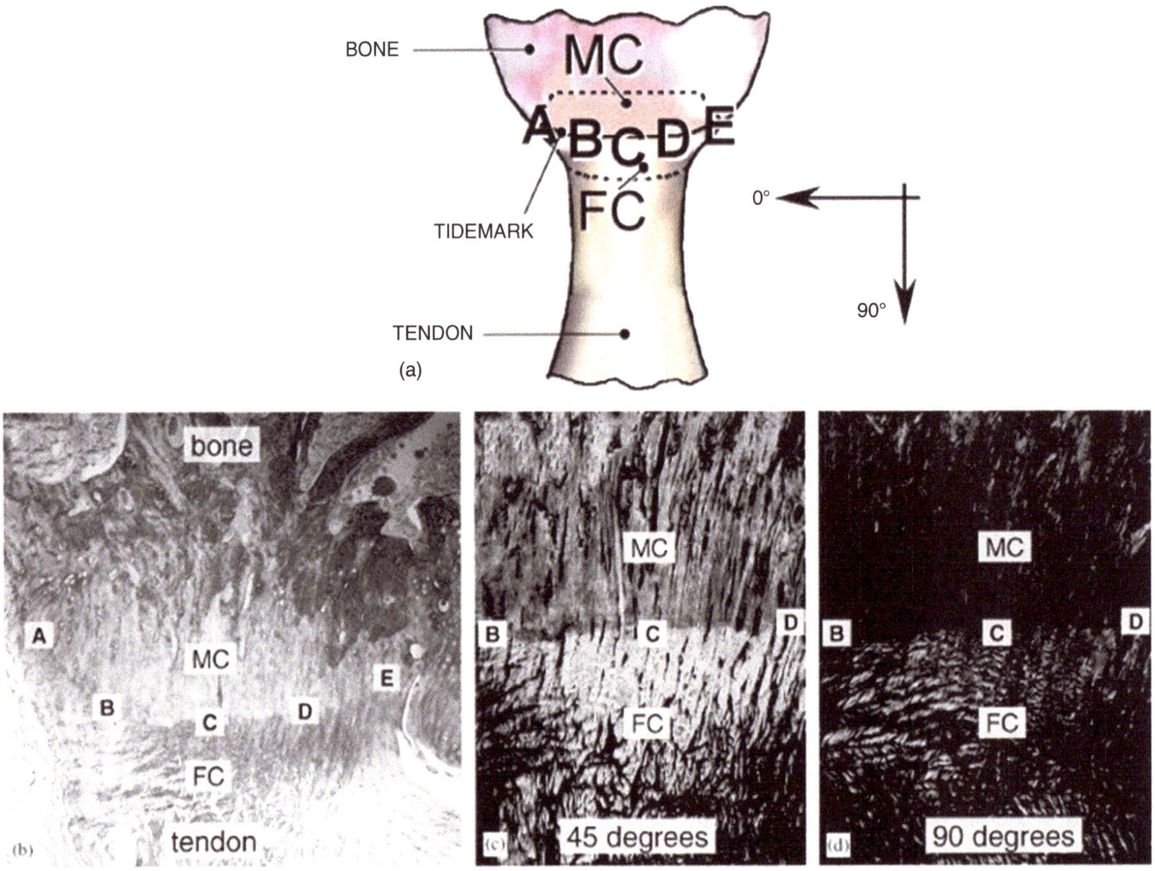

Fig. 9.1 (**a**) Tendon-to-bone insertion region. (**b**) Image of the insertion site under brightfield microscopy (10× objective). (**c**) The specimen at a higher magnification (20× objective) under polarized light microscopy. The tendon is oriented 45° relative to the orientation of the polarizer. (**d**) The tendon aligned with orientation of the polarizer [2]

Birefringence materials (i.e. anisotropic materials) have different refractive indices in different principal axes. Due to strong alignment of collagen fibers that commonly exists in soft connective tissue, the tendon-to-bone insertions have a linear birefringence response. The polarized light microscopy (PLM) technique can be used to image the collagen fiber (birefringence slow axis) alignment angles [15] due to the birefringence response allowing us to visualize the collagen fibers indirectly. The PLM technique is employed to image the collagen fiber realignment during the experiments due to its advantage of fast speed, high sensitivity, and high contrast of generated images.

The application of fast harmonic analysis method on PLM technique used to generate collagen fiber alignment maps was developed and demonstrated by Tower and Tranquillo [15, 16]. The optical setup is first described by Hauge [17], which uses a rotating quarter wave plate and a linear polarizer as the polarization state generator. A fast harmonic analysis method was used to generate alignment direction maps and polarization light retardation [15] when linear birefringence dominates the optical behavior. The tissue samples were modeled as linear birefringence materials due to strong alignment of collagen fibers that commonly exists in soft connective tissue. The advantage of using fast harmonic analysis method is its low computational cost [15]. The method had been demonstrated for a porcine heart valve leaflet [15].

The long term goal of this work is to better understand the tendon-to-bone insertion injury due to medium strain rate impact (e.g. sports activity), through a series of controlled dynamic experiments. Specifically, we imaged collagen fiber realignment during impact, to investigate the ability of the tendon-to-bone insertion to survive harsh impact events and understand the realignment process. The experimental data will provide important information regarding insertion failures and associated collagen fiber realignments.

9.2 Methods

The optic train diagram of the PLM setup used in this work is shown in Fig. 9.2 [15, 18]. The tissue sample is illuminated by a circular polarized light beam at changing azimuths, which is produced by rotating the polarizer. Due to the birefringence property of the tissue specimen, the transmitted light beam is elliptically polarized light. A fixed circular analyzer will result in elliptically polarized light, yielding a sinusoidal oscillation in the intensity of light passing through the circular analyzer. The oscillation's phase and amplitude are related to the tissue specimen's collagen fiber alignment direction (birefringence slow axis) and retardation, respectively. Harmonic image analysis techniques allow for the rapid generation of alignment direction and retardation images.

The incident non-polarized light is first polarized using a vertical linear polarizer. The polarized light is then fed through a quarter wave plate which causes the light to become elliptically polarized. The polarization state of the elliptically polarized light beam passed through the specimen is changed due to the birefringence properties of the specimen. Another quarter wave plate (slow axis at 45°) and a horizontal linear polarizer served as polarization state analyzer were placed at the other side of the sample. Images were acquired in transmission mode with a Grasshopper 2 CCD camera.

A PLM setup was built in the lab and quasi-static testing (Fig. 9.3) was used to monitor the birefringence property changes of a known material under changing stress conditions. In this setup the polarizer is rotated to 15 positions evenly spaced over 180° to acquire images. A series of images is then recorded as the sample is incrementally loaded. After taking the images they are passed through a MATLAB script which generates fiber alignment and retardation maps from the image sequence.

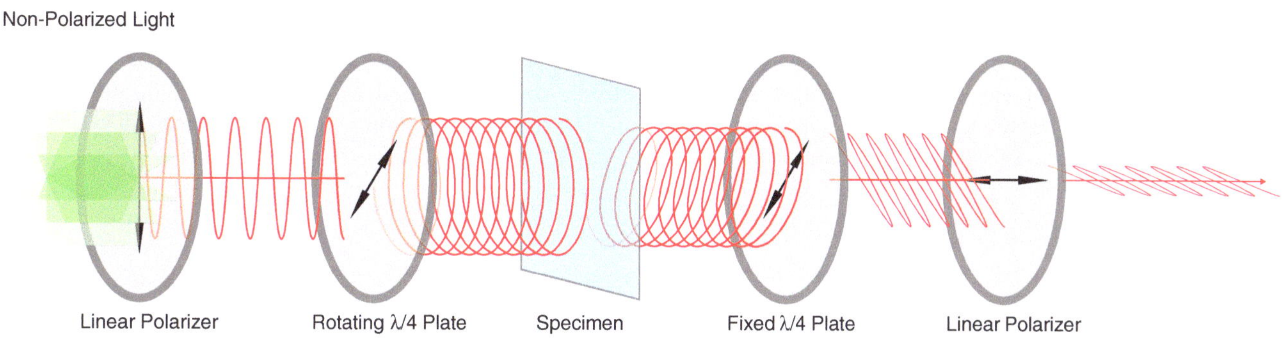

Fig. 9.2 Polarized light microscopy setup optic train diagram

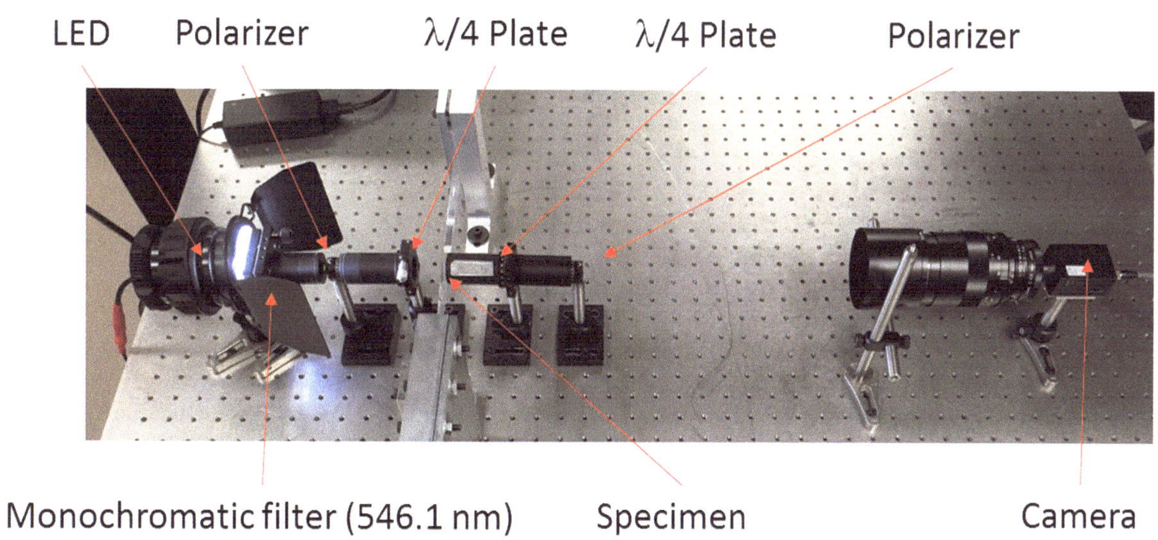

Fig. 9.3 Quasi-static test setup and self-built polarized light microscopy experimental setup

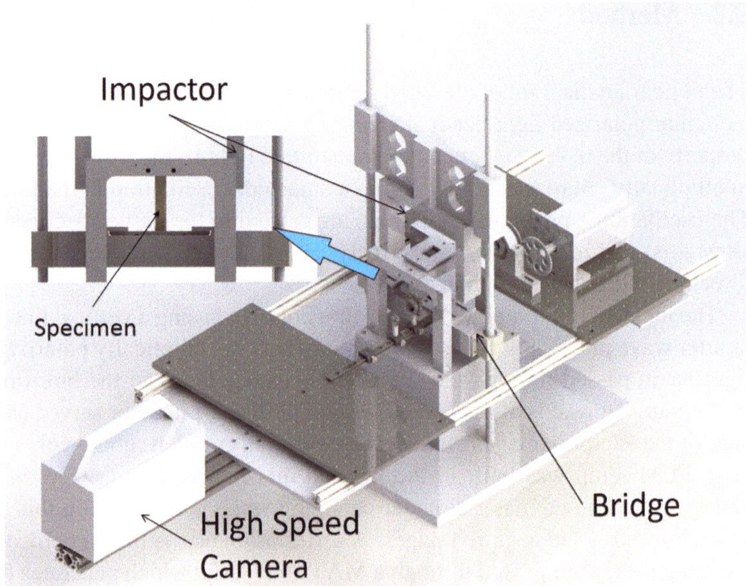

Fig. 9.4 Dynamic polarized light microscopy setup with optical elements incorporated into drop tower

To perform dynamic experiments a drop weight tower was modified for medium strain rate testing (10–100 %/s). Experiments are performed on porcine tendons to investigate their dynamic response and deformation. The PLM setup has been modified to operate at high frequency and synchronized with a high-speed camera, see Fig. 9.4, allowing for dynamic visualization of the tissue during dynamic deformation. The dynamic alignment maps will be correlated with the drop tower impacting load, which is measured using an accelerometer.

9.3 Results and Discussion

A 0.381 mm thick, 38.1 mm wide polycarbonate dogbone specimen was cut to test. A 1.651 mm diameter hole was cut at the center of the polycarbonate specimen to create a non-uniform stress field when the specimen was loaded on the test instrument. The specimen was mounted on the quasi-static testing instrument and loaded. The PLM setup was operated to generate an alignment image and retardation image for loaded polycarbonate dogbone specimens. Then, the stress applied on the polycarbonate specimen was increased slightly. The PLM setup was then operated to generate an alignment image and retardation image for the specimen under the new stress condition. These process was repeated for several times.

A series of images were recorded as the sample was incrementally loaded. Figure 9.5a, b show the alignment and retardation maps generated from one stage of loading. The vectors overlapped on top of the alignment maps were generated by averaging the corresponding alignment map over 10 × 10 pixel sub areas. Based on the stress field induced by uniform tension on a plate with a hole and the optic stress law, the experiment results matched the analytical expectation, validating the experimental setup and analysis code.

For the same polycarbonate specimen, Fig. 9.5c, d shown the alignment and retardation maps generated from another stage of loading when the strain applied on the specimen was increased by 0.2 %. The change of stress field on the specimen at the new stage resulted in the birefringence property change. Figure 9.5c shows that the slow axis alignment direction at the region above and on the right side of the hole. The retardation map also shows significant change at the region around the center hole, which is caused by the thickness change of the specimen.

Several dynamic experiments have been conducted using this modified drop tower on porcine tendon specimens. The middle part of the tendon was used for the tests. Figure 9.6 shows images of high-speed deformation of a tendon tissue sample recorded using a Photron FASTCAM SA-X2 high-speed camera (20 μm pixel size CMOS sensor). The high-speed camera was run at 12,500 frames per second with a 384 × 840 pixel resolution. The figures shown in Fig. 9.6b–d were cropped to be 179 × 326 pixels. The sequence of images shows how the sample deforms with respect to time. At the tendon surface, collagen fiber failure was observed after the tests.

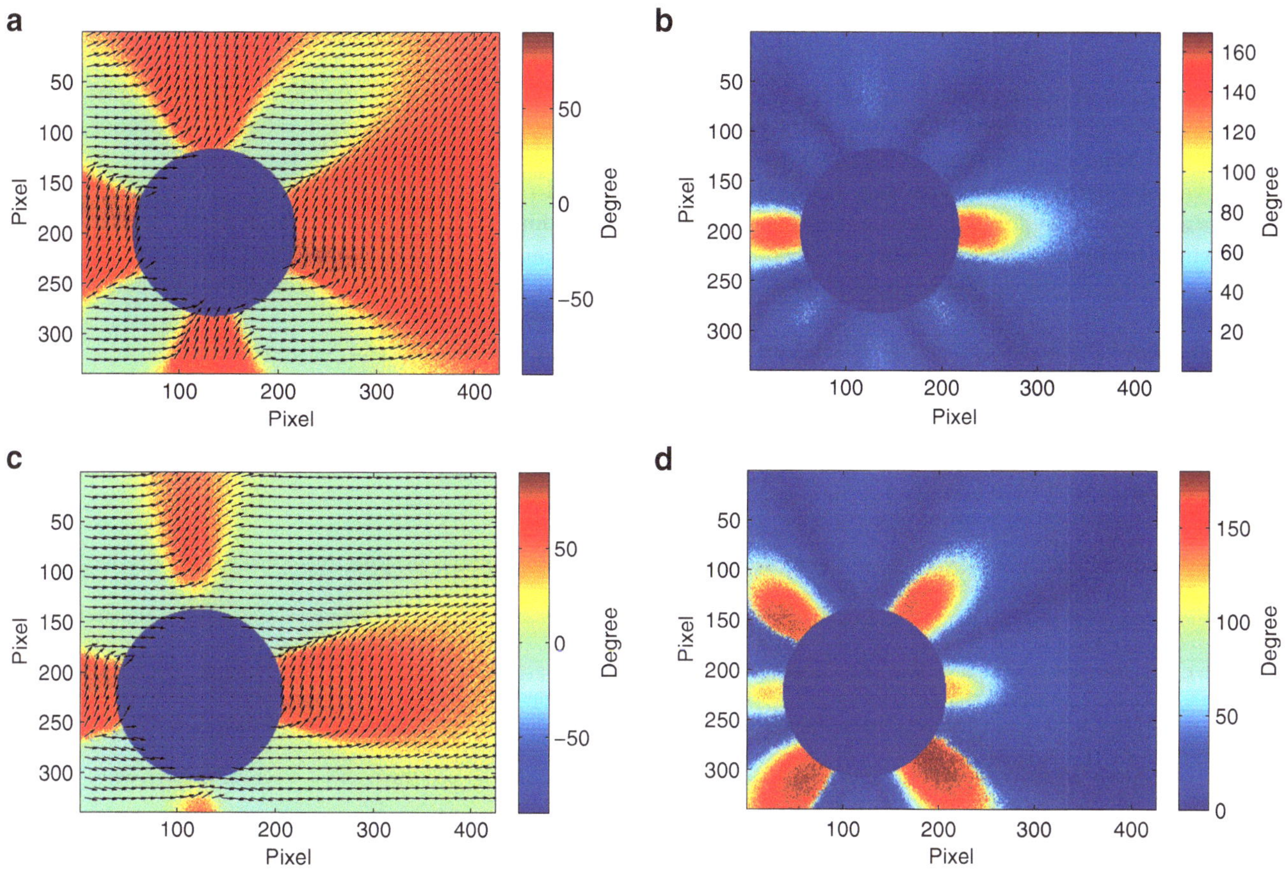

Fig. 9.5 Polarized light (**a**) alignment image and (**b**) retardation image from measurements of a polycarbonate dogbone specimen. (**c**) Alignment image and (**d**) retardation image from the same specimen but the applied stress was increased. The enter hole area was covered using a mask in the data processing

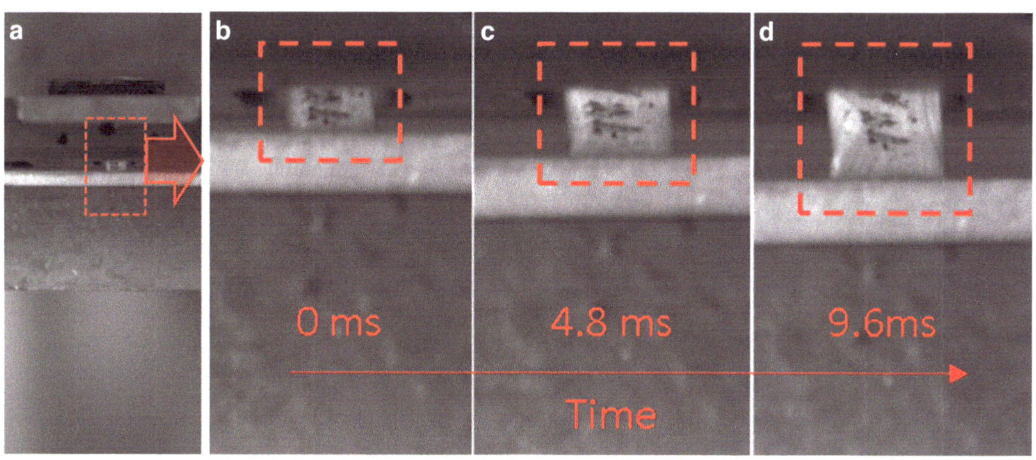

Fig. 9.6 An image captured using high-speed camera during an impact test of a tendon tissue specimen (11 mm wide, marked by *dashed boxes*) mounted in the modified drop tower. *Black marks* on the tendon specimen were made by graphite as reference. High speed camera frame rate: 12,500 FPS. (**a**) Before the impacting. (**b**) At the beginning of the impacting. (**c**) 4.8 ms after the impacting. (**d**) 9.6 ms after the impacting

9.4 Conclusions

The presented work has built the foundation for understanding dynamic deformation of collagen fiber realignment. In the future work, a high frequency PLM setup will be integrated with a high-speed camera for monitoring the tendon-to-bone insertion deformation in situ under different strain rate loadings.

Studies conducted in tendon-to-bone insertion dynamic tests will provide important information regarding insertion failures and associated collagen fiber realignments. Better understanding of the tendon-to-bone dynamic behavior would help to improve the clinical repair repairing techniques and design of engineered tissue replacement. The study on dynamic behavior of tendon-to-bone insertions will also help in designing better implants that can potentially resist against failure.

Acknowledgements The authors would like acknowledge the financial support of this project from the National Science Foundation (BMMB 1400018).

References

1. Thomopoulos, S., et al.: Variation of biomechanical, structural, and compositional properties along the tendon to bone insertion site. J. Orthop. Res. **21**(3), 413–419 (2003)
2. Thomopoulos, S., et al.: Collagen fiber orientation at the tendon to bone insertion and its influence on stress concentrations. J. Biomech. **39**(10), 1842–1851 (2006)
3. Butler, D.L., Juncosa, N., Dressler, M.R.: Functional efficacy of tendon repair processes. Annu. Rev. Biomed. Eng. **6**, 303–329 (2004)
4. Awad, H.A., et al.: Repair of patellar tendon injuries using a cell–collagen composite. J. Orthop. Res. **21**(3), 420–431 (2003)
5. Dressler, M.R., et al.: A potential mechanism for age-related declines in patellar tendon biomechanics. J. Orthop. Res. **20**(6), 1315–1322 (2002)
6. Lewis, G., Shaw, K.M.: Tensile properties of human tendo achillis: effect of donor age and strain rate. J. Foot Ankle Surg. **36**(6), 435–445 (1997)
7. Tohyama, H., Yasuda, K.: Extrinsic cell infiltration and revascularization accelerate mechanical deterioration of the patellar tendon after fibroblast necrosis. J. Biomech. Eng. **122**(6), 594–599 (2000)
8. Wren, T.A.L., et al.: Mechanical properties of the human Achilles tendon. Clin. Biomech. **16**(3), 245–251 (2001)
9. Yamamoto, N., Hayashi, K.: Mechanical properties of rabbit patellar tendon at high strain rate. Biomed. Mater. Eng. **8**(2), 83–90 (1998)
10. Ng, B.H., et al.: Strain rate effect on the failure properties of tendons. Proc. Inst. Mech. Eng. H **218**(3), 203–206 (2004)
11. Apostolakos, J., et al.: The enthesis: a review of the tendon-to-bone insertion. Muscles Ligaments Tendons J. **4**(3), 333–342 (2014)
12. Thomopoulos, S., Genin, G.M., Galatz, L.M.: The development and morphogenesis of the tendon-to-bone insertion what development can teach us about healing. J. Musculoskelet. Neuronal Interact. **10**(1), 35 (2010)
13. Harryman 2nd, D.T., et al.: Repairs of the rotator cuff. Correlation of functional results with integrity of the cuff. J. Bone Joint Surg. **73**(7), 982–989 (1991)
14. Galatz, L.M., et al.: The outcome and repair integrity of completely arthroscopically repaired large and massive rotator cuff tears. J. Bone Joint Surg. Am. **86**(2), 219–224 (2004)
15. Tower, T.T., Tranquillo, R.T.: Alignment maps of tissues: II. Fast harmonic analysis for imaging. Biophys. J. **81**(5), 2964–2971 (2001)
16. Tower, T.T., Tranquillo, R.T.: Alignment maps of tissues: I. Microscopic elliptical polarimetry. Biophys. J. **81**(5), 2954–2963 (2001)
17. Hauge, P.S.: Mueller matrix ellipsometry with imperfect compensators. JOSA **68**(11), 1519–1528 (1978)
18. Tower, T.T., Neidert, M.R., Tranquillo, R.T.: Fiber alignment imaging during mechanical testing of soft tissues. Ann. Biomed. Eng. **30**(10), 1221–1233 (2002)

Chapter 10
Self-Shifting Neutral Axis and Negative Poisson's Ratio in Hierarchical Structured Natural Composites: Bamboo

Shaowen Xu, Aniruddha Mitra, Stephen Migues, Jacob Mayfield, Michael Shinall, Bessenbacher Derek, Davis Linley, and Spratlin Russell

Abstract Bamboo is a light weight, high strength natural composite. Although it has been wildly used in the world for thousand years, the mechanism of material enhancement and microstructure/property relationships has not been well understood yet. In this study, triple points bending tests were conducted for mechanical behavior characterization of natural composite-bamboo. Digital Image Correlation was utilized to determine the deformation and strain of a bamboo beam. It was found that (a) the material responses to the applied load had both a linear and a non-linear range, (b) the beam was stiffer when it had node at two ends, (c) the neutral axis shifted toward the tensile side of the beam during loading process, (d) when samples were reloaded, the material responses to the load were different, (e) negative Poisson's ration was found in the high fiber density of the beam.

Keywords Neutral axis • Hierarchical structure • Self-shifting • Negative poisson's ratio • Digital image correlation

10.1 Introduction

In nature, biological materials have evolved for over eons to fit their environments, resulting in various multi-scale hierarchical structures with unique properties. For example, bamboo is organized from simple organic building blocks, which are as same as that of a tree, at various length scales to maximize strength, toughness and flexibility using minimal mass. Characterizing the mechanical behavior and understanding structure/property relationships of these hierarchical structures is fundamental for developing new advanced composites with superior properties. Although bamboo has been wildly used for 1000 years, the mechanism of material enhancement and microstructure/property relationships, especially the interaction of reinforced fiber and porous matrix, have not been well studied due to the complexity of the microstructure and lacking of available technologies.

In this study, triple point bending tests and microstructure analysis were conducted to characterize the deformation behavior of bamboo. In the test, Digital Image Correlation [1–7] was utilized to determine deformation and strain and understand the mechanism of material enhancement and structure/property relationships of bamboo. The effect of node structure of bamboo on material properties was also analyzed. The goals of this study were to (a) investigate material response of multi-scale hierarchical structures to the applied load, (b) determine the contributions of node structures in bamboo to bulk properties; (c) build knowledge base for research in bio-inspired advanced engineering material development.

10.2 Experiment

The bamboos used in this study were harvested from a local bamboo forest in southeastern Georgia. Structure analysis were performed in both transverse and longitudinal cross-sections. The samples used in the tests were cut along a longitudinal axis of bamboo. To investigate the effect of the bamboo node on material behavior, both samples with and without node were studied. All the samples were prepared using a same section of a bamboo so that the samples could be identical (Fig. 10.1). The wall thickness of the bamboo section was 7.10 mm. Shown in Table 10.1 were dimensions of the samples.

S. Xu (✉) • A. Mitra • S. Migues • J. Mayfield • M. Shinall •
B. Derek • D. Linley • S. Russell
Department of Mechanical Engineering, Georgia Southern University, Statesboro, GA 30458, USA
e-mail: shaowenxu@georgiasouthern.edu

C.S. Korach et al. (eds.), *Mechanics of Biological Systems and Materials, Volume 6*,
Conference Proceedings of the Society for Experimental Mechanics Series, DOI 10.1007/978-3-319-41351-8_10

Fig. 10.1 Photos of tested samples, (**a**) side view, (**b**) top view

Table 10.1 Dimension of bamboo samples for bending tests

Test number	Sample number	Height (mm)	Width (mm)	Length (mm)	Loading condition
2a	2	6.636	6.31	155.21	First time to be loaded
2b	2	6.636	6.31	155.21	Re_loaded couple day later
3a	3 (with node)	6.670	6.463	197.97	First time to be loaded
3c	3 (with node)	6.670	6.463	197.97	Re_loaded couple day later

Fig. 10.2 Experimental set up of the tests

In this study, research was focused on material responses bamboo beam when subjected to bending load. Triple point bending tests were adapted for material behavior characterization. The distance between two lower supports was 120.0 mm. In the test, Digital Image Correlation was utilized to determine the full field of displacement and strain on the side of the bending samples during the loading process (Fig. 10.2).

10.3 Results and Discussion

Bamboo is a natural composite. It has periodic appearing nodal structures along the longitudinal axis (Fig. 10.3a). The macro-fibers are parallel with each other along the longitudinal axis between two nodes. The macro-fibers are continuing between the two nodes. The macro-fibers in one section tangle with the fibers in nearby section to form the a node (Fig. 10.3c). Bamboo consists of hierarchical structures of a couple of length scale. Shown in Fig. 10.3b, d, e are the structures of the bamboo on a transverse cross section. At macro-scale, the bamboo is a fiber reinforced composite.

Fig. 10.3 Hierarchical structures of bamboo: (**a**) macro-structure of bamboo, (**b**) micro-structure of porous matrix of bamboo, (**c**) macro-fiber orientation and structure of nodal structure of bamboo; (**d**) micro-structures on transverse cross section, (**e**) micro-structures on transverse cross section near the outside surface

A macro-fiber consist of hundreds of microfibers. Each microfiber has a co-shell multilayer structure. The matrix of the bamboo is not solid. It is formed of microporous structures (Fig. 10.3b). The wall of the porous structure has also layered structure. The macro-fibers are functional gradient distributed along the radius. The fiber density is high at an outside surface and lowers near the inside. The shape and dimension of the macro-fibers vary from inside to outside (Fig. 10.3d). These hierarchical structures result in unique mechanical property of bamboo.

Plotted in Fig. 10.4 were the load-displacement curves (continue lines) and the neutral axis-displacement curves (dot lines) of the samples with (sample 3) and without (sample 2) nodes. The two load-displacement curves (continue line) were similar. They both had a linear range and a non-linear range. When loads were below 80N for the sample with the node (88N for the sample without node), the structure responses were linear. The Flexure Modulus was 12.15 GPa for the sample with node and 11.83 GPa for sample without node. The sample with node structures was stiffer. Its load-displacement was above that of without node. The sample with node needed higher load to be deformed. It was indicated that the nodes could reinforce the material and made it stiffer. It was interested to find (a) the neutral axis of the bamboo beam was not at center of the cross-section of the beam, it was offset to the outside surface of the bamboo; (b) the neutral axis continuously shifted toward high fiber density side when load increased. In linear range, the shifting was not obvious for the sample without node but little for the sample with node. The neutral axis tended to stay in its original location. When the load further increased into the non-linear range, the neutral axis started shifting toward the high fiber side (tensile side), the shifting rate was almost same for both sample with/without nodes. Shown in the Fig. 10.4, the initial location of neutral axis of the sample with nodes was more close to the center of the beam than that of the sample without the nodes when the samples subjected same load in linear range. In other words, when the deformations of the samples were the same, larger load was required and the shift of the neutral axis was smaller for the sample with the nodes.

Shown in Fig. 10.5 were load-displacement curves and neutral axis-displacement curves of the samples with (sample 3) and without (sample 2) nodes when the samples were reloaded. The responses of the material were different with when the

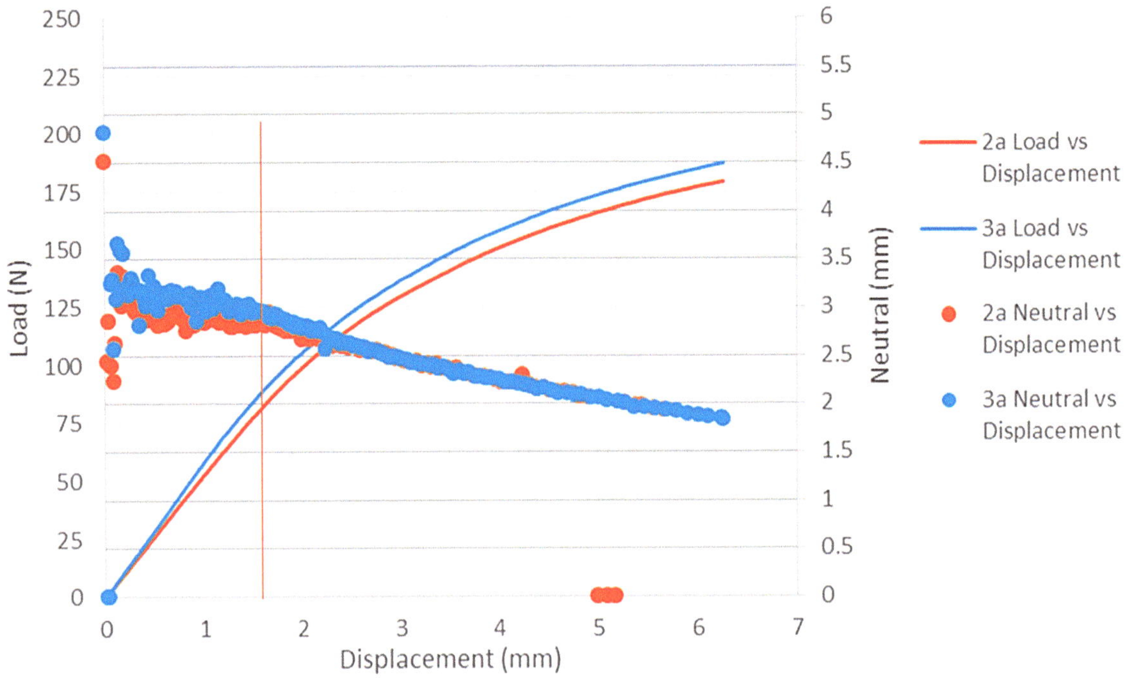

Fig. 10.4 Load vs displacement curves and location of neutral axis of the sample with node and without node

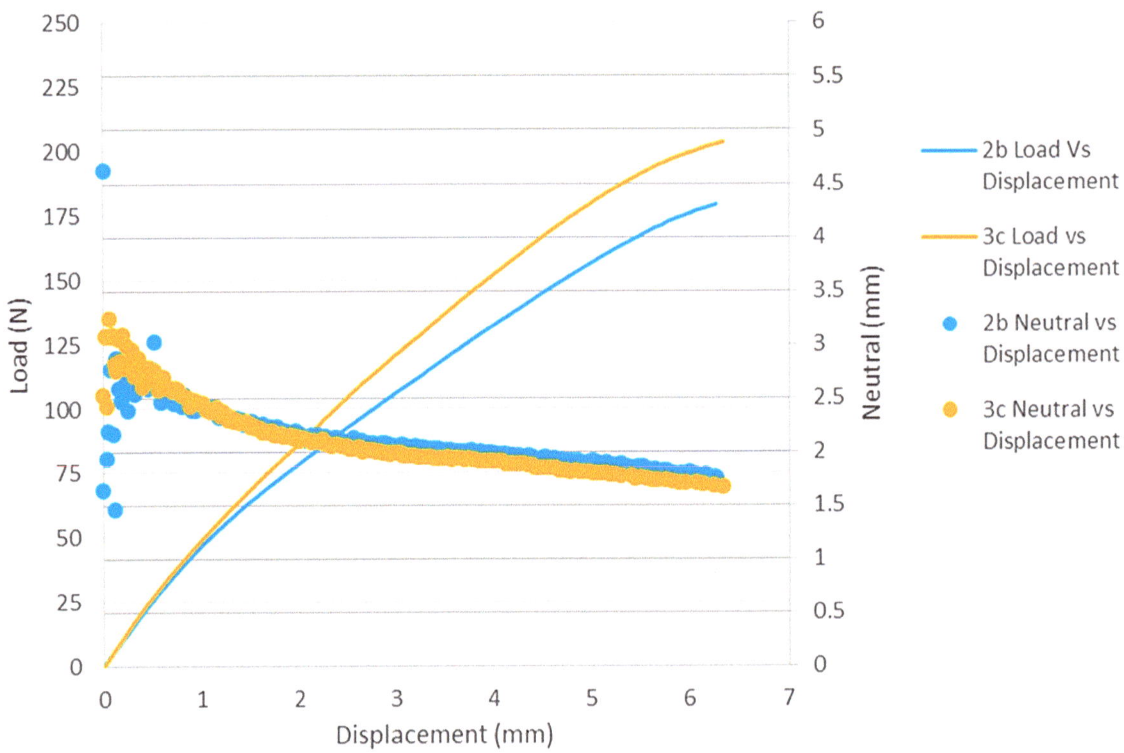

Fig. 10.5 Load vs displacement curves and location of neutral axis of the samples with node and without node on reloading conditions

samples were loaded first time. Instead of linear-nonlinear pattern, the material responded to the load in linear-nonlinear-linear-nonlinear pattern. When the load was below 30N and above 55N, the material responses were linear. When the load was between 30N and 55N, non-linear transfer ranges were observed. When the load was over 175N, non-linear responses appeared again. The sample with nodes still appeared stiffer than without nodes. When examining the neutral axis, it was found that the neutral axis shifted toward high fiber density (tensile) side with almost same shifting rate for both samples.

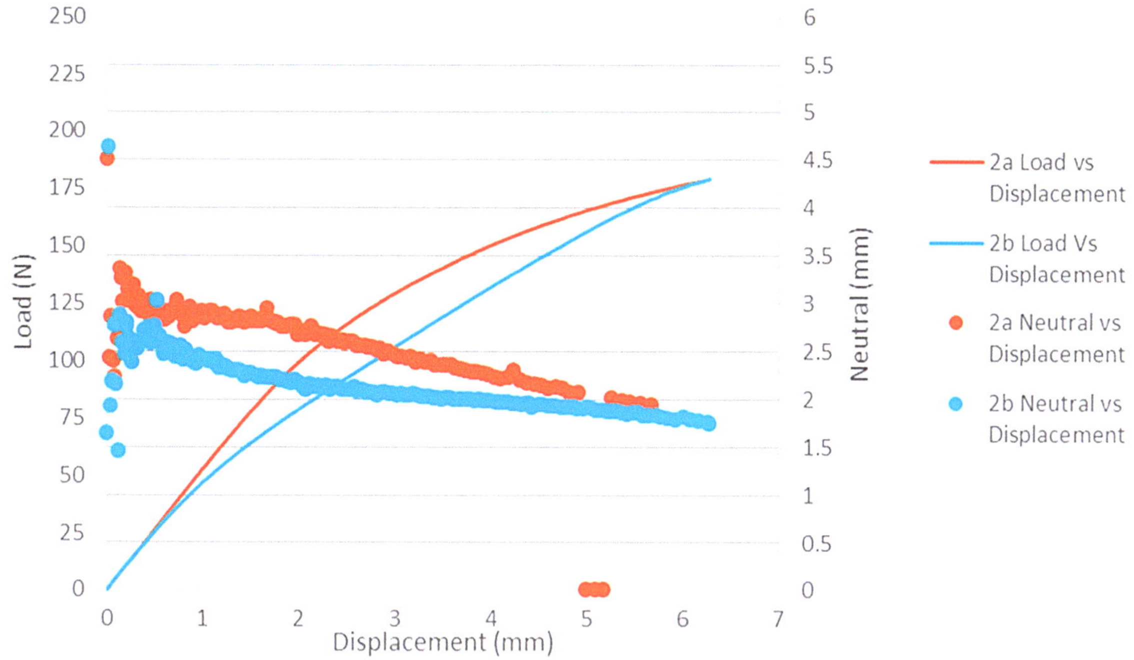

Fig. 10.6 Load vs displacement curves and location of neutral axis of the sample without node

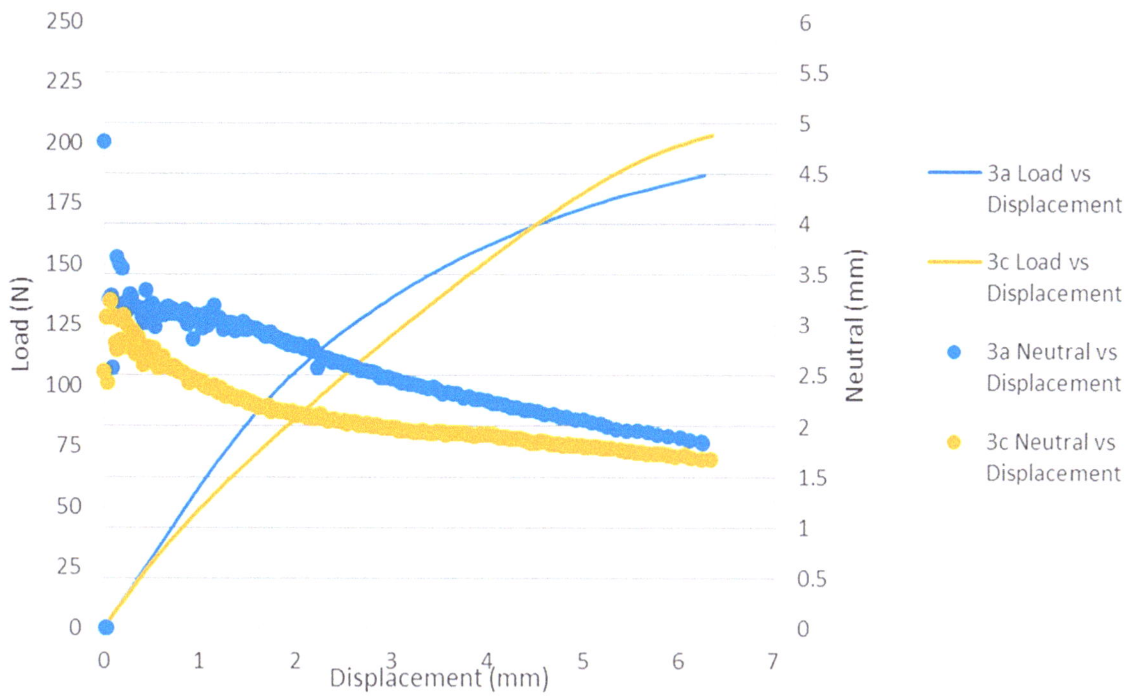

Fig. 10.7 Load vs displacement curves and location of neutral axis of the sample with node

Comparing with neutral axis shifting when the samples were loaded at first time, the axis shifted in different manner. When the load was below 88N, shifting rate of the neutral axis was high. The shifting rate became slow when load was over 88N.

Shown in Figs. 10.6 and 10.7 were the material behavior comparisons on load and reloaded conditions. Both deformation behavior and neutral axis shifting were different for load and reloaded conditions. The structure stiffness was softer, and the neutral axis quickly shifted to the high fiber density side when the samples were reloaded. The neutral axis tended to locate in the positions when it was loaded at the end of the first loading cycle. Carefully examining the neutral axis, it was found that

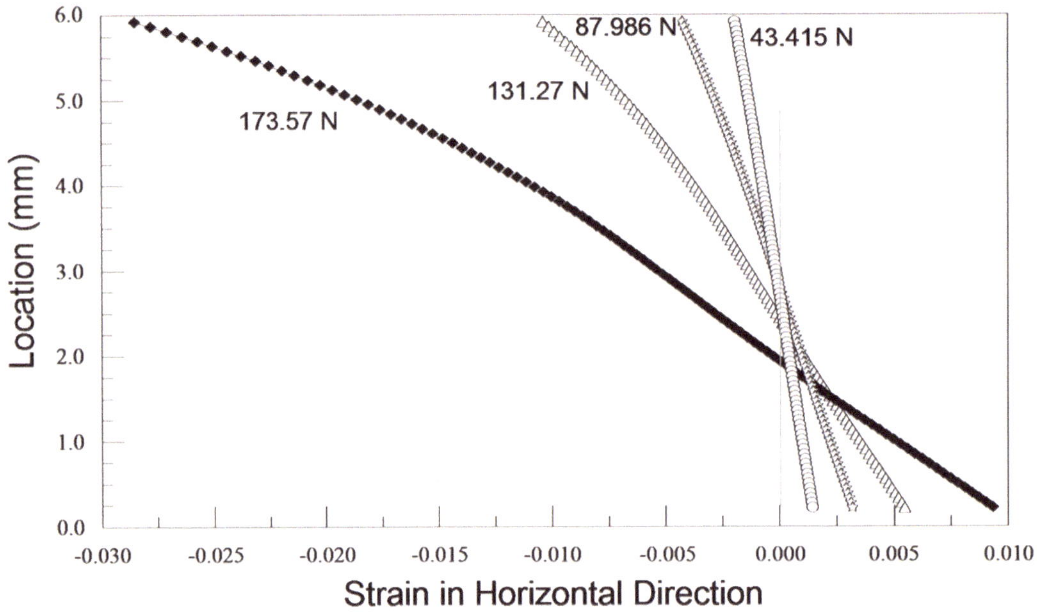

Fig. 10.8 Horizontal strain distributions on cross section at different load steps

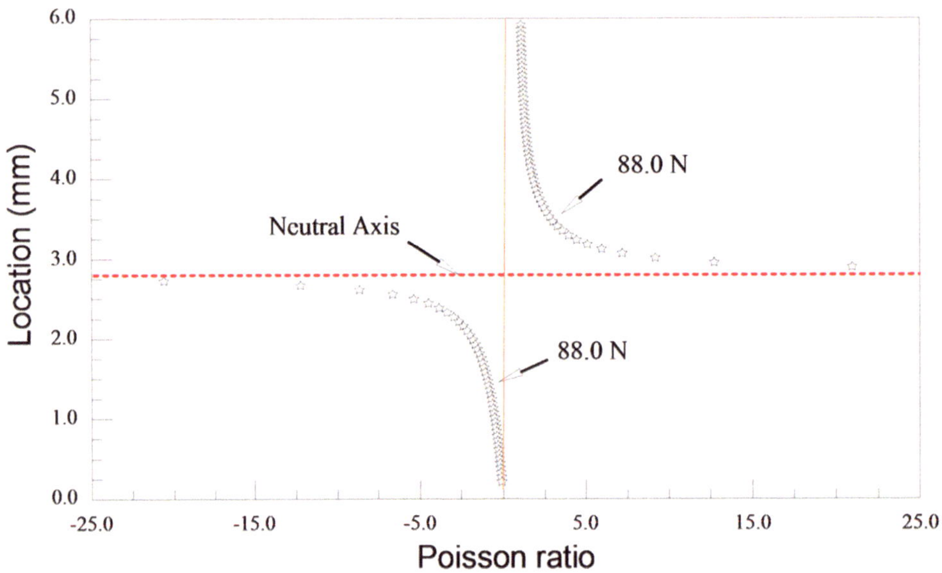

Fig. 10.9 Poisson's ratio distribution along beam thickness at the end of the linear loading range

the neutral axis shifted toward outside surface a little bit more when it was reloaded for both samples with and without nodes. Even though the neutral axis shifted in different manner for load and re-load condition, at the end, the axis tended to shift to same location. It was noted that the sample with nodes became stiffer when load over 165N.

Plotted in Fig. 10.8 were the distributions of horizontal normal strain along the thickness direction at the center of the beam. Shown in the figure, the strain distribution was linear when load was low and in linear range. It suggest the plane assumption—"the plane cross sections remain plane after bending" is still valid for this heterogeneous material. When the load was increased to nonlinear range (large then 88N), non-linear strain has developed in the high porosity side of the beam. In contract, the strain distributions were still appearing in linear manner in the high fiber density side. If carefully examining the location of zero strain (the location of the neutral axis), it was found that neutral axis tended to stay in initial location in linear range of the load, but the axis started to shift towards to high fiber density when the load increased above 88N.

Shown in Fig. 10.9 was the profile of Poisson's ratio on the cross section at the center of the beam, when the load was 88N. It was noticed the Poisson's ration was not constant on the cross section. Its profile looked like a reciprocal function. It was divided into two parts by the neutral axis. The sign of the Poisson's ratio was opposite on the two sides of the axis. It was positive above the axis, and negative below it. More close to the neutral axis, the magnitude of Poisson's ratio was larger.

10.4 Conclusion

Triple points bending tests were conducted for natural composite-bamboo. Digital Image Correlation was utilized to determine the deformation and strain of the bamboo beam. The effect of node structures on the material behavior was studied. It was found that (a) the material responses to the applied load had both a linear and a non-linear range, (b) the beam was stiffer when it had node at two ends, (c) the neutral axis shifted toward the tensile side of the beam during loading process, (d) when samples were reloaded, the material responses to the load were different, (e) negative Poisson's ration was found in the high fiber density of the beam.

References

1. Sutton, M.A., Wolters, W.J., Peters, W.H., Ranson, W.F., McNeill, S.R.: Determination of displacements using an improved digital correlation method. Image Vis. Comput. **4**(3), 143–150 (1983)
2. Bruck, H.A., McNeill, S.R., Sutton, M.A., Peters, W.H.: Digital image correlation using Newton Raphson method of partial differential correction. Exp. Mech. **29**, 261–267 (1989)
3. Sutton, M.A., Bruck, H.A., Chae, T.L., Turner, J.L.: Development of computer methodology for the analysis of surface deformations in magnified images. In: Vander Voort, G.F. (ed.) MICON-90, Advances in Video Technology for Microstructural Evaluation of Materials, ASTM STP-1094, vol. 109, American Society for Testing and Materials, Philadelphia (1990)
4. Ning, J., Xu, S., Wang, Y., Lessner, S.M., Sutton, M.A., Anderson, K., Bischoff, J.E.: Deformation measurements and material property estimation of mouse carotid artery using a microstructure-based constitutive model. J. Biomech. Eng. **132**(12), 121010 (2010)
5. Tiwari, V., Sutton, M.A., McNeill, S.R., Xu, S., Deng, X., Fourney, W.L., Bretall, D.: Application of 3D image correlation for full-field transient plate deformation measurements during blast loading. Int. J. Impact. Eng. **36**(6), 862–874 (2009)
6. Xu, S., Tiwari, V., Deng, X., Sutton, M.A., Fourney, W.L.: Identification of interaction pressure between structure and explosive with inverse approach. Exp. Mech. **51**(6), 815–830 (2011)
7. Xu, S.: Failure detection of temporary structures with digital image correlation for construction safety applications. In: 1st International Digital Image Correlation Society Conference & Workshop, Columbia SC, USA, (Nov. 3–5, 2015)

Chapter 11
High-Speed Holography for In-Vivo Measurement of Acoustically Induced Motions of Mammalian Tympanic Membrane

Payam Razavi, Jeffrey Tao Cheng, Cosme Furlong, and John J. Rosowski

Abstract Measurements of human Tympanic Membrane (TM, eardrum) motions require nanometer and microsecond spatio-temporal resolutions while maintaining a field-of-view of about one centimeter. Previously, we have developed holographic methods to successfully measuring continuous and transient responses of post-mortem TM's. To expand our capabilities to in-vivo measurements, it is necessary to overcome such challenges as the influence of submillimeter physiological motions as well as the confined location of the TM. We are developing novel High-speed Digital Holographic Methods (HDHM) in combination with recently developed image and data processing algorithms to overcome such challenges. Our developments have unique capabilities that utilize the full spatio-temporal resolution of high-speed cameras (i.e., >147,000 points at >42,000 fps) to measure nanometer-scale TM motions in the audible range (0.02–20 kHz). We present preliminary holographic measurements made on an anesthetized chinchilla in a controlled anechoic chamber in-vivo and in-vitro. To the best of our knowledge, these data are reported for the first time and establish the potential of HDHM as a hearing research and clinical tool to further expand our understanding of the human hearing processes.

Keywords High-speed digital holography • In-vivo measurements • Middle ear mechanics • Transient acoustic excitation • Tympanic membrane

11.1 Introduction

The transformation of sound energy in mammalian middle ear to mechanical vibrations in inner ear is not yet well understood, while it is essential for understanding the hearing processes and diagnosis of middle ear diseases [1, 2]. TM is the interface between the middle and outer ear and transforms sound energy into mechanical vibrations of the ossicles. It is confined at the end of the ear canal and has a conical shape of about 1–3 mm depth and about 8 mm diameter in humans. TM is thin (50–150 μm) and consists of three distinct layers: mucosal, epidermis, and fibrous [2, 3]. Understanding the transformation of sound energy requires accurate measurement of the full-Field Of View (FOV) spatio-temporal motions, shape, and ultrastructure [3]. In addition, measurements of the TM motions are challenging because of required nano-meter displacement and micro-second temporal resolutions, while maintaining a FOV of about six orders of magnitude larger than its motions (~8 mm) [4].

Available quantitative clinical TM diagnostics tools such as tympanometry and Laser Doppler Vibrometry (LDV) lack full-field measuring capabilities. Holographic methodologies are powerful nondestructive testing tools that have been used in the past to enable full-FOV measurements of the vibrating patterns of the TM surface [5–7]. Løkberg et al. used electronic

P. Razavi (✉)
Center for Holographic Studies and Laser micro-mechaTronics (CHSLT), Worcester Polytechnic Institute, Worcester, MA, USA

Mechanical Engineering Department, Worcester Polytechnic Institute, Worcester, MA, USA
e-mail: prazavi@wpi.edu

J.T. Cheng • J.J. Rosowski
Eaton-Peabody Laboratory, Massachusetts Eye and Ear Infirmary, Boston, MA, USA

Department of Otology and Laryngology, Harvard Medical School, Boston, MA, USA

C. Furlong
Center for Holographic Studies and Laser micro-mechaTronics (CHSLT), Worcester Polytechnic Institute, Worcester, MA, USA

Mechanical Engineering Department, Worcester Polytechnic Institute, Worcester, MA, USA

Eaton-Peabody Laboratory, Massachusetts Eye and Ear Infirmary, Boston, MA, USA

Department of Otology and Laryngology, Harvard Medical School, Boston, MA, USA

C.S. Korach et al. (eds.), *Mechanics of Biological Systems and Materials, Volume 6*,
Conference Proceedings of the Society for Experimental Mechanics Series, DOI 10.1007/978-3-319-41351-8_11

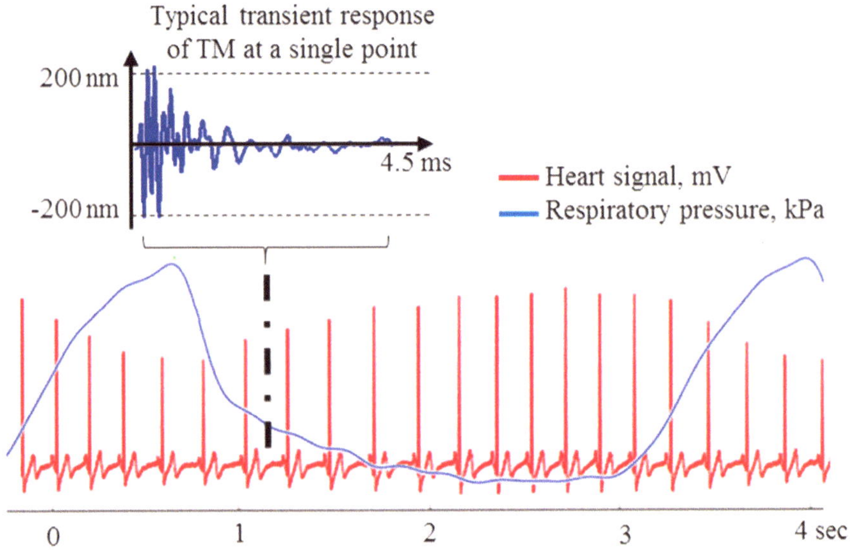

Fig. 11.1 Typical timing diagram of heartbeat and respiratory signals together with typical transient response of TM measured at a single point and produced by an acoustic click

speckle pattern interferometry (ESPI) to obtain preliminary qualitative measurements of vibration of human TM in-vivo [8]. Previously, we have developed various holographic methods to measure steady state and transient responses of post-mortem TM's [4, 9–13]. However, measurements in live ears are adversely affected by physiological in-vivo motions due to respiration, blood circulation, and muscle motion [8]. To expand our capabilities to in-vivo measurements, it is necessary to overcome the following challenges:

- The influence of submillimeter physiological motions (i.e., breathing, heartbeat, tremor etc.) while measuring motions with nano-meter resolution, as illustrated in Fig. 11.1.
- Access to and orientation of the TM within the external ear necessitating a large depth of focus.
- Technological challenges, such as increases in the speed of measurements without compromising spatial resolution and measuring sensitivity.

11.2 Methods

To overcome the challenges mentioned above we advanced our holographic tools and developed High-speed Digital Holographic Methods (HDHM) to capture the transient motion of the TM at high frequencies (i.e., >10 kHz) nearly free from low frequency physiological motions. We have developed and implemented high-speed 2 + N frame acquisition methods based on hybrid spatio-temporal Local Correlation (LC) phase sampling approach that allows for quantification of the TM transient deformations by utilizing two reference frames, $I_{ref}, I_{ref+\pi/2}$, and N consecutive deformed frames, $(I_{def})_i$, $i \in 1, 2 \ldots N$, recorded before and throughout the evolution of an event [9], shown in Fig. 11.2.

The timing diagram of the events occurring during measurements is shown in Fig. 11.2a. By use of continuous phase shifting method and developed phase quality optimization algorithm, two frames with $\pi/2$ phase shift is chosen from hundreds of the reference holograms to extract the exact undeformed state of the TM. Next, the sample desired excitation occurs. At the same time, while keeping the PZT at its $\pi/2$ position, a series of high-speed (e.g., >40 kHz) holograms $(I_{def})_i$, $i \in 1, 2..N$, measure the TM motions [11]. The corresponding spatio-temporal variation of the double-exposure optical phase change, $\phi(m, n, t)$, is quantified by correlating the intensities of each deformed and reference frames with

$$\phi(m,n,t) = tan^{-1}\left[\frac{\rho\left(I_{ref}(m,n),\ I_{def}(m,n,t)\right)}{\rho\left(I_{ref+\pi/2}(m,n),\ I_{def}(m,n,t)\right)}\right] \tag{11.1}$$

where ρ is the Pearson's correlation coefficient for finite discrete sets having the recoded intensities as arguments and computed based on a spatial kernel (e.g., with dimensions of 3 × 3 or 5 × 5 pixels) centered around each measurement point, (m, n), of a pair of reference and deformed frames [9, 14]. Measuring accuracy of HDHM has been verified with Laser Doppler Vibrometry [9].

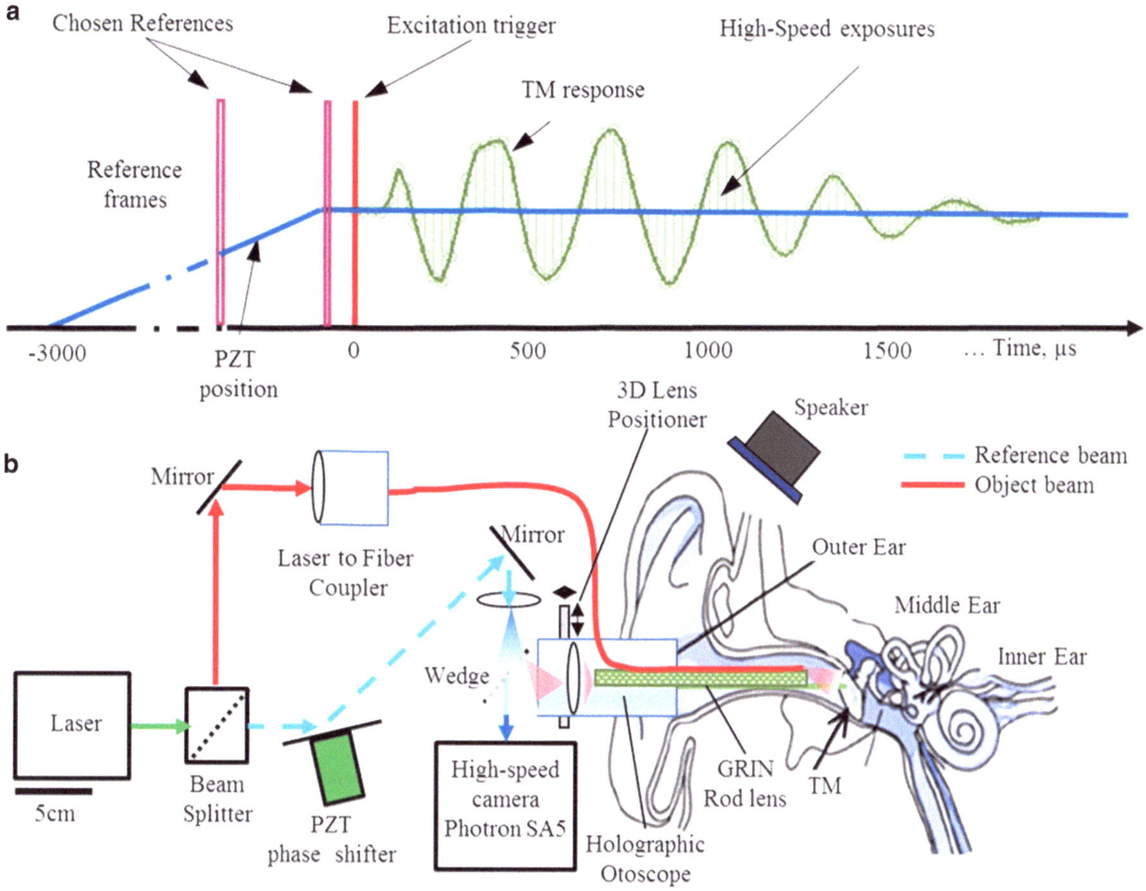

Fig. 11.2 Schematic representation of the HDHS and the acquisition timing diagrams: (**a**) timing diagram demonstrating the principle of operation of the high-speed 2 + *N* frame acquisition method [7]; and (**b**) schematic representation of the developed high-speed holographic system equipped with a custom-made otoscope. Illumination and observation inside the ear canal is performed by fiber optics and a GRadient INdex (GRIN) rod lens, respectively. The otoscope has a relatively large depth of focus, which is sufficient to measure motions of the cone-shaped TM

11.3 Experimental Setup

A portable setup with remotely controlled capabilities was realized at Worcester Polytechnic Institute (WPI) and moved inside a double-walled sound booth located in the facilities of the Massachusetts Eye and Ear Infirmary (MEEI). The HDHM setup includes a high-speed camera (Photron SA-5), a CW laser (532 nm, 50 mW), variable beam splitter, beam combiner wedge, mirrors, beam expanders, piezoelectric phase shifter, laser to fiber coupler and a holographic otoscope, as shown in Fig. 11.3. The laser beam is split into reference and illumination beams by use of a beam splitter. The illumination beam, after being coupled into a fiber optic waveguide, irradiates the TM through the earcanal. The object beam and the reference beam are combined at the CCD camera by use of the wedge. The sample is excited by a speaker (placed within 15 cm of the TM surface) with 50 μs square pulses that produce an acoustic click. The sound pressure variations at the surface of the TM are measured by a probe microphone. During the measurements, cooling fans associated with the laser and camera systems are switched off. All the measurements were repeated thrice consecutively with 100 ms intervals to investigate the repeatability of the motions under various environmental conditions.

All animal procedures were approved by the Massachusetts Eye and Ear Animal Care Committee. Chinchillas were anesthetized and remained so throughout the in-vivo measurements. The cartilaginous and boney part of ear canals were resected to expose the majority of the TM surface and the middle ear cavity was vented, but otherwise intact. The TM was painted with a thin layer of ZiO solution for improved reflectivity. After a series of in-vivo measurements, the animal was euthanized and measurements repeated without repositioning the animal.

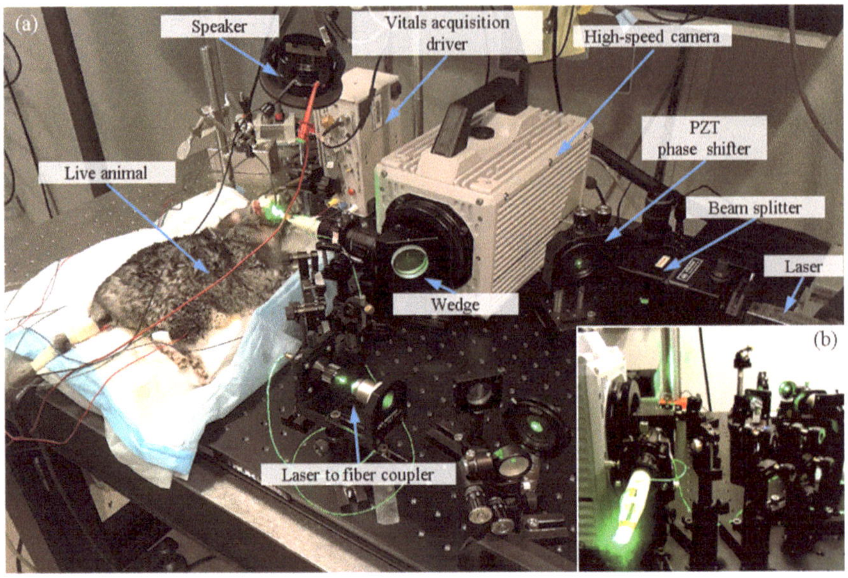

Fig. 11.3 HDHM setup during live animal measurements: (**a**) opto-mechanical components inside a double-walled sound booth; and (**b**) view of the otoscope head

11.4 Representative Results

We report preliminary results of acoustically induced transient response of a chinchilla under two excitation amplitudes and two conditions: in-vivo and post-mortem.

11.4.1 Quantitative Full-Field-of-View In-Vivo Measurements

Two sets of three consecutive experiments were performed at two excitation levels for each of the pre- and post-mortem conditions. Figure 11.4 shows eight temporal instances of the normalized mean ($N = 6$) of these measurements together with the motions at six specific locations across the TM. Despite large physiological motions, highly repeatable measurements were recorded. In addition, a linear response of the TM under different loading conditions was observed.

11.4.2 Comparison of Pre- and Post-mortem Response of the TM in the time domain

We observed similarities and differences between live and post-mortem measurements, including:

- The motions in both live and post-mortem cases are very similar, including the size and location of motion maxima at the beginning of the response to the click (see the 143 μs panels in Fig. 11.5a), but the patterns diverge somewhat as time progresses (see the differences in magnitude and number of local maxima and minima at 1171 μs).
- Similar conclusion can be drawn looking at the temporal response of selected location on the TM surface, Fig. 11.5b.
- Similar comparison of the time waveform of the displacements at different TM locations suggest the average decay time of the displacement is increased postmortem.
- The large increases in the decay time of the displacements seen post-mortem are consistent with post-mortem changes in the damping provided by the cochlear load.
- Maximum peak-to-peak motions of the membrane surface is increased by about 20 % in post-mortem condition.

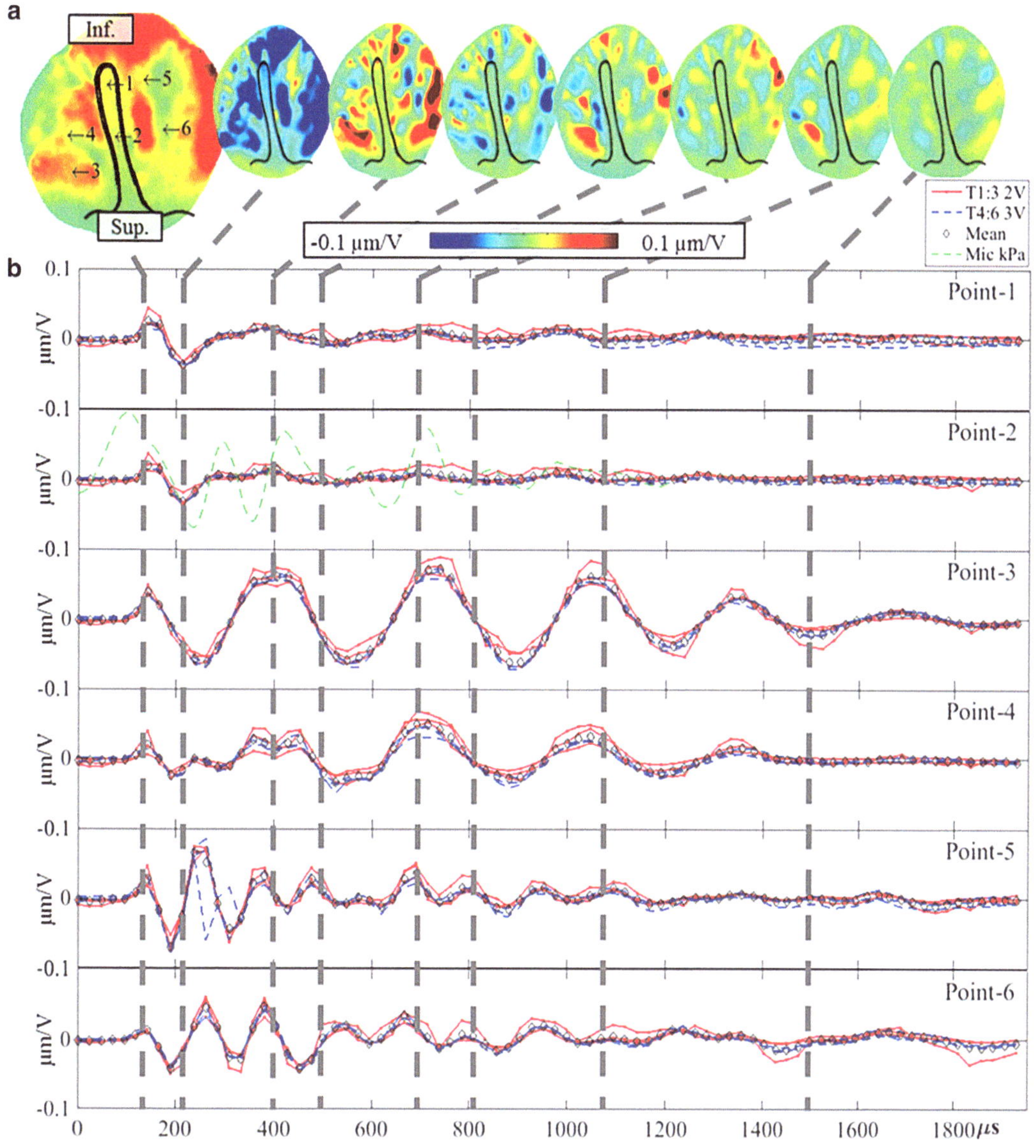

Fig. 11.4 In-vivo measurement results up to 1.8 ms; (**a**) mean ($N = 6$) displacement maps taken at eight temporal instances between 143 and 1500 μs. The color bar codes instantaneous displacements between ±0.1 μm/V. The map at 143 μs shows the umbo peak motion and the numbers and *arrows* indicate six locations where displacement versus time waveforms are plotted in (**b**). Point-1 is at the umbo, and the manubrium is outlined in *black*; (**b**) displacement waveforms at six locations normalized by the stimulus voltage are illustrated. At each location, six individual measurements at two different stimulus voltages (T1-3: 2 V and T4-6: 3 V) are plotted together with the corresponding mean, shown by *open diamond*. The sound pressure waveform recorded by the microphone is shown in *green* in the panel of point-2

11.4.3 Comparison of Pre- and Post-mortem Response of the TM in the Frequency Domain

FFT analyses were performed on the displacement waveforms at each point on the TM surface to define maps of the displacement magnitude, as shown in Fig. 11.6, at the frequencies with prominent spectral peaks obtained from the mean of the FFTs, as marked by circles in Fig. 11.7. After analysis of the results, the following was observed:

- The displacement maps show more spatially complex shapes with lower amplitudes in the in-vivo condition.
- The similarities of the frequencies of peak displacements in pre- and post mortem conditions suggest minimal changes in the mass and stiffness of the middle ear.

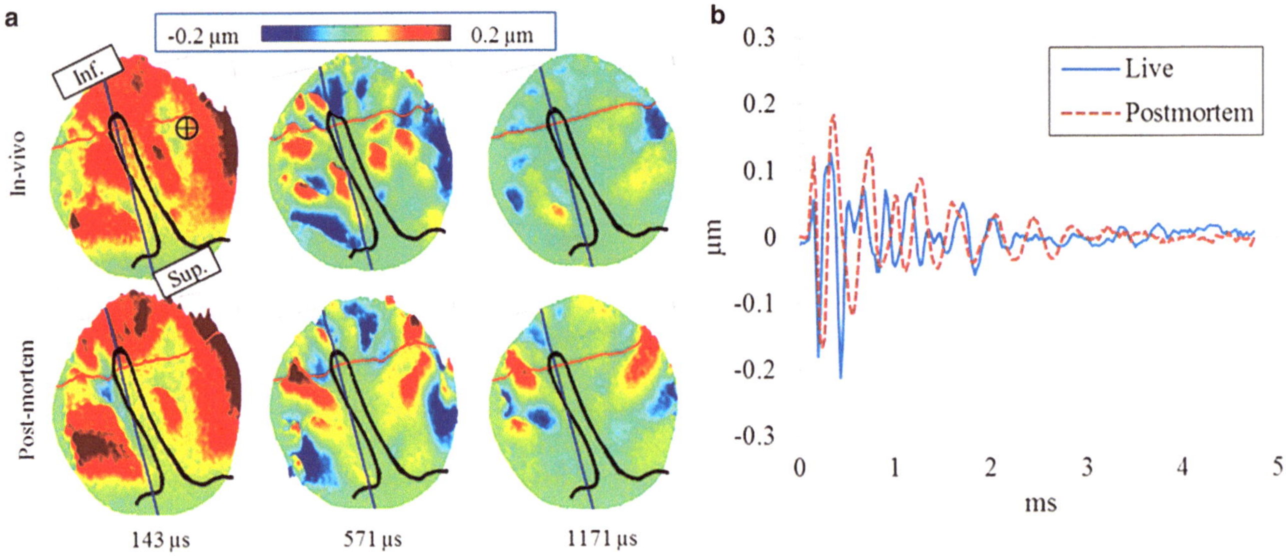

Fig. 11.5 Comparison of live and post-mortem spatio temporal evolution of motions in a chinchilla TM: (**a**) three temporal instances of full-FOV motions of the sample; and (**b**) temporal comparison of motions at the location shown with *circled plus* in (**a**)

Fig. 11.6 Maps of the displacement magnitudes at the spectral peaks isolated from Fig. 11.7 and identified by *open circle*

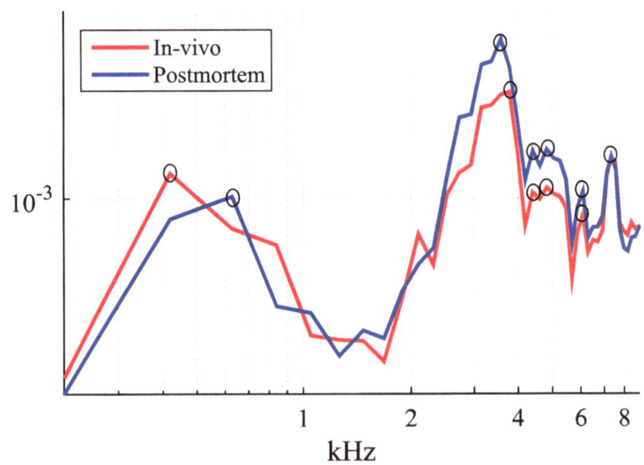

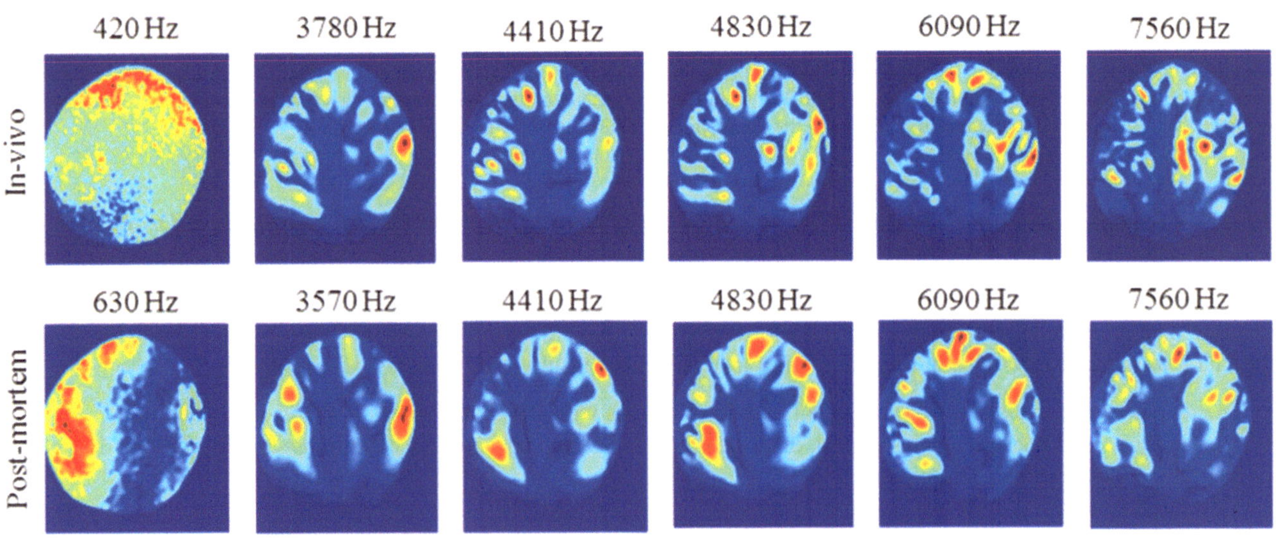

Fig. 11.7 Comparison of live and post-mortem spatio-temporal mean of the FFTs of the displacement waveform at each point. Selected spectral peaks are shown by *open circle*

11.5 Conclusions and Future Work

We report preliminary quantitative measurements of TM motions in full-FOV in a live animal. Our techniques provide high temporal (i.e., <23.8 μs) and spatial (i.e., >100 k data points) resolutions with nano-meter (i.e., <15 nm) displacement sensitivity in the presence of physiological and environmental noises. We observed similarities and differences in the response of the TM in live and post-mortem conditions. The preliminary results establish the potential of HDHM as a hearing research and clinical tool to expand our understanding of human hearing. Future work includes the analysis together with the interpretation of the live and post-mortem measurements for further understanding of hearing function and potential medical applications. Fundamental research work will be focused on the initial transient dynamics of the TM, its relationship to the energy transfer into the middle ear, and its connection to previous steady-state dynamics research done by our group and others. In addition, further improvements of the opto-mechanical design of the holographic otoscope will be performed.

Acknowledgements This work has been funded by the National Institute on Deafness and Other Communication Disorders (NIDCD), the National Institute of Health (NIH), the Massachusetts Eye and Ear Infirmary (MEEI), NSF MRI program award: CMMI-1428921, and the Mittal Fund. The authors would like to acknowledge contributions of Nima Maftoon, Melissa McKinnon, and Michael E. Ravicz from the MEEI and Ivo Dobrev and Koohyar Pooladvand from the CHSLT at the Worcester Polytechnic Institute.

References

1. Rosowski, J.J., Cheng, J.T., Ravicz, M.E., Hulli, N., Hernandez-Montes, M., Harrington, E., Furlong, C.: Computer-assisted time-averaged holograms of the motion of the surface of the mammalian tympanic membrane with sound stimuli of 0.4–25 kHz. Hear. Res. **253**, 83–96 (2009)
2. Decraemer, W., Funnell, W.: Anatomical and mechanical properties of the tympanic membrane. In: Chronic Otitis Media: Pathogenesis-Oriented Therapeutic Management, pp. 51–84. Kugler Publications, The Netherlands (2008)
3. Van der Jeught, S., Dirckx, J.J.J., Aerts, J.R.M., Bradu, A., Podoleanu, A.G., Buytaert, J.A.N.: Full-field thickness distribution of human tympanic membrane obtained with optical coherence tomography. J. Assoc. Res. Otolaryngol. **14**, 483–494 (2013)
4. Khaleghi, M., Cheng, J.T., Furlong, C., Rosowski, J.J.: In-plane and out-of-plane motions of the human tympanic membrane. J. Acoust. Soc. Am. **139**, 104–117 (2016)
5. Khanna, S.M., Tonndorf, J.: Tympanic membrane vibrations in cats studied by time-averaged holography. J. Acoust. Soc. Am. **51**, 1904–1920 (1972)
6. Tonndorf, J., Khanna, S.M.: Tympanic-membrane vibrations in human cadaver ears studied by time-averaged holography. J. Acoust. Soc. Am. **52**, 1221–1233 (1972)
7. del Socorro Hernández-Montes, M., Furlong, C., Rosowski, J.J., Hulli, N., Harrington, E., Cheng, J.T., Ravicz, M.E., Santoyo, F.M.: Optoelectronic holographic otoscope for measurement of nano-displacements in tympanic membranes. J. Biomed. Opt. **14**, 034023-034023-9 (2009)
8. Løkberg, O., Hagmoen, K., Gundersen, T.: Vibration measurement of the human tympanic membrane-in vivo. Acta Otolaryngol. **89**, 37–42 (1980)
9. Dobrev, I., Furlong, C., Cheng, J.T., Rosowski, J.J.: Full-field transient vibrometry of the human tympanic membrane by local phase correlation and high-speed holography. J. Biomed. Opt. **19**, 96001 (2014)
10. Razavi, P., Ravicz, M.E., Dobrev, I., Cheng, J.T., Furlong, C., Rosowski, J.J.: Response of the human tympanic membrane to transient acoustic and mechanical stimuli. Hear. Res. (2016). doi:10.1016/j.heares.2016.01.019
11. Razavi, P., Dobrev, I., Ravicz, M., Cheng, J.T., Furlong, C., Rosowski, J.J.: Transient response of the eardrum excited by localized mechanical forces. In: Tekalur, S.A., Zavattieri, P., Korach, C.S. (eds.) Mechanics of Biological Systems and Materials, Volume 6, pp. 31–37. Springer, Cham (2016)
12. Khaleghi, M., Furlong, C., Cheng, J., Rosowski, J.: Characterization of acoustically-induced forces of the human eardrum. In: Tekalur, S.A., Zavattieri, P., Korach, C.S. (eds.) Mechanics of Biological Systems and Materials, Volume 6, pp. 147–154. Springer, Cham (2016)
13. Khaleghi, M., Furlong, C., Cheng, J., Rosowski, J.: Thin-shell behavior of mammalian tympanic membrane studied by digital holography. In: Barthelat, F., Korach, C., Zavattieri, P., Prorok, B.C., Grande-Allen, K.J. (eds.) Mechanics of Biological Systems and Materials, Volume 7, pp. 19–25. Springer, Cham (2015)
14. Georgas, P.J., Schajer, G.S.: Modulo-2pi phase determination from individual ESPI images. Opt. Lasers Eng. **50**, 1030–1035 (2012)

Chapter 12
Rheology of Soft and Rigid Micro Particles in Curved Microfluidic Channels

Jia Liu, Yuhao Qiang, Michael Mian, Weihe Xu, and E. Du

Abstract We investigated the rheological behavior of micro particles in inertial flow in a curved microfluidic channel. Different from the typical microfluidic regime operating at low Reynolds number, inertial flow provides hydrodynamic manipulation, namely inertial focusing of particles at high flow speeds. Primary influences of inertial flow on particle motions are several: repulsive force from the wall due to a pressure buildup in the constriction between the wall and the particle, shear gradient lift force due to the parabolic flow profile at microscale, and secondary drag force in the cross-sectional direction due to channel curvature. These forces result in particle moving across the streamlines to certain predictable equilibrium positions in the flow. With regard to soft particles, their flow behavior and equilibrium positions may deviate from the theoretical predictions based on rigid particles. This study provides a proof-of-concept of inertial focusing-based separation of particles with different deformability. We demonstrated its capability by separating yeast cells and polystyrene particles of similar sizes in a double spiral channel.

Keywords Biological cell • Inertial focusing • Hydrodynamics • Deoxygenation • Particle separation

12.1 Introduction

Sickle cell disease (SCD) is caused by an inherited abnormal sickle hemoglobin (HbS), where the glutamic acid (E/Glu) in the normal β-chain is substituted by valine [1, 2]. This leads to intracellular HbS polymerization at low partial oxygen pressure (pO_2), and consequent morphological deformation, e.g. into its characteristic "sickle" shape. Diagnosis of SCD relies on detection of the presence of significant quantities of HbS by isoelectric focusing, cellulose acetate electrophoresis, high-performance liquid chromatography, or DNA analysis [3, 4]. A potential alternative method is label-free detectoin and counting fraction of red blood cells (RBCs) that can sickle under low pO_2 levels. This is supported by a comprehensive microfluidics study of sickle RBCs using a hypoxic microfluidic chip and a microvascular occlusion model [5]. It was found that the fraction of in vitro hypoxia-induced cell sickling is strongly correlated with the mean HbS concentrations. Cellular rheological behavior evaluated by mean cell velocity and capillary obstruction ratio, is closely associated with oxygen concentration and patients' hematological measures. In this case, a simple, effective yet relatively low-cost sample preparation, i.e. sorting rigid sickled cells from deformable unsickled cells, could benefit SCD diganosis and treatment.

Microfluidics has been broadly used in cell level operations. Separation of different cell types has been a subject of intense study. Among various microfluidic approaches, a promising method is inertial focusing, which was first implemented experimentally by Di Carlo et al. [6]. The physics behind inertial focusing and its applications are reviewed previously [7]. Briefly, in a straight microchannel, when Reynolds number is relatively high, competition between viscous dag and inertial life-forces exerted on a particle determines its equilibrium position. Such phenomenon has been used to focus and concentation of particles. In a curved microchannel, particles experience a secondary drag force in the cross-section, resulting in a better separation of one particle type from another. In such case, a marked feature of inertial focusing is high throughput. In addition, it has been demonstrated to be able to separate varied biological cells based on their size, shape and mechanical deformability [7–10]. Representative applications were found in separating different blood components from whole blood [11], circulating tumor cells from peripheral blood [12].

J. Liu • Y. Qiang • M. Mian • E. Du (✉)
Department of Ocean and Mechanical Engineering, Florida Atlantic University, Boca Raton, FL 33431, USA
e-mail: edu@fau.edu

W. Xu
National Synchrotron Light Source II, Brookhaven National Laboratory, Upton, NY 11973, USA

C.S. Korach et al. (eds.), *Mechanics of Biological Systems and Materials, Volume 6*,
Conference Proceedings of the Society for Experimental Mechanics Series, DOI 10.1007/978-3-319-41351-8_12

Previous theoretical study showed that elastic particles experienced an additional force which tends to push the particle away from the wall of channel [8]. Therefore, we hypothesize that rigid sickled RBCs should have an equilibrium position different from those softer, unsickled cells.

In this paper, we report the rheological behavior of microparticles in a double spiral channel, which was initially developed by Sun et al. [13] and utilized to separate tumor cells (MCF-7 and Hela cells) from diluted whole blood [14]. We explored the feasibility of this technique in separation of rigid particles from soft particles that are of similar sizes, and tested our hypothesis with a mixture of polystyrene particles and yest cells. Secondly, in order to apply this technique to sickle cells, we demonstrated a deoxygenation apparatus, as an essential component of sickle cell sample preparation to induce cell sickling before cell suspension is introduced into the spiral channel for separation. These preliminary studies indicated that this approach is promising to separate rigid sickled cells from soft unsickled ones.

12.2 Materials and Method

The inertial focusing for separation between soft and rigid cells consists of two components, including a hypoxia chamber for sample deoxygenation and a spiral channel for particle focusing and separation (Fig. 12.1). The spiral channel was fabricated using standard soft-lithography techniques. A 10:1 ratio mixture of (Polydimethylsiloxane) PDMS base and agent was thoroughly mixed and centrifuged to remove air bubbles. Then the mixture was poured over the mold and cured for 2 h in a convection oven at 70 °C. The cured PDMS was then peeled off from the mold and bonded to a glass slide with an air plasma treatment. A suspension of cells/particles was firstly loaded in a gas permeable silicone tubing (0.5 mm ID, 2.1 mm OD, 50 cm long, Cole-Parmer) by a syringe pump (New Era Pumps Systems Inc.). In order to fully deoxygenate the suspension, silicone tubing was placed in a plastic vial exposed to a hypoxic gas mixture (5 % CO_2 balanced with N_2). A PEEK tubing (Cole-Parmer) was used to connect silicone tubing and the spiral channel in order to prevent re-oxygenation of cell suspension. After the particle/cell suspension was completely deoxygenated, it was pumped into the spiral channel for separation.

To determine the actual oxygen concentration of the deoxygenated cell suspension in the spiral channel, an oxygen-sensitive dye, tris(2,2′-bipyridyl dichlororuthenium) hexahydrate (Ru(bpy), Sigma-Aldrich) was added into the medium and tested before working on particles. This fluorophore presents a linear relationship between dissolved oxygen concentration and fluorescence intensity [15]. We confirmed that after a 10-min deoxygenation in silicone tubing, the oxygen concentration of the medium pumped into the sprial channel was less than 3 % (Fig. 12.2). Hence, we expect this setup can be readily applied to sickle cells, since an oxygen concentration below 5 % is close to the in vivo pO_2 in venous blood and was sufficient to induce cell sickling in vitro [5, 16].

To demonstrate the capability of the spiral channels in inertial focusing and separation between rigid and soft particles, fluorescent polystyrene particles (Nile red, 5.0–5.9 μm, Spherotech) and yeast cells were mixed at a ratio of 1:5 and diluted into deionized water to a final total concentration of 1×10^6 particles per milliliter. Since the mean diameter of the fluorescent particles and the yeast cells are very close, we assume that the difference in particle deformability would be a

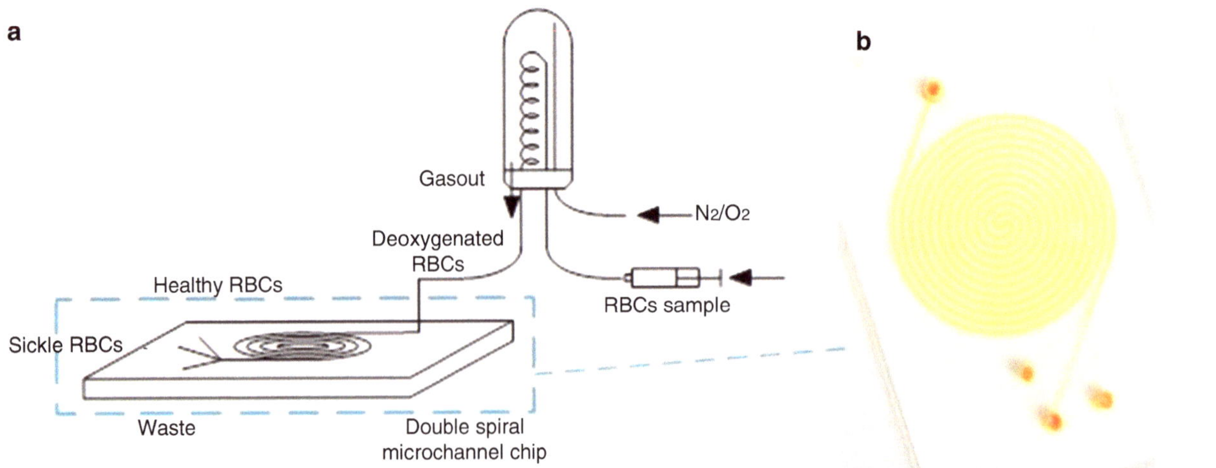

Fig. 12.1 (**a**) Schematic of inertial focusing of deoxygenated cell/particle suspension. (**b**) A photograph of the double spiral microfluidic channel Yellow dye was added to the channel for a visualization

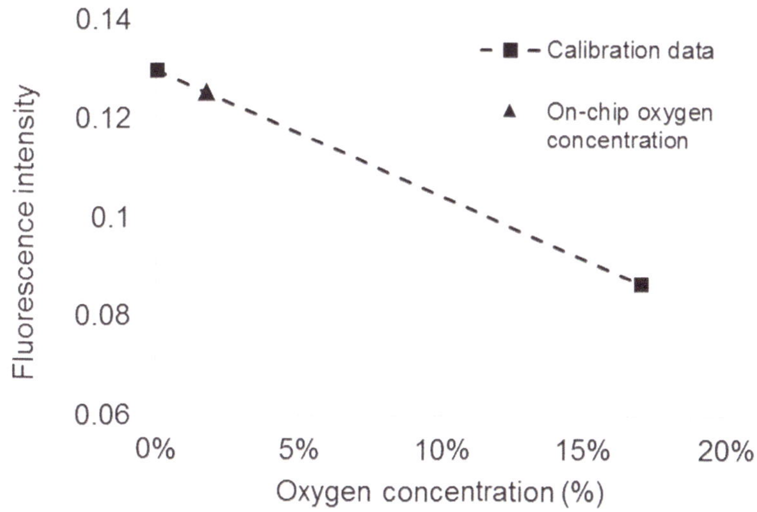

Fig. 12.2 Linear relationship between fluorescence intensity and oxygen concentration

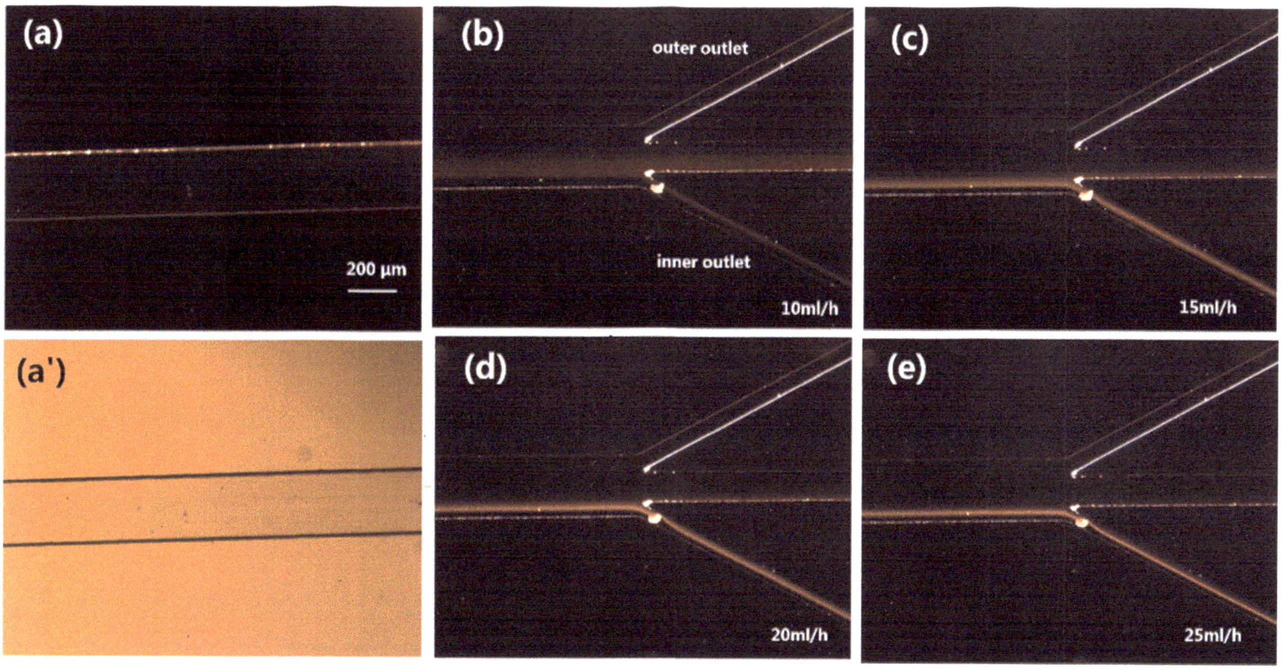

Fig. 12.3 (**a**) Fluorescence image and (**a′**) bright field image of uniformly distributed particles near the inlet. (**b–e**) Fluorescence images of particle focusing at outlets under various flow rates

primary factor that led to the different rheological behaviors of these two types of particles. The double spiral microchannel was initially filled with deionized water, followed by loading the suspension of particle mixture through the syringe pump. Flow rate was set to be 10, 15, 20 and 25 ml/h.

12.3 Results and Discussion

Particles were uniformly distributed at the entrance near the inlet (Fig. 12.3a, a′) and became focused at the outlets. When flow rate increases, the width of the focusing band decreased (Fig. 12.3b–e). It was found that at all the flow rates we tested, fluorescent particles were found to be focused along the inner side of the inner outlet.

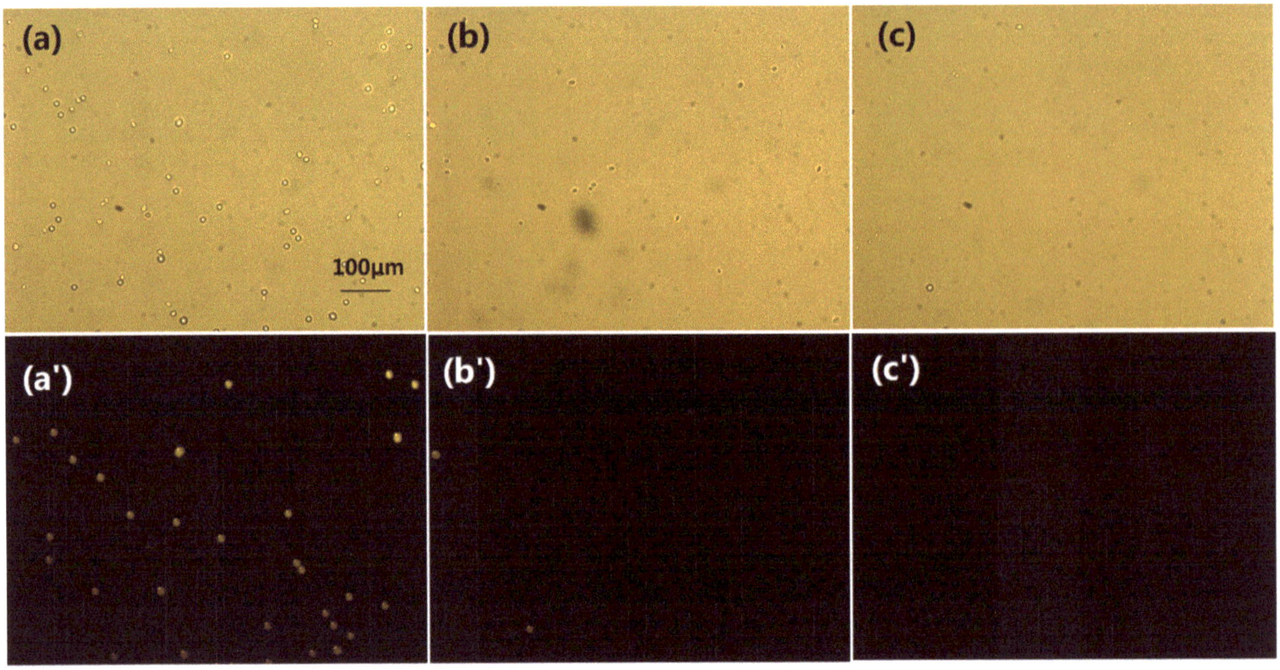

Fig. 12.4 (**a–c**) Bright field images and (**a′–c′**) fluorescent images of particles collected at inner outlet, middle outlet and outer outlet, respectively, at a flow rate of 15 ml/h. Samples were adjusted to a same target volume

Fig. 12.5 Fractions of fluorescent particles measured at the inner outlet and fraction of yeast cells measured at the middle outlet

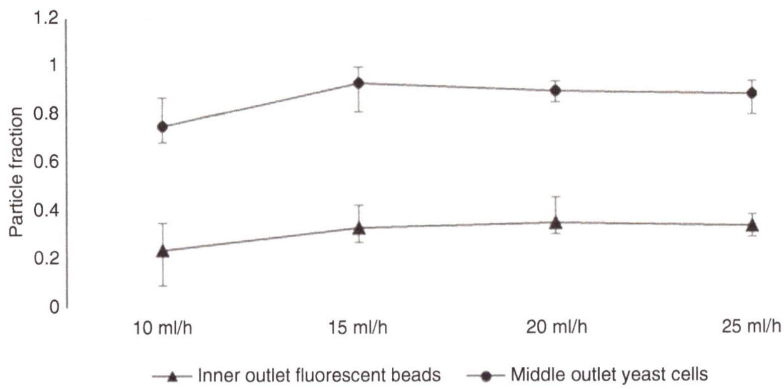

Under a flow rate of 15 ml/h, fractions of cells collected at the inner and middle outlets were quite similar (Fig. 12.4a, b). In contrast, few particles, including both yeast cells and fluorescent particles were found in the outer outlets (Fig. 12.4c, c′). Comparing the fluorescence images and bright field images, we found fluorescent particles were enriched at the inner outlet (Fig. 12.4a, a′) while yeast cells were enriched at the middle outlet (Fig. 12.4b, b′). These results indicate that this approach is capable of separating between rigid and soft particles and can be potentially utilized to separate rigid sickled cells from soft unsickled cells.

To evaluate the separation efficiency, five samples were obtained at each outlet and cell concentrations were counted (Fig. 12.5). The triangular points represent the fractions of fluorescent particles to the total amount of two kinds of particle came out from the inner outlets and the circular points represent the fractions of yeast cell to the total amount of two kinds of particle came out from the middle outlet. As the original ratio of the concentration of fluorescent particles to yeast cells is 1:5. The best performance in enrichment of the fluorescent particles and the yeast cells was found at the flow rate of 15 ml/h, which may be associated with the center-directing effect from the deformability-induced lift force.

12.4 Conclusion

This paper presented an inertial focusing method to separate rigid particles based on their deformability. We demonstrated such capability using yeast cells and polystyrene particles of similar sizes. A deoxygenation setup was calibrated and tested to be effective in deoxygenating sickle RBCs, which will facilitate generation of two separate populations of RBCs in sickle blood. In combination with this deoxygenation setup, we expect the double spiral channel will have the potential to separate rigid sickled cells from soft unsickled cells.

Acknowledgement This material is based upon work supported by the National Science Foundation under Grant No. 1464102.

References

1. Pauling, L., Itano, H.A., Singer, S.J., Wells, I.C.: Sickle cell anemia, a molecular disease. Science **110**, 543–548 (1949)
2. Ingram, V.M.: A specific chemical difference between the globins of normal human and sickle cell anaemia hemoglobin. Nature **178**, 792–794 (1956)
3. Bender, M.A., Douthitt Seibel, G.: Sickle cell disease. In: Pagon, R.A., et al. (eds.) GeneReviews®. University of Washington, Seattle, WA (1993)
4. National Heart LaBI: The management of sickle cell disease. NIH Publication. 02–2117 (2002)
5. Du, E., Diez-Silva, M., Kato, G.J., Dao, M., Suresh, S.: Kinetics of sickle cell biorheology and implications for painful vasoocclusive crisis. Proc. Natl. Acad. Sci. U. S. A. **112**, 1422–1427 (2015)
6. Di Carlo, D., Irimia, D., Tompkins, R.G., Toner, M.: Continuous inertial focusing, ordering, and separation of particles in microchannels. Proc. Natl. Acad. Sci. U. S. A. **104**, 18892–18897 (2007)
7. Di Carlo, D.: Inertial microfluidics. Lab Chip **9**, 3038–3046 (2009)
8. Hur, S.C., Henderson-MacLennan, N.K., McCabe, E.R.B., Di Carlo, D.: Deformability-based cell classification and enrichment using inertial microfluidics. Lab Chip **11**, 912–920 (2011)
9. Mach, A.J., Di Carlo, D.: Continuous scalable blood filtration device using inertial microfluidics. Biotechnol. Bioeng. **107**, 302–311 (2010)
10. Hur, S.C., Brinckerhoff, T.Z., Walthers, C.M., Dunn, J.C.Y., Di Carlo, D.: Label-free enrichment of adrenal cortical progenitor cells using inertial microfluidics. PLoS One **7**(10), e46550 (2012)
11. Jäggi, R.D., Sandoz, R., Effenhause, C.S.: Microfluidic depletion of red blood cells from whole blood in high-aspect-ratio microchannels. Microfluid. Nanofluid. **3**, 7 (2007)
12. Nagrath, S., et al.: Isolation of rare circulating tumour cells in cancer patients by microchip technology. Nature **450**, 1235–1239 (2007)
13. Sun, J., et al.: Size-based hydrodynamic rare tumor cell separation in curved microfluidic channels. Biomicrofluidics **7**, 11802 (2013)
14. Sun, J., et al.: Double spiral microchannel for label-free tumor cell separation and enrichment. Lab Chip **12**, 3952–3960 (2012)
15. Chang-Yena, D.A., Galea, B.K.: An integrated optical oxygen sensor fabricated using rapid-prototyping techniques. Lab Chip **3**, 5 (2003)
16. Du, E., Mendelsohn, L., Nichols, J.S., Dao, M., Kato, G.J.: Quantification of anti-sickling effect of Aes-103 in sickle cell disease using an in vitro microfluidic assay. Blood **124**, 2699 (2014)

Chapter 13
Microfluidic Approaches for Biomechanics of Red Blood Cells

E. Du

Abstract Studies of cellular mechanics have led to important advances in our understanding of the physiological processes of human diseases as well as provided adjunct biophysical markers for disease diagnosis and therapeutic evaluations. Microfluidic assays allow novel manipulations of single cells quickly and effectively under a physiologically appropriate environment. This paper briefly summarizes our two recently developed microfluidic techniques for probing mechanics of diseased human red blood cells (RBCs). In particular, the in vitro microvascular model allowed characterization of blood micro-rheology in terms of cellular velocity and capillary obstructions under transient hypoxic conditions. These analyses suggested novel mechanics indicators of disease severity and drug treatment in sickle cell anemia. The second microfluidic technique allowed rapid characterization of electrically coupled mechanics of cell membranes from a dielectrophoresis-induced uniaxial stretching of single RBCs. The obtained force-deformation curves agreed well with the independent studies reported in literature for both healthy RBCs and those influenced by *Plasmodium falciparum*. These findings clearly demonstrated the capabilities of microfluidics in quantitative, mechanical measurements of RBCs and can be potentially applied to other cell types.

Keywords Microfluidics • Cellular mechanics • Red blood cell • Sickle cell anemia • Malaria

13.1 Introduction

Normal red blood cells (RBCs) have a biconcave disc shape, about 8 μm in diameter and 2 μm in thickness; they must undergo severe passive deformation while maintaining mechanical stability during circulation, where they are subjected to intensive mechanical stimulation in blood flow or when squeezing through micro capillaries and splenic sinusoids [1]. The remarkable deformability of RBCs largely attributes to the mechanical properties of the membrane, its only structural component. During the repeated passages of several hundred 1000 times in their 120-day normal lifespan, RBCs lose their membrane integrity, become rigid, and are degraded by the mononuclear phagocyte system [2]. Various pathological and physiological conditions can accelerate this process, such as hereditary hemolytic anemias [3], extracorporeal circulation for hemodialysis [4], cardiopulmonary bypass [5], and cold storage of RBC products for transfusion purposes [6, 7].

Many engineering techniques, such as micropipette aspiration [8], atomic force microscopy [9, 10], optical tweezers [11–13], diffraction phase microscopy [14–16], magnetic twisting cytometry [17], have enabled high-precision mechanical measurements of individual cells and cellular components. These studies have demonstrated critical roles of cellular mechanics in various pathophysiological processes and provided novel biophysical markers for disease diagnosis and drug efficacy testing, in malaria [18–20], sickle cell disease [16, 21, 22], and cancer [23–25]. These techniques have advantages in high resolutions of force and displacement for probing single-cells and subcellular components. Despite the intense studies in cellular mechanics, limited measurements have yet been obtained at the single-cell level. It remains a challenging task to measure single cells under physiological conditions close to their in vivo microenvironments. However, it is crucial to measure large quantities of cells in order to account for the statistical dispersion in nature among the biological cells. Implementation of such measurements will lead to a better understanding of cellular dynamics and intercellular heterogeneity in their native biological context.

Microfluidics has been demonstrated to be an versatile tool for high throughput single-cell studies [20, 26–30], as well as intercellular communications in response to mechanical stimuli [31]. In order to recreate the specific factors that cells are exposed to in vivo, microfluidic systems can be designed to incorporate controls of flow, temperature, gas, chemical

E. Du (✉)
Department of Ocean and Mechanical Engineering, Florida Atlantic University, 777 Glades Road, Boca Raton, FL 33431, USA
e-mail: edu@fau.edu

C.S. Korach et al. (eds.), *Mechanics of Biological Systems and Materials, Volume 6*,
Conference Proceedings of the Society for Experimental Mechanics Series, DOI 10.1007/978-3-319-41351-8_13

gradients, and surface molecules etc. Our work has focused on development of microfluidic assays for cellular mechanics measurements in the context of human blood diseases. For instance, a fundamental mechanism of vascular occlusion in sickle cell disease has been recognized as a result of deoxygenated intracellular sickle hemoglobin. Sickle hemoglobin forms into fibers at low partial pressure of oxygen, causing rigid sickle cells that can severely impact blood flow in narrow vessels when they interact with other cell types in a complex manner [32]. Since many factors can be associated with deteriorated cellular mechanical properties, such as molecular level-variations in hemoglobin types and concentrations as well as local micro environmental factors, a testing platform with high flexibility will be useful to elucidate the role of each individual factors.

13.2 Microfluidic Techniques

13.2.1 Cellular Rheology Under Hypoxia

We developed an in vitro microvascular model for sickle cell disease. This model allowed rheological measurements of single cells under controlled transient hypoxia through a double-layer polydimethylsiloxane (PDMS) structure. Development and initial applications of this model have been reported in our previous publications [33, 34]. The key idea of this model is to control the oxygen level in cell suspension by diffusing gas mixtures of specific oxygen concentrations through a gas-permeable PDMS film, while forcing individual sickle cells moving through a series of periodic micro gates under a differential pressure at body temperature (Fig. 13.1). Initially, sickle cells were deformable and can move freely through the micro gates in exposure to a gas mixture with atmospheric oxygen levels. When a hypoxic gas mixture was introduced, oxygen concentration in the cell channel reduced to the same level as the hypoxic gas due to a diffusion equilibrium. As a result, the deoxygenated sickle hemoglobin inside cells form into rigid fibers, turning initially deformable cells into rigid sickled cells.

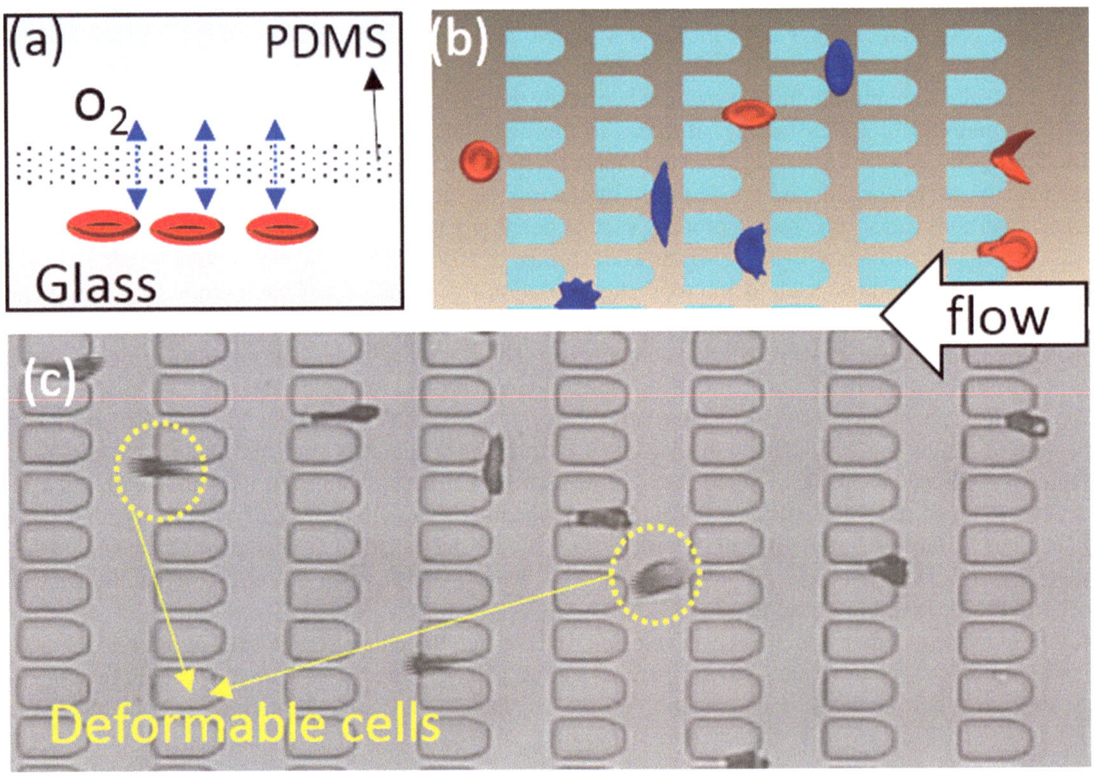

Fig. 13.1 In vitro microvascular model for sickle cell occlusion under transient hypoxia. (**a**) Schematic double-layer PDMS structure for gas diffusion. (**b**) Schematic of variations in cellular morphological sickling and obstructions (*blue*—rigid cells; *red*—deformable cells). (**c**) Experimental observation of single-cell rheology in microfluidic mimic of microvasculature. Highlighted cells are deformable while other cells are unable to pass through micro grates

Sickle cell rheological behavior was quantified by mean cellular velocity and capillary obstruction ratio. We observed marked variations in cellular morphology and rheological behavior in response to oxygen changes. To access whether changes in cellular rheological behavior are associated with the cell sickling process, we performed independent test on cellular morphological sickling in a static condition at high magnifications. This experiment allowed a direct observation of morphological deformation caused by intracellular polymerization of deoxygenated sickle hemoglobin. The fraction of sickled cells has been found to be strongly correlated with the fraction of capillary obstruction, indicating intracellular sickle hemoglobin polymerization was the primary factor that caused rigid sickled cells. Besides intracellular polymers, cellular shape also determined the rheological behavior of a cell passing through the micro gates. For instance, not all sickle-shaped cells will block the flow, they may conform their shapes by folding before the sickle hemoglobin fibers are formed, or they are narrow enough to pass through the confined apertures in alignment with the flow streamlines even if after cell sickling. Strong correlation has also been found between the in vitro rheological factors and patients' clinical measures, such as disease severity level, hematological parameters, and therapeutic treatment [34].

13.2.2 Electrically Coupled Cell Mechanics

We have established a microfluidic technique for the study of mechanical deformability of multiple single cells simultaneously through an Electrokinetic phenomenon, called dielectrophoresis [35]. This approach utilizes electrical forces exerted on the electrically induced dipole at the cell/medium interface to create a uniaxial stretching on single cells. The electric dipole is assumed to have equal magnitude but opposite signs that are separated by a distance equivalent to the major axes of the stretched RBCs. In this case, proper electrical frequencies are required to create positive dielectrophoresis with respect to the polarizabilities of cell and surrounding medium.

One side of the cell is held tightly at the electrode edges of maximum gradient of electric strength, which can be assumed as a fixed condition. The other side of the cell is subjected to an electric force acting in the opposite direction, resulting in a unidirectional stretching. We were able to probe multiple cells simultaneously through interdigitated thin-film electrode arrays embedded in a PDMS microfluidic chamber (Fig. 13.2). All cells were stretched in response to the alternating current electrical field. Some cells were stretched with their dimple in the plane of the stretching force and other cells were observed to spin around their major axes, which may be attributed to a torque exerted on the cells due to their geometrical asymmetry. This approach has been applied to measure the mechanical deformation of RBCs under electrical excitations at varied frequencies. Force-deformation profiles of RBCs indicated that those influenced by *Plasmodium falciparum* are more rigid than their healthy counterparts, in agreement with the independent studies by optical tweezers [13, 36]. The nonlinear force-deformation relationship can be used to determine shear modulus, using the closed forms of linear shear modulus derived from numerical simulations of optical tweezers stretching [36].

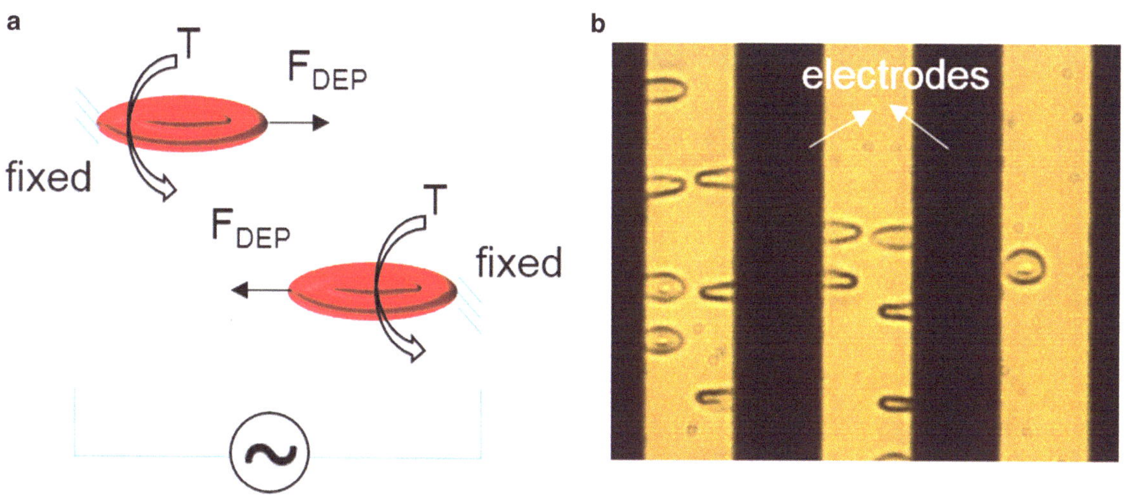

Fig. 13.2 (**a**) Schematic top view of electrically coupled mechanics of RBCs. (**b**) Experimental observation of unidirectional stretched RBCs in the electrical field at radio frequency generated by the interdigitated electrode arrays

13.3 Discussion

We have demonstrated that microfluidics provides a flexibility in the study of cellular mechanics at a relatively high throughput compared to existing methods. The sickle cell microvascular model has been designed to achieve specific measurements under physiological conditions mimicking cellular environment in vivo, such as flow, oxygen level and temperature. Relationships between in vitro microfluidics measurements and cellular disease states have been established. This model can be further improved for more comprehensive study of single cells and cell-cell interactions. For instance, in the model of vascular occlusion in sickle cell disease, we demonstrated the cellular velocity and capillary obstructions in response to oxygen changes. Additional factors, such as a miniature fluidic network mimicking blood vasculature, endothelial lining of microchannel, and other blood components can be added to the current model to elucidate the interferences between these biological factors and single-cell, intercellular mechanics. The electrically coupled mechanics method provided an alternative method for probing single cell mechanics. The electrical excitations can be further programmed to study the dynamics of cells in response to different loading rates.

Acknowledgement This work was supported by the NIH grant (R01HL094270 and U01HL114476).

References

1. Mohandas, N., Gallagher, P.G.: Red cell membrane: past, present, and future. Blood **112**, 3939–3948 (2008)
2. Lang, F., Lang, E., Foller, M.: Physiology and pathophysiology of eryptosis. Transfus. Med. Hemother. **39**, 308–314 (2012)
3. Barabino, G.A., Platt, M.O., Kaul, D.K.: Sickle cell biomechanics. Annu. Rev. Biomed. Eng. **12**, 345–367 (2010)
4. Polaschegg, H.D.: Red blood cell damage from extracorporeal circulation in hemodialysis. Semin. Dial. **22**, 524–531 (2009)
5. Black, K.L., Jones, R.D.: The discocyte-echinocyte transformation as an index of human red cell trauma. Ohio J. Sci. **75**, 225–230 (1976)
6. Hendrickson, J.E., et al.: Rapid clearance of transfused murine red blood cells is associated with recipient cytokine storm and enhanced alloimmunogenicity. Transfusion **51**, 2445–2454 (2011)
7. Bosman, G.J.C.G.M., et al.: The proteome of red cell membranes and vesicles during storage in blood bank conditions. Transfusion **48**, 827–835 (2008)
8. Chien, S.: Micropipette method for studying red and white blood-cells. Clin. Hemorheol. **5**, 655 (1985)
9. Lulevich, V., Zink, T., Chen, H.Y., Liu, F.T., Liu, G.Y.: Cell mechanics using atomic force microscopy-based single-cell compression. Langmuir **22**, 8151–8155 (2006)
10. Carvalho, P.A., Diez-Silva, M., Chen, H., Dao, M., Suresh, S.: Cytoadherence of erythrocytes invaded by *Plasmodium falciparum*: quantitative contact-probing of a human malaria receptor. Acta Biomater. **9**, 6349–6359 (2013)
11. Huruta, R.R., et al.: Mechanical properties of stored red blood cells using optical tweezers. Blood **92**, 2975–2977 (1998)
12. Dao, M., Lim, C.T., Suresh, S.: Mechanics of the human red blood cell deformed by optical tweezers. J. Mech. Phys. Solids **51**, 2259–2280 (2003)
13. Mills, J.P., Qie, L., Dao, M., Lim, C.T., Suresh, S.: Nonlinear elastic and viscoelastic deformation of the human red blood cell with optical tweezers. Mech. Chem. Biosyst. **1**, 169–180 (2004)
14. Park, Y.K., et al.: Refractive index maps and membrane dynamics of human red blood cells parasitized by *Plasmodium falciparum*. Proc. Natl. Acad. Sci. U. S. A. **105**, 13730–13735 (2008)
15. Park, Y.K., et al.: Metabolic remodeling of the human red blood cell membrane. Proc. Natl. Acad. Sci. U. S. A. **107**, 1289–1294 (2010)
16. Byun, H., et al.: Optical measurement of biomechanical properties of individual erythrocytes from a sickle cell patient. Acta Biomater. **8**, 4130–4138 (2012)
17. Puig-De-Morales-Marinkovic, M., Turner, K.T., Butler, J.P., Fredberg, J.J., Suresh, S.: Viscoelasticity of the human red blood cell. Am. J. Phys. Cell Phys. **293**, C597–C605 (2007)
18. Rey, J., et al.: Reduced erythrocyte deformability associated with hypoargininemia during *Plasmodium falciparum* malaria. Sci. Rep. **4**, 3767 (2014)
19. Hosseini, S.M., Feng, J.J.: How malaria parasites reduce the deformability of infected red blood cells. Biophys. J. **103**, 1–10 (2012)
20. Bow, H., et al.: A microfabricated deformability-based flow cytometer with application to malaria. Lab Chip **11**, 1065–1073 (2011)
21. Barodka, V.M., et al.: New insights provided by a comparison of impaired deformability with erythrocyte oxidative stress for sickle cell disease. Blood Cell Mol. Dis. **52**, 230–235 (2014)
22. Itoh, T., Chien, S., Usami, S.: Effects of hemoglobin concentration on deformability of individual sickle cells after deoxygenation. Blood **85**, 2245–2253 (1995)
23. Zhang, Z.F., Xu, J., Hong, B., Chen, X.L.: The effects of 3D channel geometry on CTC passing pressure—towards deformability-based cancer cell separation. Lab Chip **14**, 2576–2584 (2014)
24. Gabor, H., Weiss, L.: Perturbations in cancer cell deformability and resistance to shear forces. Invas. Metast. **6**, 166–179 (1986)
25. Cohen, M.H.: Red-blood-cell deformability (Rbcd) influences chemotherapy results in small cell lung-cancer (Sclc) patients (Pts). Proc. Am. Assoc. Cancer Res. **22**, 501 (1981)
26. Huang, S., et al.: Dynamic deformability of *Plasmodium falciparum*-infected erythrocytes exposed to artesunate in vitro. Integr. Biol. (Camb.) **5**, 414–422 (2013)

27. Dudani, J.S., Gossett, D.R., Tse, H.T.K., Di Carlo, D.: Pinched-flow hydrodynamic stretching of single-cells. Lab Chip **13**, 3728–3734 (2013)
28. Guido, I., Jaeger, M.S., Duschl, C.: Dielectrophoretic stretching of cells allows for characterization of their mechanical properties. Eur. Biophys. J. Biophys. Lett. **40**, 281–288 (2011)
29. Chen, J., et al.: Electrodeformation for single cell mechanical characterization. J. Micromech. Microeng. **21**, 054012 (2011)
30. Shevkoplyas, S.S., Yoshida, T., Gifford, S.C., Bitensky, M.W.: Direct measurement of the impact of impaired erythrocyte deformability on microvascular network perfusion in a microfluidic device. Lab Chip **6**, 914–920 (2006)
31. Polacheck, W.J., Li, R., Uzel, S.G., Kamm, R.D.: Microfluidic platforms for mechanobiology. Lab Chip **13**, 2252–2267 (2013)
32. Embury, S.H.: The not-so-simple process of sickle cell vasoocclusion. Microcirculation **11**, 101–113 (2004)
33. Li, X., et al.: Patient-specific blood rheology in sickle-cell anaemia. Interface Focus **6**, 20150065 (2016)
34. Du, E., Diez-Silva, M., Kato, G.J., Dao, M., Suresh, S.: Kinetics of sickle cell biorheology and implications for painful vasoocclusive crisis. Proc. Natl. Acad. Sci. U. S. A. **112**, 1422–1427 (2015)
35. Du, E., Dao, M., Suresh, S.: Quantitative biomechanics of healthy and diseased human red blood cells using dielectrophoresis in a microfluidic system. Extreme Mech. Lett. **1**, 35–41 (2014)
36. Dao, M., Li, J., Suresh, S.: Molecularly based analysis of deformation of spectrin network and human erythrocyte. Mat. Sci. Eng. C **26**, 1232–1244 (2006)

Chapter 14
Custom Indentation System for Mechanical Characterization of Soft Matter

Chelsey Simmons, Andres Rubiano, Daniel Stewart, and Brandey Andersen

Abstract Soft, hydrated samples of biomaterials and live tissues are hard to characterize with standard equipment designed for engineering materials. To accommodate millimeter-scale samples with sub-kPa moduli, we designed a cantilever-based indentation system using piezoelectric actuators and capacitive sensors compatible with a variety of probe tips and sample holders. Embryonic mouse hearts, rat gastrointestinal tissues, human pancreas, and rat brain tissue have all been successfully characterized using this system.

Keywords Experimental apparatus • Tissue mechanics • Viscoelasticity • Indentation • Compression testing

14.1 Introduction

Common techniques to characterize soft matter include macroscale compression and tension of centimeter-scale tissue samples and nanoscale indentation by atomic force microscopy or nanoindentation instruments. However, the wide variety in indentation methods and material models obfuscate comparisons among the data and do not accommodate millimeter-scale samples like embryonic mouse hearts or excised human tumors. To characterize the material properties of a variety of biomaterials and isolated tissues, we designed a cantilever-based indentation system for soft matter. We then fit the indentation data to the common standard linear solid viscoelastic material model using a modified Hertz contact model, which allows us to calculate a steady-state elastic modulus (E_{ss}) and viscosity, as well as an instantaneous (E_a) and a total elastic modulus for comparison to existing data in the literature. We have used this system to experimentally characterize synthetic soft matter, biomaterials, and excised tissue.

14.2 Methods

Force-Displacement Quantification with Custom Indenter. We modified a tool designed to measure friction forces to create a custom Multi-Scale Indenter (MSI, Fig. 14.1) [1, 2]. The cantilever-based probe (Fig. 14.1b2) is displaced vertically using a software-controlled piezoelectric stage (Fig. 14.1b1, P-628.1CD, Physik Instrumente). A custom program in LabVIEW (National Instruments) was used to control indentation profile and to read deflection of cantilever tip with capacitive sensor (Fig. 14.1b3, C8S-3.2-2.0 and compact driver CD1-CD6, Lion Precision) through a data acquisition card system (NI 9220 and cDAQ-9171, National Instruments). The probe with a 4 mm-diameter rigid tip (Fig. 14.1b4) is brought

C. Simmons (✉)
Department of Mechanical and Aerospace Engineering, University of Florida, PO Box 116250, Gainesville, FL, USA

J. Crayton Pruitt Family Department of Biomedical Engineering, Herbert Wertheim College of Engineering, University of Florida, PO Box 116250, Gainesville, FL, USA

Division of Cardiovascular Medicine, Department of Medicine, College of Medicine, University of Florida, PO Box 116250, Gainesville, FL, USA
e-mail: css@ufl.edu

A. Rubiano
Department of Mechanical and Aerospace Engineering, University of Florida, PO Box 116250, Gainesville, FL, USA

D. Stewart • B. Andersen
J. Crayton Pruitt Family Department of Biomedical Engineering, Herbert Wertheim College of Engineering, University of Florida, PO Box 116250, Gainesville, FL, USA

C.S. Korach et al. (eds.), *Mechanics of Biological Systems and Materials, Volume 6*,
Conference Proceedings of the Society for Experimental Mechanics Series, DOI 10.1007/978-3-319-41351-8_14

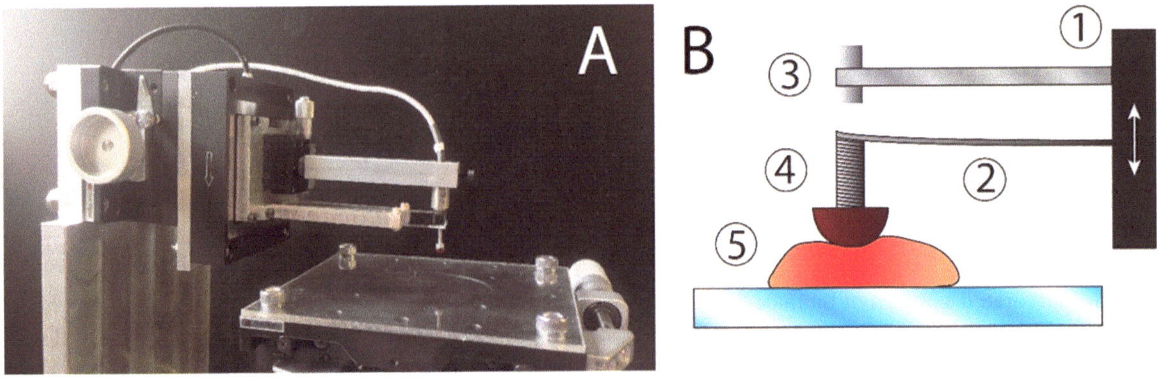

Fig. 14.1 (**a**) Custom indentation system for characterization of soft matter. (**b**) A rigid probe (*4*) indents tissue or soft matter sample (*5*) located on top of an acrylic surface. The probe is mounted directly to a cantilevered flexure (*2*), and capacitance probes (*3*) measure the displacement of a target on the calibrated flexure relative to movement of piezoelectric stage (*1*), yielding forces

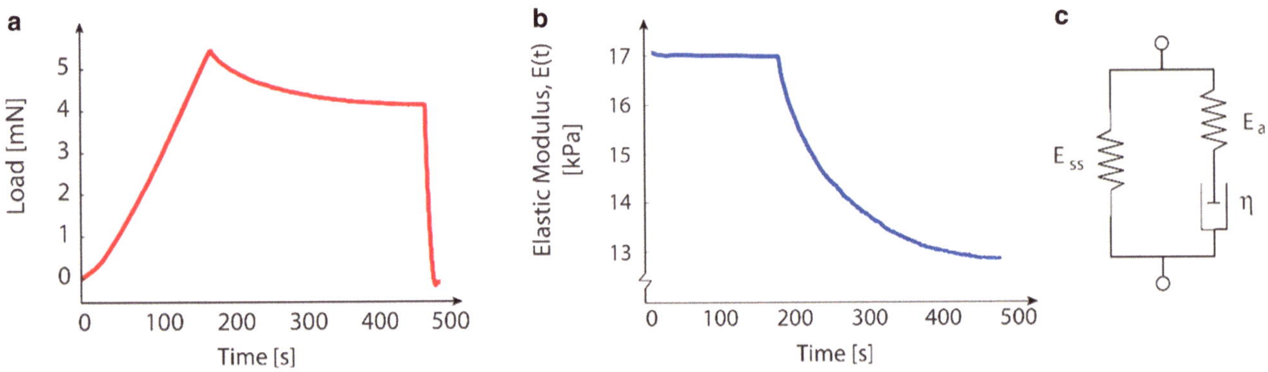

Fig. 14.2 Representative data for indentation of a slice of heart tissue from a rat. (**a**) Acquired force over time throughout an indent-hold-retract cycle. (**b**) Plotted elastic modulus as a function of time using the Hertz JKR contact model for spherical tip in contact with flat sample. (**c**) Springs and dashpot representation of the Standard Linear Solid model

into contact with the tissue sample following a time-dependent displacement profile set by the user in custom LabVIEW code. The reacting normal force of the tissue bends the titanium cantilever, and the relative displacement of the cantilever tip is measured by capacitive probe. The stiffness of the cantilever, calibrated by suspending weights and measuring deflection, is utilized by the LabVIEW code to determine the normal load throughout the indentation cycle.

Contact Mechanics and Viscoelastic Relaxation Model. Force-displacement data from the original indentation is converted into stress relaxation-like modulus as a function of time using the Hertz contact model for a spherical tip (Eq. 14.1):

$$E(t) = \frac{3 \cdot F(t) \cdot (1 - \nu^2)}{4 \cdot \sqrt{R} \cdot \delta(t)^{\frac{3}{2}}} \tag{14.1}$$

where $E =$ effective modulus, $F =$ normal force calculated by MSI through LabVIEW, $\nu =$ Poisson's Ratio, $R =$ tip radius, and $\delta =$ indentation depth [3].

We then fit modulus data to a standard linear solid (SLS) model, a common viscoelastic constitutive model composed of representative springs (linear elasticity) and a dashpot (viscosity) (Fig. 14.2). This simple combination of linear elastic and viscous components has a straightforward physical interpretation, making this model particularly useful for interdisciplinary investigations with collaborators from broad backgrounds. The strain rate-dependent and "liquid-like" properties of the material are represented by viscosity, η, while $E_{ss} + E_a$ is the effective modulus of the tissue during initial compression. As the dashpot η relaxes, E_a trends to zero and E_{ss} can be interpreted as the steady-state or equilibrium modulus.

To determine these parameters, we fit the relaxation portion of the calculated *E(t)* versus time plot to the following equation [4]:

$$E_t(t) = E_{ss} + E_a \cdot e^{-t/\tau} \tag{14.2}$$

where E_t = total effective elastic modulus as a function of time, E_{ss} = steady state elastic modulus after complete relaxation, E_a = apparent strain-rate-dependent contribution to initial elastic modulus, and $\tau = \eta/E_a$ with η = viscosity. Special care needs to be taken when selecting an appropriate relaxation time for the exponential regression to result in positive, real number values for the three properties. Long relaxation times can cause viscosity values to be imaginary, while insufficient time can return inaccurate values for both elastic moduli. Additionally, all data points used to obtain SLS coefficients must reside within the relaxation portion of the elastic modulus function for the viscosity to be accurate and positive.

Soft Matter Handling and Testing. A variety of soft matter samples have been used with this system, including synthetic hydrogels like polydimethylsiloxane and polyacrylamide, biomaterials like collagen gels and decellularized tissue scaffolds, and excised tissue from animals and humans. Maintaining in vitro sample stability and hydration and quantifying sample thickness are paramount concerns for successful characterization. Tissue samples are typically received from the operating room 1–2 h after excision, during which time they are stored on ice. Upon receipt, we transfer to a humid container lined with water-soaked tissues to maintain hydration and continue to store on ice. Characterization of fully submerged samples is complicated by buoyancy forces on the large probe tip, so for hydrogels, we typically constrain sample in custom silicone wells or on high friction surface (e.g. sandpaper) and surround sample with room temperature saline.

Consistency of indentation data relies on indents approximately 10 % of sample thickness, so quantification of sample thickness is also important. For large (centimeter-scale) samples, we typically section them to known thickness with a matrix slicer (e.g. Ted Pella or Zivic). For smaller or awkwardly shaped samples, we measure the distance between the surface holding the sample and the top of the sample by using the encoded piezostage for thin (<800 μm) sections and the manual z-stage vernier scale for thicker samples.

14.3 Results

Operation of MSI. Through appropriate cantilever design, we are able to apply normal loads of less than 10 μN. Our chosen cantilever and capacitive probe combination can measure forces with uncertainties ~25 μN and resolution ~0.5 μN. Real-time monitoring using a graphical user interface in LabVIEW allows visualization of indentations and estimation of uncertainty in calculated force (Fig. 14.3). Our current configuration has enabled characterization of the steady-state modulus for hydrogels (0.1–100 kPa), human pancreas (500 Pa–20 kPa), rat small intestine and colon (1–10 kPa), rat, mouse, and human brain tissue (500 Pa–4 kPa), and rat hearts (5–50 kPa). Details on select results are below.

While we have chosen to fit our force-displacement indentation data to a viscoelastic model, the system can be configured for unconfined or porous-platen compression and fit a variety of different material models. By controlling the piezoelectric stage through LabVIEW, arbitrary profiles can be created to satisfy test conditions for many constitutive models.

Cross-Linking of Collagen Hydrogels. In the body, collagen concentration and the mechanical properties of tissues are correlated. By creating a tunable range of moduli over different collagen concentrations, it is possible to isolate one variable from the other and study the effects of the mechanical microenvironment on cell behavior. In our system, stiffness can be controlled using different conditions of collagen concentration and glutaraldehyde crosslinking that result in overlapping values of elastic moduli (Fig. 14.4). Figure 14.4 also demonstrates custom silicone holder designed for efficient handling of soft biomaterials.

Mechanical Properties of Normal and Malignant Brain Tissue. This indentation system has been used to characterize changes in health and disease of tissues [2]. For example, brain tissue from mice with xenograft human tumors demonstrated significant differences in mechanical properties between normal and tumor regions (Fig. 14.5). Results from such characterizations are critical to designing relevant biomimetic microenvironments in which to test cancer cells for proliferation, migration, and chemoresistance.

14.4 Conclusion

While many groups have characterized the elastic properties of common tissues such as hearts, pancreas, and intestine, they vary widely in indentation methods. These variations in methods and material models obfuscate comparisons among the data. Reported data on the elastic modulus of heart muscle, for example, range from common accepted values for brain tissue

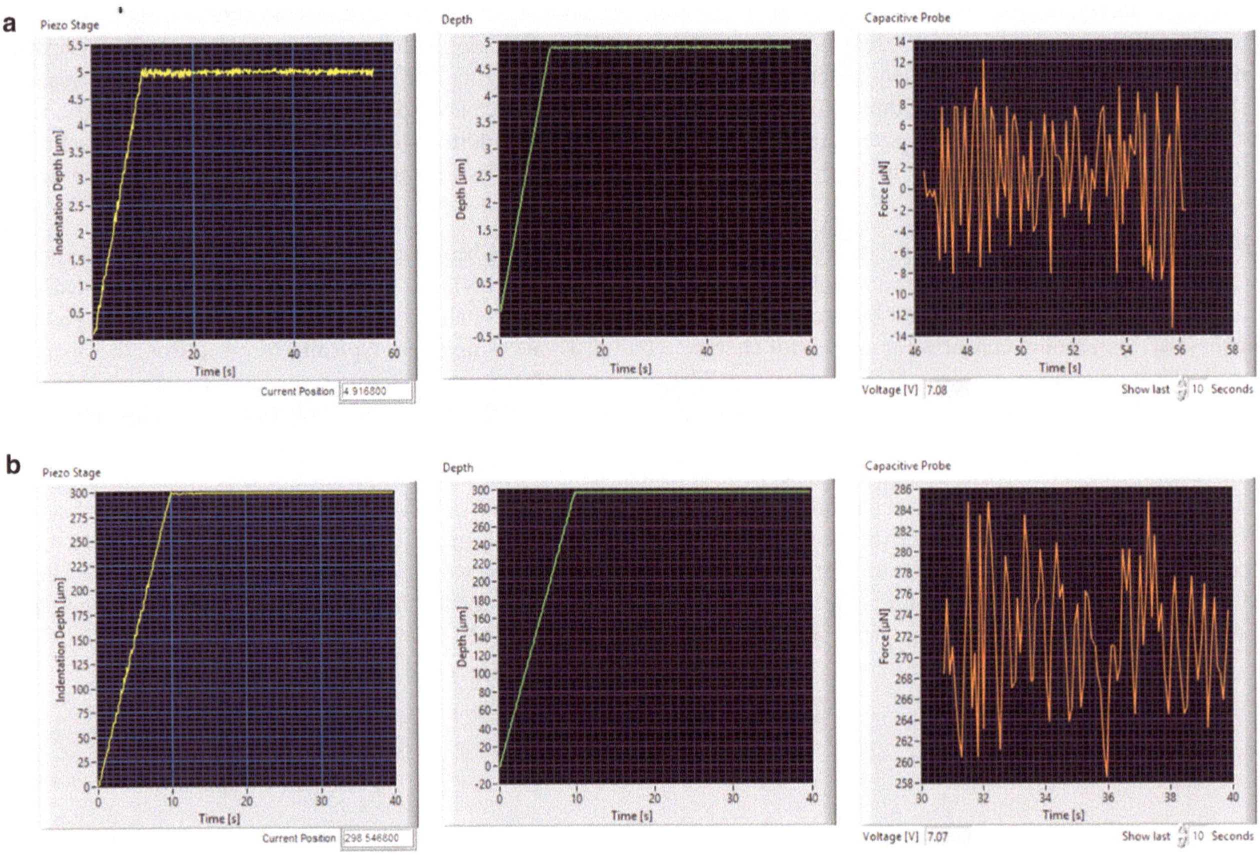

Fig. 14.3 Screen shots of LabVIEW graphical user interface during example indentations (**a**) in air and (**b**) in contact with a polyacrylamide gel. *Left panels* show measured displacement at tip of cantilever by capacitive probe; *center panels* show actual displacement of piezoelectric stage and base of cantilever; *right panels* show fluctuations in instantaneous calculated force

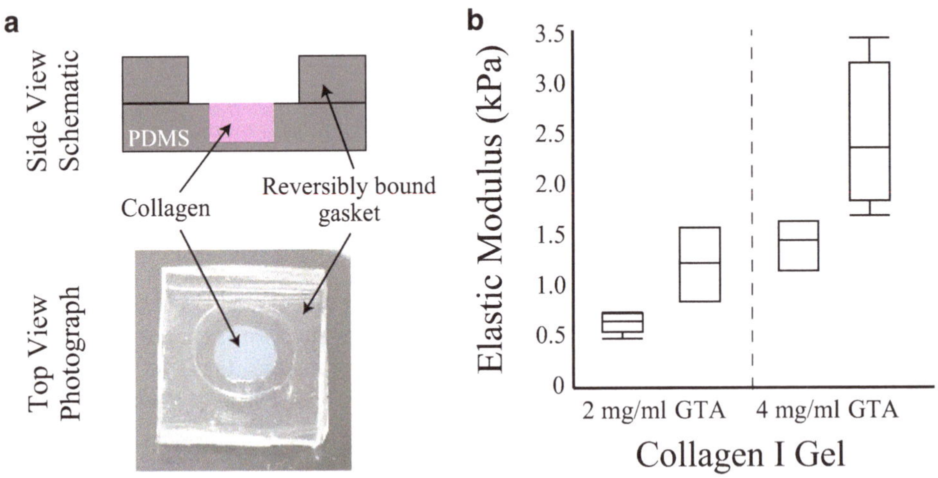

Fig. 14.4 (**a**) Silicone-based constructs are custom fabricated for collagen gel experiments to be compatible with cell culture and indentation characterization. Three mm-thick sheets of polydimethylsiloxane (PDMS) are punched with 7 mm-diameter biopsy punch and permanently bound to a 250 µm-thick silicone sheet. Corresponding top view photograph shows polymerized collagen gel and a silicone gasket reversibly bound to the top to hold media. (**b**) Treatment with 1 % glutaraldehyde (GTA) for 4 h increases the modulus of collagen gels, enabling future mechanotransduction experiments independently investigating collagen density and modulus effects

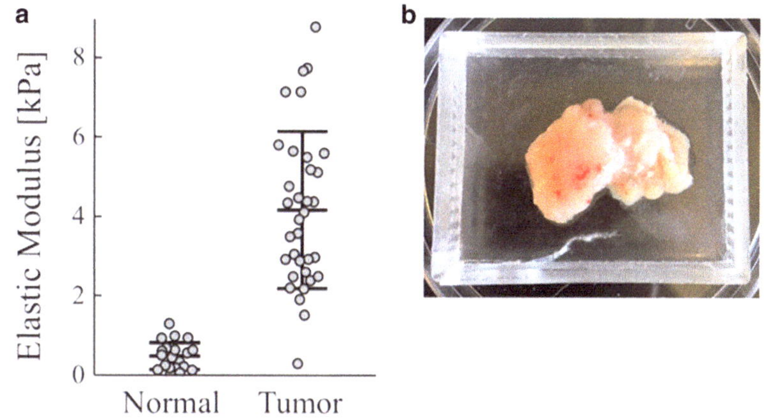

Fig. 14.5 Tumor region of mouse brain displays significantly greater elastic modulus than normal region in same brain ($p < 0.0001$) when characterized with MSI. (**a**) Steady-state elastic modulus was found to be significantly different between normal mouse brain ($n = 47$ indentations, $E_P = 0.62 \pm 0.07$ kPa) and the tumors ($n = 43$ indentations, $E_P = 3.69 \pm 0.33$ kPa). (**b**) Xerograph human cancer cells were implanted into mouse brains (*right portion of tissue*) and formed large tumors (*left portion of tissue*). Tissue was indented 3–4 h after isolation from euthanized animals

(~1 kPa) to pre-calcified bone (100–1000 kPa) depending on method and model used to approximate elastic modulus [2]. We have demonstrated a custom-built indentation apparatus that effectively quantifies force-displacement data on arbitrarily shaped hydrogels and tissues. By fitting the force-displacement data to a modified Hertz contact model and a Standard Linear Solid viscoelastic model, we are able to effectively characterize and compare a variety of soft matter materials.

Acknowledgements The authors would like to acknowledge research support from Medtronic (CSS) and the UF Institute for Cell & Tissue Science and Engineering (AR).

References

1. Krick, B.A., Vail, J.R., Persson, B.N.J., Sawyer, W.G.: Optical in situ micro tribometer for analysis of real contact area for contact mechanics, adhesion, and sliding experiments. Tribol. Lett. **45**(1), 185–194 (2011)
2. Rubiano, A., Qi, Y., Guzzo, D., Rowe, K., Pepine, C., Simmons, C.: Stem cell therapy restores viscoelastic properties of myocardium in rat model of hypertension. J. Mech. Behav. Biomed. Mater. **59**, 71–77 (2015)
3. Hayes, W.C., Keer, L.M., Herrmann, G., Mockros, L.F.: A mathematical analysis for indentation tests of articular cartilage. J. Biomech. **5**(5), 541–551 (1972)
4. Moore, S.W., Keller, R.E., Koehl, M.A.: The dorsal involuting marginal zone stiffens anisotropically during its convergent extension in the gastrula of *Xenopus laevis*. Development **121**(10), 3131–3140 (1995)

Chapter 15
Experimental Evaluation of Blast Loadings on the Ear and Head with and Without Hearing Protection Devices

Tim J. Walilko, Ryan D. Lowe, Ted F. Argo, G. Doug Meegan, Nathaniel T. Greene, and Daniel J. Tollin

Abstract This work will describe the experimental evaluation of blast loadings on post mortem human surrogates instrumented to obtain pressure and strain measurements at various locations as a precursor to studies on biological models to investigate neurological damage to the auditory cortex. The evaluation was conducted to validate improvements to auditory hazard models, and investigate the potential of bone conduction as a mechanism for propagating energy into the ear/head at levels sufficient to cause damage to components of the auditory system. The effectiveness of various hearing protection devices were investigated in a large shock tube at blast overpressures warfighters are typically exposed. Pressure measurements were made in the outer ear, middle ear, and within the cochlea using fiber optic pressure gages. Strain was monitored on the frontal and temporal bones. This work was supported by the U.S. Army Medical Research and Material Command under Contract No. W81XWH-15-2-0002.

Keywords Blast overpressure • Auditory injury • Hearing protection devices • Insertion loss • Electro-acoustic models

15.1 Introduction

In a combat environment damage to the peripheral auditory system (ear) is typically a result of exposure to high intensity impulsive pressure waves (blast) generated by an explosive device, shoulder launched rocket, mortar round, and/or weapons fire. While the devices that generate blast exposures may differ, all of the blast waves generated have a similar shape, a discontinuous rise in pressure over ambient conditions followed by an exponential decay. To investigate how hearing protection devices (HPDs) can be used to mitigate damage to the ear, the pressure-time traces from various weapons and explosive devices have been captured in the combat environment and duplicated in the laboratory using a shock tube. The following is a brief introduction on how the blast waves were generated and applied to a post mortem human surrogate (PMHS) to characterize the neurological damage induced to the ear. This work is important since symptoms of hearing damage (headache, sleep disturbances, avoidance/withdrawal from social settings) mimic symptoms associated with traumatic brain injury (TBI). Through a better understanding of the contribution of auditory dysfunction to these symptoms, better mitigation, diagnosis, and treatment protocols can be developed. The first step in this process is understanding how the pressure wave interacts with the structures of the head and how protective equipment such as hearing protection devices mitigate or enhance the interaction.

15.2 Methods

Data from 120 shock tube tests using eight different PMHS specimens were conducted to characterize the change in pressure at various locations in ear with and without hearing protection. The peak pressure for the four pressure waves used in the effort were 7, 28, 55 and 83 kPa which covers the range of pressures service members are typically exposed to in combat. The pressure wave was generated by a mobile shock tube (Fig. 15.1) built and designed by Applied Research Associates (Littleton, CO). The mobile shock tube consists of a 50-ft trailer which houses an 18 in. diameter shock tube (Fig. 15.1),

T.J. Walilko • R.D. Lowe (✉) • T.F. Argo • G.D. Meegan
Applied Research Associates, Inc. Rocky Mountain Division, 7921 Shaffer Parkway, Littleton, CO 80127, USA
e-mail: rdlowe@ara.com

N.T. Greene • D.J. Tollin
Department of Physiology and Biophysics, University of Colorado School of Medicine, 13001 E 17th Place, Aurora, CO 80045, USA

C.S. Korach et al. (eds.), *Mechanics of Biological Systems and Materials, Volume 6*,
Conference Proceedings of the Society for Experimental Mechanics Series, DOI 10.1007/978-3-319-41351-8_15

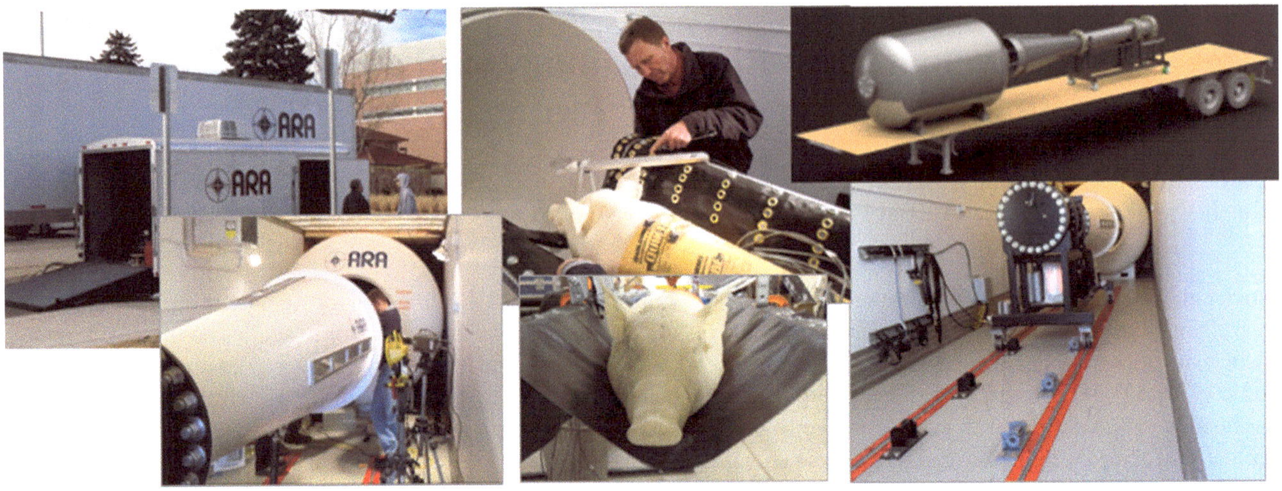

Fig. 15.1 Mobile Shock Laboratory is capable of generating a 3 ms in duration pressure wave in the excess of 480 kPa (70 psi). This is equivalent to standing 2.5 m from a detonating M107 105 mm Artillery round. Longer duration pressure pulses can be generated with the shock tube at lower peak pressure levels

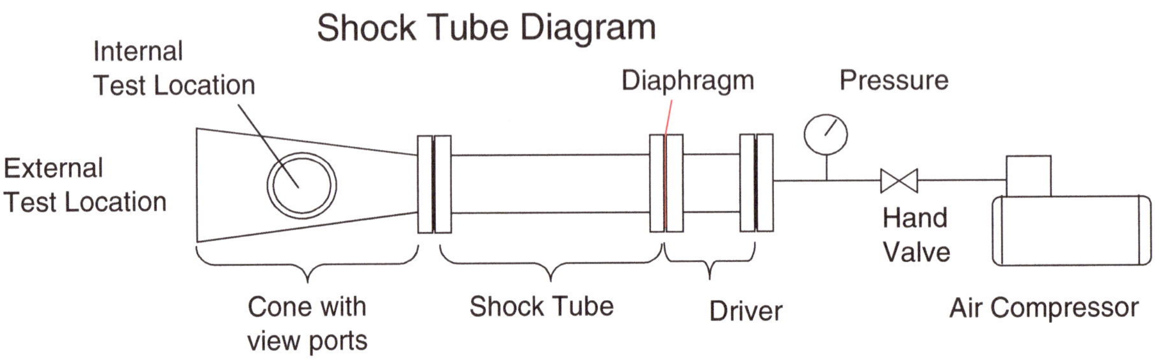

Fig. 15.2 Schematic of the 18 inch-diameter Shock Tube with 8-foot Expansion Cone. To operate the tube the driver pressure section is filled with compressed gas until the diaphragm ruptures, thereby sending a shock wave down the shock tube section and into the cone. The burst pressure is controlled by the thickness of the diaphragm material used

acoustic suppressors to minimize external impulse noise during testing, and an 8-ft expansion cone (Fig. 15.2) which enables generation of high impulse loads with blast wave durations in excess of 5 ms. The blast loads generated by the shock tube have been verified against free field blasts of the same magnitude (Fig. 15.3). The mobile laboratory was selected for this study since it can be moved to the site of testing to minimize the necessary transportation of biological subjects between the blast testing facility and support and imaging centers.

To measure the response of the head, 26 sensors were implanted into the head including some sensors placed in the auditory system and others placed on the skull surface. The purpose of the sensors was to collect the data required for characterizing the pressure inside the external auditory canal with and without the insertion of hearing protection. The sensors selected for this test series are shown in Table 15.1. The instrumentation includes fiber optic pressure gauges inside the ear (Fig. 15.4), pressure gauges on the surface of the skull near the ear, and strain gauges that directly measured the flexure/vibrations of the temporal and frontal bones resulting from the blast exposure. All of the data was digitized at a sampling rate 200 kHz with an anti-aliasing filter of 50 kHz.

To determine the non-linear parameters within the Auditory Hazard Assessment Algorithm for Humans (AHAAH) hearing protection module, four different types of HPD were evaluated. For performance comparisons, the response measured using each type of protection was compared against the empty ear. Table 15.2 is a brief description of the HPD that were used in the PMHS testing.

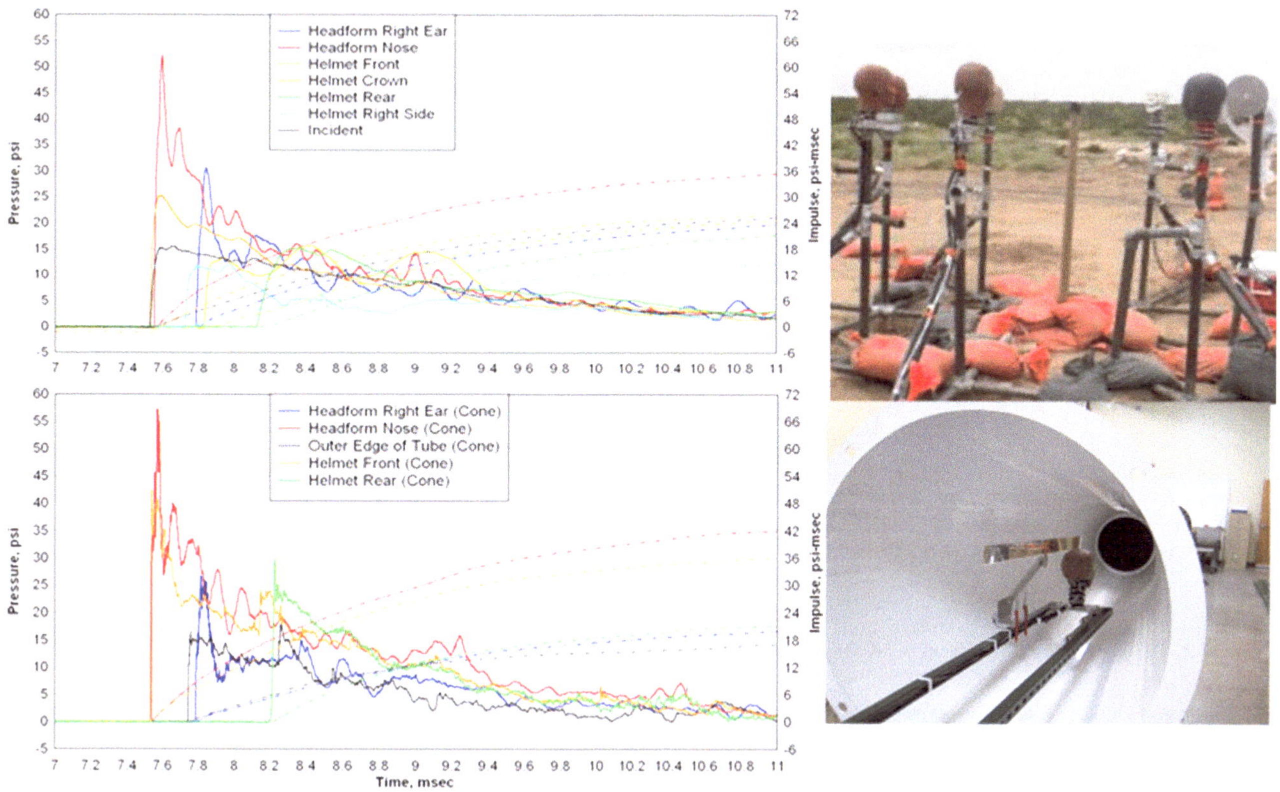

Fig. 15.3 Comparison of Free-field and shock tube blast loads on a Hybrid III anthropometric test device (ATD) headform show that the pressure wave and impulse energy generated by the shock tube engulfs the headform in a similar manner as a free filed pressure wave. Data demonstrates that this method is appropriate for the evaluation of hearing protection devices

Table 15.1 The suite of instrumentation used to evaluate the response of PHMS and pressure in the outer middle and inner ear during blast overpressure exposure with and without HPD

Location	Type	Manufacturer	#	Purpose
Skull surface	Pressure gauges	Endevco	5	Reference pressure on skull surface
	Strain gauge	Omega	5 × 3 each	Quantify frequency content generated by skull flexure
Outer ear	Pressure gauges	FISO	2	Pressure behind hearing protection
Middle ear	Pressure gauges	FISO	2	Pressure in middle ear space
Inner ear	Pressure gauges	FISO	2	Pressure in cochlea
Free field	Pressure gauges	Entran	2	Reference pressure inside expansion cone

15.3 Results

To quantify the performance of the four HPD evaluated the PMHS data collected was processed to determine the insertion loss for each HPD. Insertion loss is the difference between the maximum estimated pressure for the open-ear condition and the maximum pressure measured when a hearing protector is placed for the same condition on the same PMHS. To calculate the insertion loss the pressure time wave in Fig. 15.5 was decomposed into its frequency components Fig. 15.6 for both the empty unconcluded ear and for the case when the ear canal was occluded with a HPD. The insertion loss was calculated by subtracting the occluded frequency components from the empty ear case paired by the pressure at the surface of the head. This process was completed for each hearing protection all four pressure conditions (Fig. 15.7).

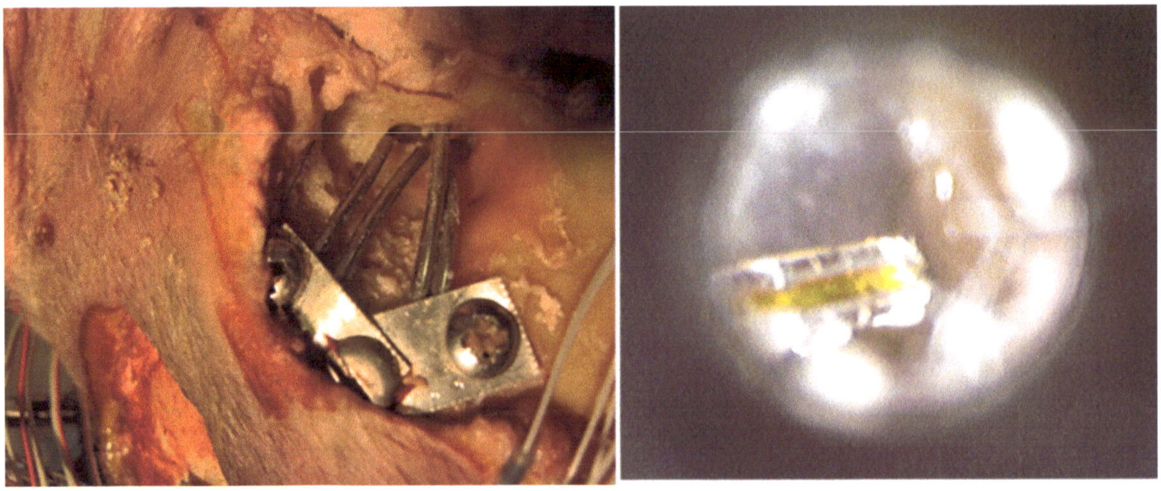

Fig. 15.4 To measure pressures behind the HPD fiber optic pressure gages were inserted into the ear canal (*right*) through a mastectomy and held in place using stainless steel tubing (*left*). Fiber optic pressure gages were also used to measure the pressure in the middle and inner ear (cochlea)

Table 15.2 Description of the four types of hearing protection devices evaluated in the effort, including in the ear (ITE) and outside the ear (OTE) configurations

<table>
<tr><th>Equipment</th><th>Sound Attenuation/Frequency Response at Non-Blast level</th><th>Picture</th></tr>
<tr><td>ITE-1 = 3M E-A-R Classic Roll-Down Foam Earplug</td><td><table>
<tr><td>Frequency [Hz]</td><td>63</td><td>125</td><td>250</td><td>500</td><td>1000</td><td>2000</td><td>4000</td><td>8000</td></tr>
<tr><td>Mean Attenuation [dB]</td><td>22.3</td><td>23.3</td><td>24.6</td><td>26.9</td><td>27.4</td><td>34.1</td><td>41.6</td><td>40.4</td></tr>
<tr><td>Standard Deviation [dB]</td><td>5.4</td><td>5.3</td><td>3.6</td><td>5.4</td><td>4.8</td><td>3.1</td><td>3.5</td><td>6.4</td></tr>
</table></td><td></td></tr>
<tr><td>ITE-2 = 3M Combat Arms Earplug in Open (Nonlinear) Position</td><td><table>
<tr><td>Frequency [Hz]</td><td>63</td><td>125</td><td>250</td><td>500</td><td>1000</td><td>2000</td><td>3150</td><td>4000</td><td>8000</td></tr>
<tr><td>Mean Attenuation [dB]</td><td>4.1</td><td>4.5</td><td>11</td><td>18.7</td><td>24.9</td><td>29.8</td><td>25.8</td><td>18.7</td><td>26.5</td></tr>
<tr><td>Standard Deviation [dB]</td><td>2.7</td><td>2.8</td><td>3.9</td><td>3.2</td><td>3.3</td><td>2.7</td><td>3.3</td><td>3.6</td><td>3.3</td></tr>
</table></td><td></td></tr>
<tr><td>ITE-3 = ARA ShotShields in Open (Nonlinear) Position</td><td><table>
<tr><td>Frequency [Hz]</td><td>63</td><td>125</td><td>250</td><td>500</td><td>1000</td><td>2000</td><td>3150</td><td>4000</td><td>8000</td></tr>
<tr><td>Mean Attenuation [dB]</td><td>25.9</td><td>25.3</td><td>24.2</td><td>20.7</td><td>27.4</td><td>18.1</td><td>24.3</td><td>21.5</td><td>24.2</td></tr>
<tr><td>Standard Deviation [dB]</td><td>0.8</td><td>0.8</td><td>0.9</td><td>1.1</td><td>1.4</td><td>0.9</td><td>0.8</td><td>0.7</td><td>0.7</td></tr>
</table></td><td></td></tr>
<tr><td>OTE = 3M ComTac XP/ComTac III, no power</td><td><table>
<tr><td>Frequency [Hz]</td><td>125</td><td>250</td><td>500</td><td>1000</td><td>2000</td><td>4000</td><td>8000</td></tr>
<tr><td>Mean Attenuation [dB]</td><td>11.5</td><td>17.9</td><td>27.8</td><td>30</td><td>32.1</td><td>36.2</td><td>40.3</td></tr>
<tr><td>Standard Deviation [dB]</td><td>2.5</td><td>2.7</td><td>1.8</td><td>2.3</td><td>3</td><td>2</td><td>3.1</td></tr>
</table></td><td></td></tr>
</table>

15.4 Discussion

The experimental data collected was used to improve a hearing protection module (HPM) designed for evaluation of hearing at high intensity impulsive pressure levels that were not generated in the lab. The most common electro-acoustic model for the human auditory system is the Auditory Hazard Assessment Algorithm for Humans (AHAAH). The AHAAH model is a nonempirical model that uses electronic elements to represent biological function of the structures within the ear. For an electro-acoustic model, voltages represent pressure, currents represent volume velocity, and charges represent volume displacement. Using electric elements that include inductors, resistors, and capacitors, the pressure changes in the ear can be mimicked and represented by changes in voltage throughout the circuit. The pressures collected in this effort allowed

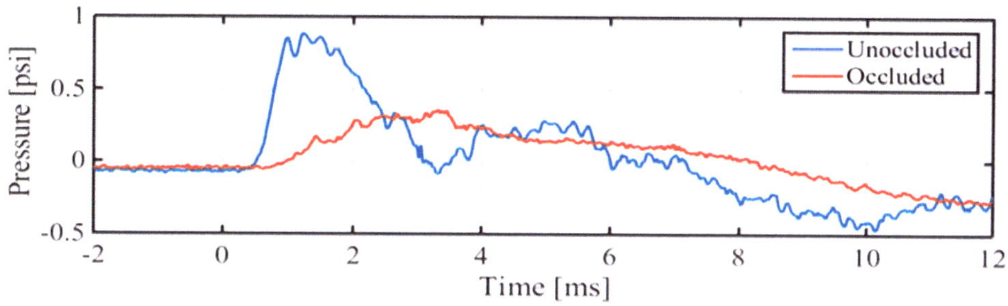

Fig. 15.5 Comparison of the pressure time trace from sensor in ear canal with and without hearing protection device

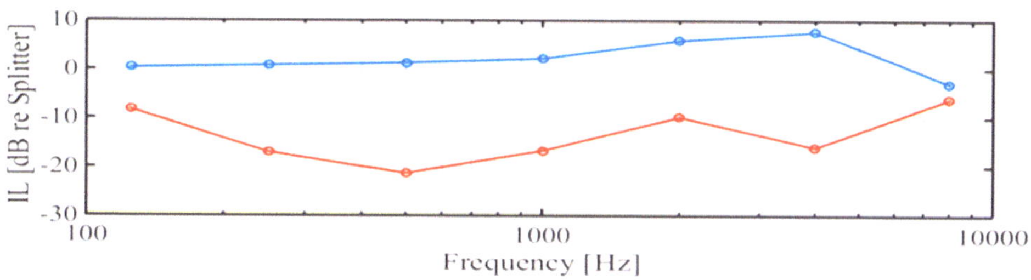

Fig. 15.6 Comparison of the pressure in ear canal with and without hearing protection device in the frequency domain

researchers to empirically validate the model using PMHS data and modify/change the parameters of the individual elements to better represent the biological function of the structures within the ear.

In this effort, the Insertion Loss (IL) curves provided by a hearing protection device (HPD) were deconstructed into a three-piston model (Fig. 15.8). Each 'piston' (Skin, Device, Air leak) occupies a different portions of the 100 Hz to 10 kHz frequency band typically associated with hearing protection. In Fig. 15.9 the IL data from the Combat arms hearing protection that was evaluated for the different pressure levels tested are represented by the blue circles. MATLAB was used to fit the three piston resonator circuit with the natural frequencies of each piston indicated by the vertical lines.

Using curves such as the one shown in Fig. 15.9 frequency bands for hearing protection improvements during impulse noise exposures generated by blast can be identified. Looking from 7 to 55 kPa the 'leak' path is always present suggesting a possible area for improvement for HPD. Also observable is the "Material" adds protection above 2 kHz as pressure increases; unfortunately, all the energy in the pressure wave is in lower frequency bands suggesting that the development of new materials for the HPD may not reduce the risk of hearing damage. The last piston 'skin' shows negligible transmission. This is most likely due to the viscoelastic properties of the tissue that dampen out the energy prior to reaching the sensate structures of the ear.

15.5 Future/Ongoing Work

Future work includes the conversion of the linear elements within the current model (Fig. 15.8) to nonlinear elements. Results using the PMHS data and linear elements suggest an over prediction of the performance of the HPD (Fig. 15.10). This over prediction of the HPD performance can potentially lead to inadvertent increase in hearing damage. Ongoing work is examining if a better approximation can be reached if the data collected at multiple pressure levels is better represented in the model with the addition of non-linear elements. Ongoing work will also investigate if the addition of a pathway for bone conduction for a more accurate prediction at higher intensity shots is required.

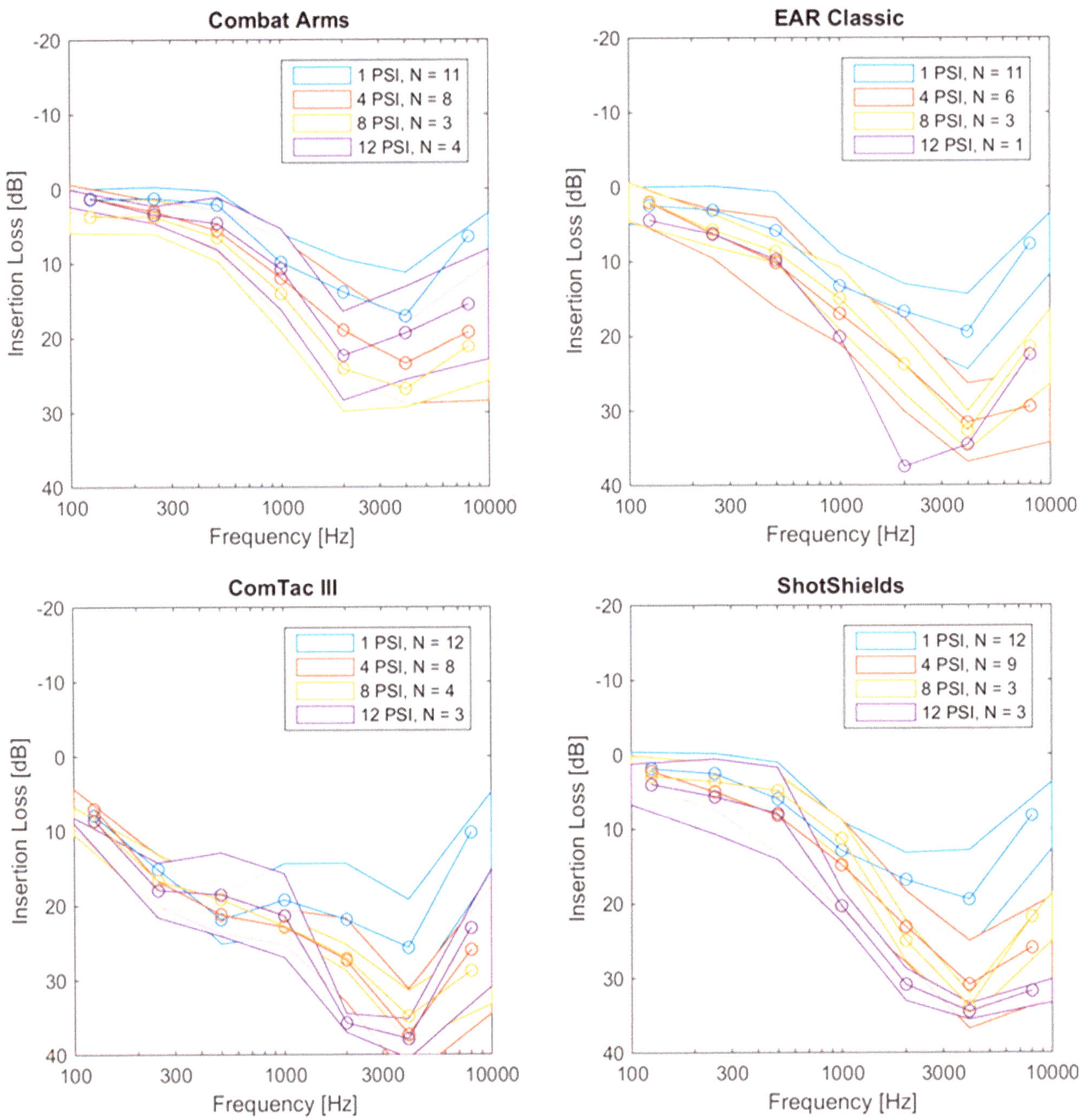

Fig. 15.7 Insertion loss curves for the four different hearing protection devices

15.6 Summary

To predict the performance of earplugs and earmuffs worn during impulse noise exposures generated by blast, such as shoulder launched rockets and weapons fire an electro acoustical model is under development. The model captures the energy flow through the ear by representing the pathways associated with the HPD as a rigid piston, leakage, energy transmitted through the skin. The model is being validated against PMHS data collected at a range of pressures to which service members are regularly exposed. This is a deviation from current methods under which the hearing protection is evaluated at lower exposure levels and the data extrapolated to the higher levels. Insertion loss curves developed under this effort show a nonlinear performance of HPD at higher pressure levels suggesting the extrapolation of data may over predict

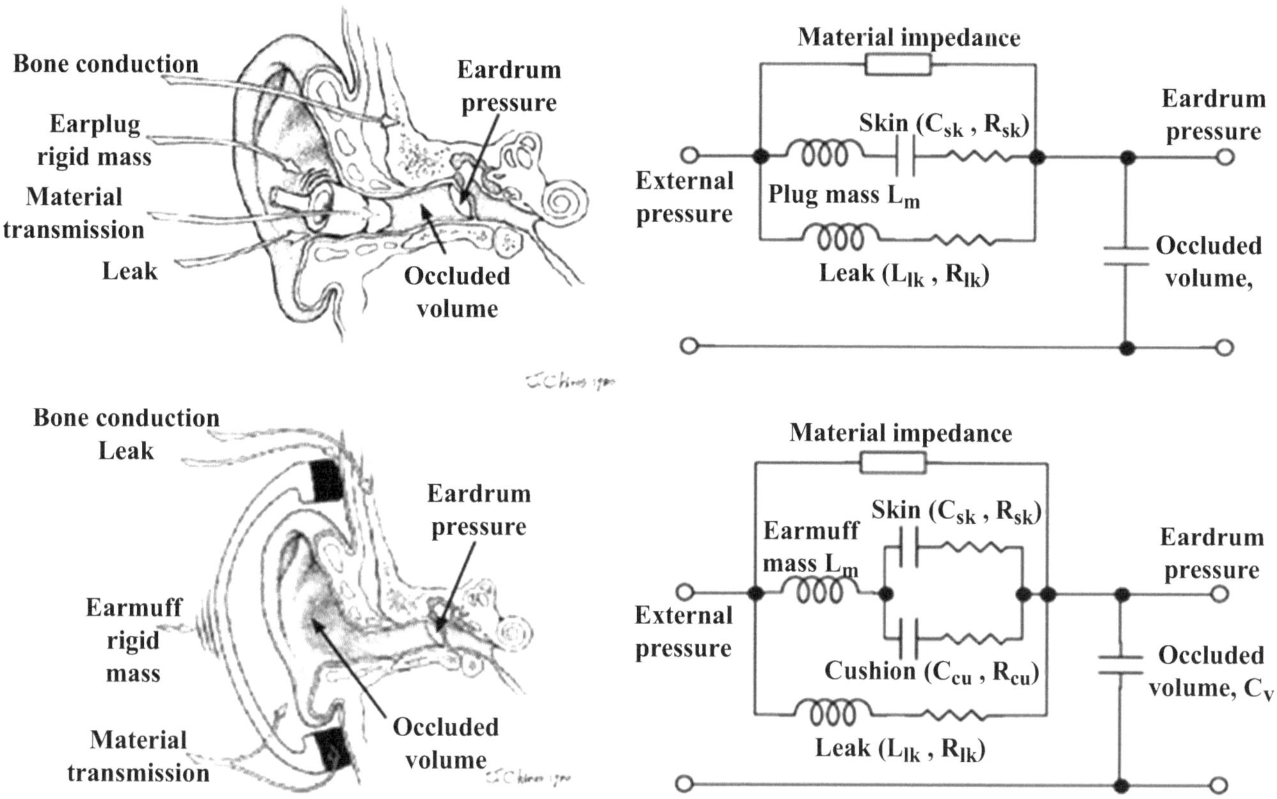

Fig. 15.8 Schematic of the three piston hearing protection module illustrating how electrical elements are used to represent the hearing protection device and the biological function of the structures within and surrounding the ear canal (reprint from Kalb (2013) [1])

the performance of the HPD leading to a greater risk of hearing loss under conditions the service members are regularly exposed. Furthermore with the better understanding of contribution of auditory dysfunction to the symptoms associated with TBI better mitigation, diagnosis and treatment protocols can be developed.

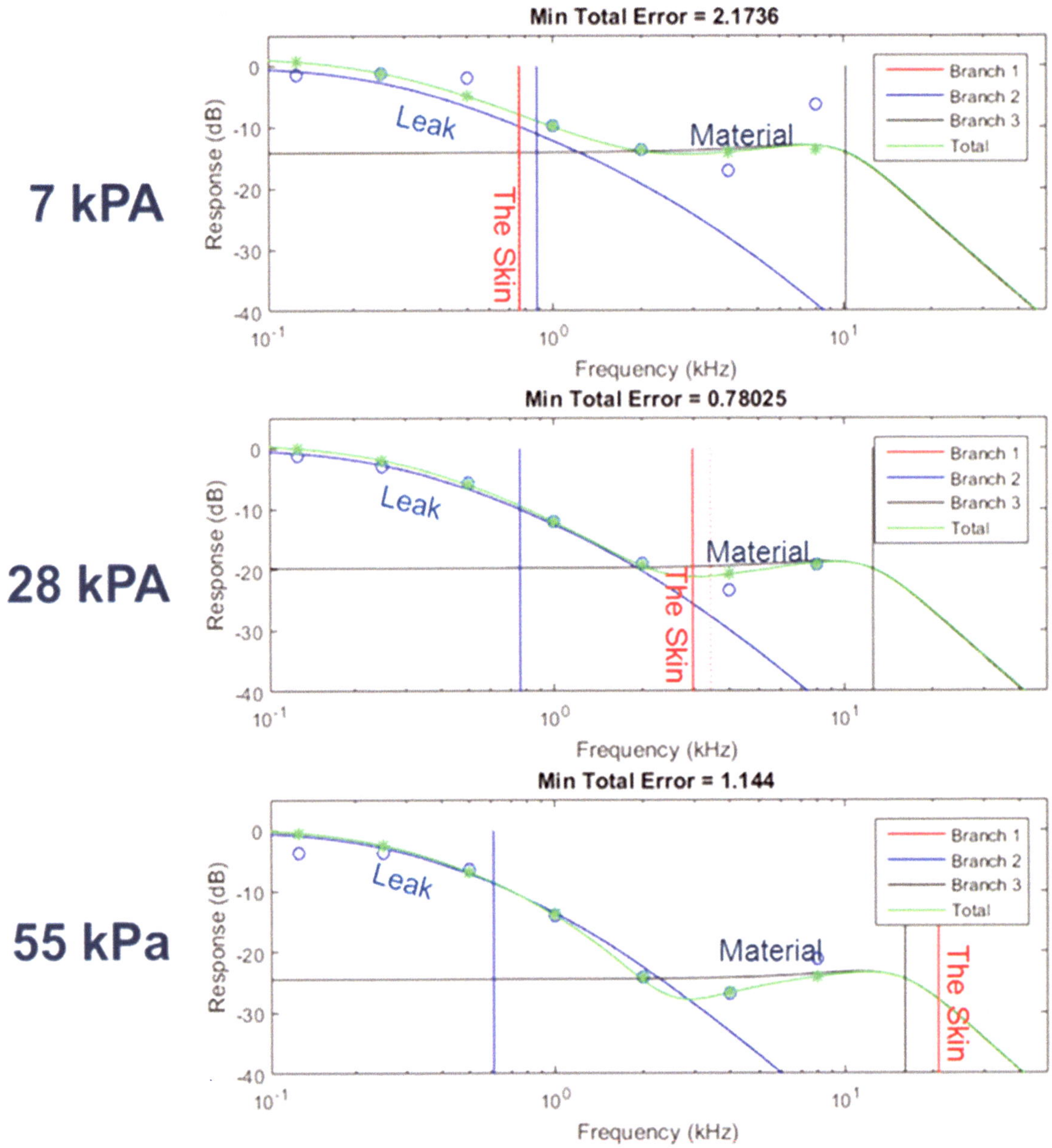

Fig. 15.9 Output in the frequency domain from the three branch resonator circuit for the Combat Arms HPD broken down by pathway illustrating how the energy is transmitted through and around the hearing protection for various levels of blast pressure

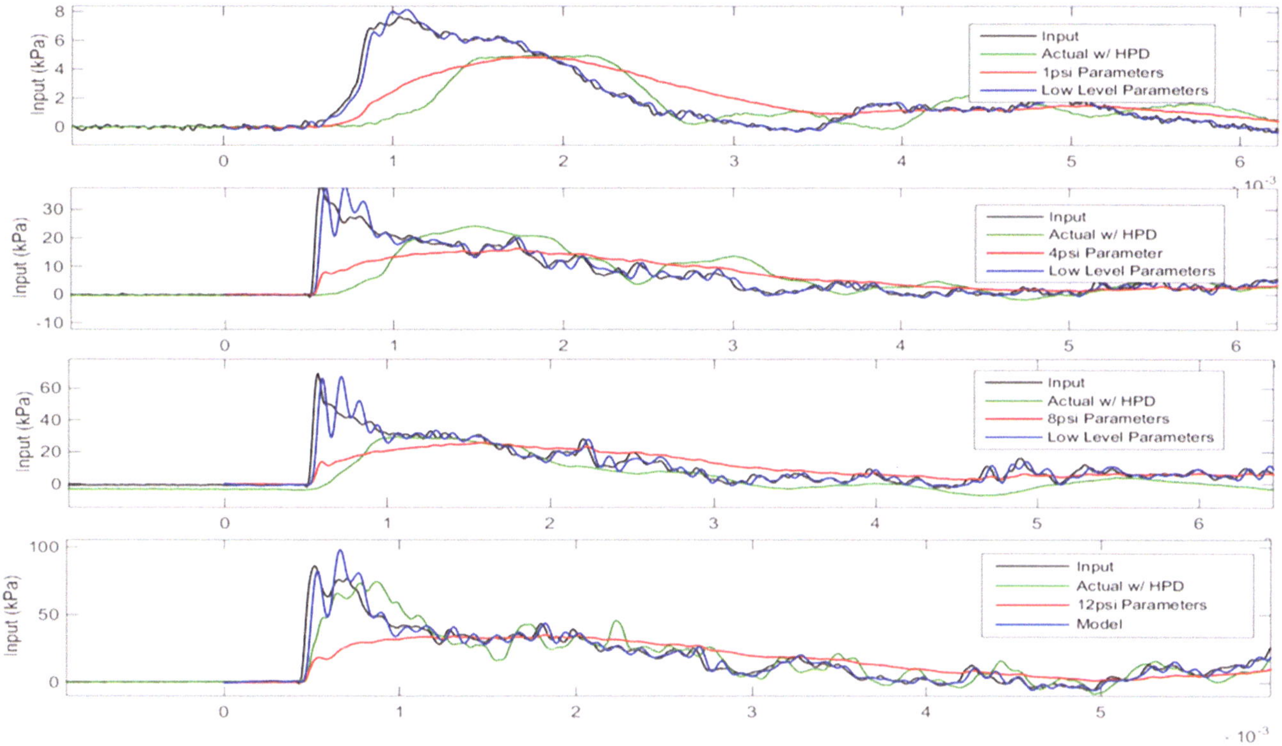

Fig. 15.10 Results using purely linear elements suggest that a better approximation can be reached if data is collected at multiple pressure levels but the addition of non-linear elements and a pathway for bone conduction are required for accurate prediction at higher intensity shots

Acknowledgements This work was supported by the U.S. Army Medical Research and Material Command under Contract No. W81XWH-15-2-0002. The authors would also like to acknowledge the contributions of Dan Welsh (ARA), Dr. Mohammed Al Hussaini (CU) and Peter Mouser at the CU Center for Surgical Innovation without which the program would not have succeeded.

Reference

1. Kalb, J.: An electroacoustic hearing protection simulator that accurately predicts pressure levels in the ear based on standard performance metrics. ARL-TR-6562. Army Research Lab Aberdeen Proving Ground MD Human Research and Engineering Directorate (2013)

Chapter 16
A Mechano-Hydraulic Model of Intracranial Pressure Dynamics

D. Evans, C. Drapaca, and J.P. Cusumano

Abstract The brain is a multi-phase mixture of cerebrospinal fluid (CSF), vasculature, and brain cells. The intracranial pressure (ICP) of the CSF can be monitored to assess patients suffering from serious neurological disorders, such as hydrocephalus and traumatic brain injury. However, the only reliable way to measure ICP is to drill a hole through the skull to insert a pressure sensor inside the brain. Therefore, it is necessary to create accurate mathematical models that couple the temporal dynamics of ICP with clinically-observable physiological parameters that can be measured non-invasively. In this paper, we present a new mechano-hydraulic mathematical model that allows one to study the dynamics arising from the coupling between the mechanics of brain, CSF circulation, and cerebral blood flow. We will study the dynamics of the model and examine domains of stability.

Keywords Intracranial pressure • Brain • Cerebrospinal fluid • Mathematical model • Stability

16.1 Introduction

Intracranial pressure (ICP) monitoring provides critical physiological information on brain damage in patients suffering from neurological disorders such as hydrocephalus and traumatic brain injury (TBI). The only reliable method of measuring ICP involves drilling a hole directly through the skull, and inserting a catheter with a pressure probe into the brain [1]. Given the invasiveness of this approach, ICP monitoring is only performed on patients in very critical conditions during life-saving neurosurgical interventions. Mathematical models that couple ICP dynamics and other physiological parameters whose monitoring is less invasive provide a powerful alternative to the direct measurement of ICP. In addition, mathematical models that also incorporate brain's regulatory mechanisms of ICP could prove essential in designing better diagnostic and therapeutic procedures for a wide range of neurological diseases. Although the potential exists to create a model that could be used to estimate ICP in real time, the current state of the art is still far from this goal. A comprehensive review of mathematical models of ICP dynamics was written by Wakeland and Goldstein [2].

One particular model proposed by Ursino and Lodi [3], predicts ICP dynamics from non-invasive measurements of arterial blood pressure, which appears to be in agreement with the ICP measured in patients with TBI. The model, which we will refer to as the Ursino-Lodi model, is a generalization of the one-compartment mathematical model of ICP dynamics pioneered by Marmarou et al. [4] that includes several blood compartments, each with variable conductances and compliances, and ICP auto-regulation due to blood vessel wall tension, and viscous forces. However, the Ursino-Lodi model does not account for the mechanics of brain cells and thus it cannot predict how dramatic changes in ICP dynamics may affect the functions of the brain which are essential to life.

In this paper we introduce a novel mathematical model that couples the hydraulic topology of the Ursino-Lodi model with the mechanics of deformation of the brain tissue as predicted by the linear viscoelastic Kelvin-Voigt model [5]. We then study the dynamics of the model and show that the model is stable and that the mechanical responses of the various components of the model happen on multiple time scales. This work lays the foundation for future exploration into necessary conditions for limit cycles to exist in the state space.

D. Evans (✉) • C. Drapaca • J.P. Cusumano
Department of Engineering Science and Mechanics, The Pennsylvania State University, University Park, PA, USA
e-mail: dje5104@psu.edu; csd12@psu.edu; jpc3@psu.edu

C.S. Korach et al. (eds.), *Mechanics of Biological Systems and Materials, Volume 6*,
Conference Proceedings of the Society for Experimental Mechanics Series, DOI 10.1007/978-3-319-41351-8_16

16.2 Physiological Considerations

The human brain is a multi-phase mixture of many different materials, including cerebrospinal fluid (CSF), brain cells, and cerebral vasculature. The majority of what is considered brain tissue consists of white matter and grey matter, which are both soft and mildly compressible materials [6]. CSF is a colorless liquid made of 99 % water that fills the brain's ventricles, the subarachnoid space (space between the two deeper meninges that envelop the brain tissue: pia mater and the arachnoid mater), and intracellular space. The presence of CSF introduces damping to the brain motion, which cushions and protects it from injury [7]. CSF is constantly formed by the choroid plexuses of the brain's ventricles from the cerebral blood, circulates through the brain tissue and spinal cord, and is constantly reabsorbed into the cerebral veins located in dura mater (above the arachnoid mater) [1]. There exists an approximately constant pressure gradient from the inside of the brain (ventricles containing CSF) to the outside (dura matter). This pressure gradient contributes to the CSF flow, allowing it to be reabsorbed by the venous system. The rate of absorption of CSF varies linearly with ICP [8]. The absorption of CSF back into the bloodstream, and hence, its removal from the intracranial compartment, is the primary natural mechanism for lowering ICP. As a result, abnormally high ICP is sometimes caused by high resistances to resorption.

In addition to the brain naturally regulating pressure via the volume of CSF, the brain also contains autoregulatory mechanisms to control the flow of blood in the cerebral arteries [9]. In order to ensure a constant artery-to-vein difference in oxygen concentration, the brain automatically regulates the cerebral blood flow (CBF) in a process known as *cerebral autoregulation*. As oxygen consumption, blood pressure, and blood viscosity change, the cerebral arteries dilate or constrict to regulate the flow of blood to the brain. The volume of the cerebral arteries, the compression of the brain tissue, and the volume of CSF within the skull are all controlled by ICP. The linked mechanisms of feedback and control through all of these processes have the potential to create instability in the steady-state behavior of the ICP, most notably being the case of Lundberg A-Waves, which sometimes arise in patients who are suffering from traumatic brain injury [1]. These waves are self-sustained low-frequency oscillations in the ICP waveform. The ICP, which is normally approximately constant, increases dramatically in oscillations with periods of approximately 5–20 min, and amplitudes of about 50–100 mmHg [1].

16.3 Mathematical Model

The schematic presented in Fig. 16.1 forms the basis for the following equations.

From the hydraulic system, the terms P_o, P_A, and P_v represent the blood pressure of the extracranial arteries, the intracranial arteries, and veins, respectively. The pressure of the cerebrospinal fluid (CSF) is P_B. The resistances R_a and R_c are fluid resistances of the arteries and capillaries as given by the Hagen-Poiseuille law. The fluid resistances R_f and R_o represent, to first-order, the pressure-dependent processes responsible for formation and resorption of CSF, respectively, and I is the constant, pressure-independent rate of CSF formation.

The volumetric fluid resistances that have just been outlined are the extent of this model's similarity to the Ursino-Lodi model. The effect of fluid storage in the arterial and CSF compartments has been previously modeled as a fluid capacitance element, corresponding to the compartment's storage capacity at a given pressure. This process is directly related to the mechanical deformation of the boundaries of each compartment, both brain tissue and artery. In order to fully describe this effect, while taking advantage of recent research contributions concerning the brain's material properties, we deem it necessary to directly include brain mechanics in our model.

The storage of fluid in both the arterial and CSF compartments is governed by the motion of the boundary of that compartment. For the CSF compartment, an increase in the amount of fluid stored is accompanied by a deformation in the surrounding brain tissue. We also allow the arterial walls to dilate and constrict, governing the amount of arterial blood is present in the brain. The Monro-Kellie doctrine is a well-known clinical idealization of the skull, which states that the skull is rigid, and therefore the volume that it contains is constant [1]. The effects of this assumption in this model is a conservation of volume between the volume of CSF, arterial blood, and brain tissue. The mechanical system consists of effective masses (m), dashpots (η), and springs (k), of the arterial wall (subscript A), and of the brain (subscript B). The relative displacement of each mass with respect to the spring's equilibrium position is u, where a positive sign indicates an expansion of the respective fluid chamber. The areas A_A and A_B are the effective contact areas that represent the boundaries of the arteries and CSF chamber, respectively.

We begin by applying the conservation of mass principle to the two fluid compartments, assuming an incompressible fluid. Equating the net flow into the vasculature chamber with the rate of change of its volume yields the following equation,

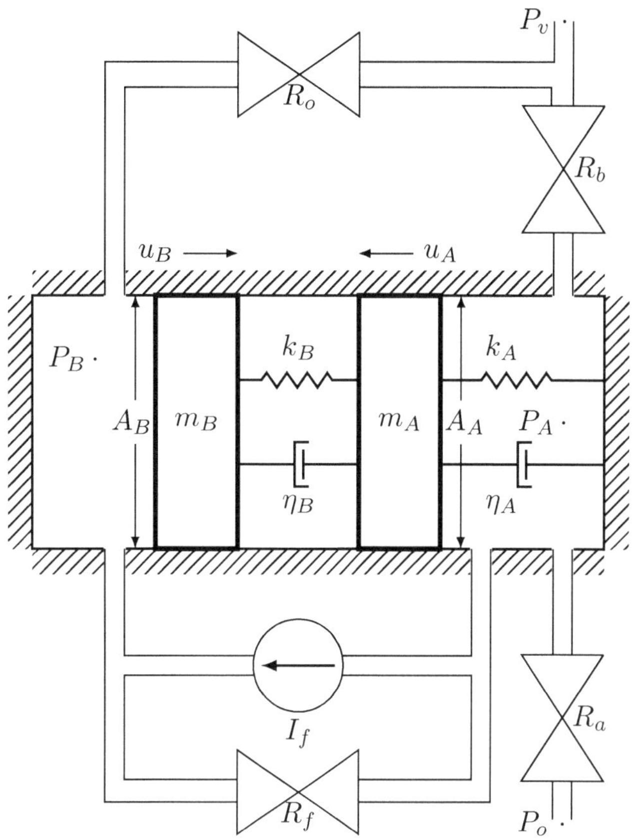

Fig. 16.1 Schematic of the proposed mechano-hydraulic system

$$\frac{P_o - P_A}{R_a} + \frac{P_v - P_A}{R_c} + \frac{P_B - P_A}{R_f} - I = A_A \dot{u}_A, \tag{16.1}$$

and applying the same principle to the CSF chamber yields

$$\frac{P_v - P_B}{R_o} + \frac{P_A - P_B}{R_f} + I = A_B \dot{u}_B, \tag{16.2}$$

where the dot ($\dot{}$) operator indicates the derivative with respect to time.

Next, we apply Newton's second law to the two masses of the system. Summing the forces acting on the mass of the arteries and setting them equal to the mass times its acceleration gives us

$$m_A \ddot{u}_A + \eta_A \dot{u}_A + \eta_B(\dot{u}_A + \dot{u}_B) + k_A u_A + k_B(u_A + u_B) = P_A A_A, \tag{16.3}$$

and for the brain,

$$m_B \ddot{u}_B + \eta_B(\dot{u}_A + \dot{u}_B) + k_B(u_A + u_B) = P_B A_B. \tag{16.4}$$

Equations (16.3) and (16.4), when written in matrix notation, assume the following form:

$$M\ddot{u} + C\dot{u} + K\text{u} = A\text{P}, \tag{16.5}$$

where $\mathbf{u} = [u_A \quad u_B]^T$ is defined as the column vector of configuration variables. However, since the pressures P_A and P_B are linear functions of the velocities $\underline{u}$, as given by Eqs. (16.1) and (16.2), the vector of pressures $\mathbf{P} = [P_A \quad P_B]^T$ can be written in the following form:

$$\mathbf{P} = \phi + E\ddot{u}, \tag{16.6}$$

where ϕ is a 2×1 vector and E is a 2×2 matrix, neither of which depend on u or its time derivatives. Let us define a new coordinate x that is the deviation from the static equilibrium point, which occurs at ($\ddot{u} = 0$, $\dot{u} = 0$).

$$x = u - K^{-1}A\phi \tag{16.7}$$

We find that we can combine Eqs. (16.5), (16.6), and (16.7) in the following form.

$$M\ddot{x} + (C - AE)\dot{x} + K\text{x} = 0 \tag{16.8}$$

16.4 Model Parameters

Many of the material properties required by our model are extremely difficult to accurately measure on living tissue. Acceptable, agreed-upon values are still being converged upon by the research community. We kept the same values for the fluid resistances as in the Ursino-Lodi model [3], and used the values reported in [11] and [5] for the spring and damping coefficients of the brain, respectively. To obtain a rough estimate of the stiffness of the arterial wall, k_A, consider an elastic hollow cylinder with initial internal radius $r_0 = 0.5$ cm, thickness $t = 0.6$ mm, effective length $L = 1$ m, and modulus of elasticity $E = 80$ kPa. With uniform internal pressure loading, the stress in the walls is

$$\sigma = \frac{pr}{t} = \frac{F}{tL} \tag{16.9}$$

while the strain in the walls is

$$\epsilon = \frac{2\pi(r - r_0)}{2\pi r_0} = \left(\frac{r}{r_0} - 1\right). \tag{16.10}$$

If we idealize the arterial wall with having a linear stress-strain relationship via

$$\sigma = E\epsilon, \tag{16.11}$$

then we can define the spring coefficient k_A to be

$$k_A = \frac{dF}{dr} = \frac{ELt}{r_0} \tag{16.12}$$

16.5 Qualitative Analysis

In our qualitative analysis, we let all the parameters in (16.8) be the same order of magnitude. The system (16.8) has a four-dimensional state-space: two state variables are the positions u_A and u_B, while the other two are the velocities $\dot{u}_A$ and $\dot{u}_B$. Therefore, a full graphical depiction of this state space would require a four-dimensional image, which is why we project this space into two-dimensional planes. In Fig. 16.2, we present the configuration space, with the velocities absent. In addition, the clinically-relevant values offered by this model are the pressures P_A and P_B, and not the state variables themselves. Rather, the pressures can be written as an algebraic function of the state variables, as seen in Eqs. (16.1) and (16.2). Thus, in Fig. 16.3, we show transient response in P_B due to the same initial conditions that are shown in Fig. 16.2.

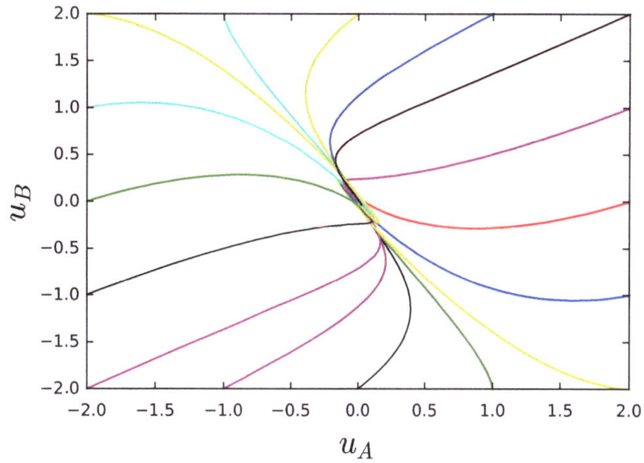

Fig. 16.2 Solution trajectories projected onto the $u_A - u_B$ plane. All velocity initial conditions are set to zero

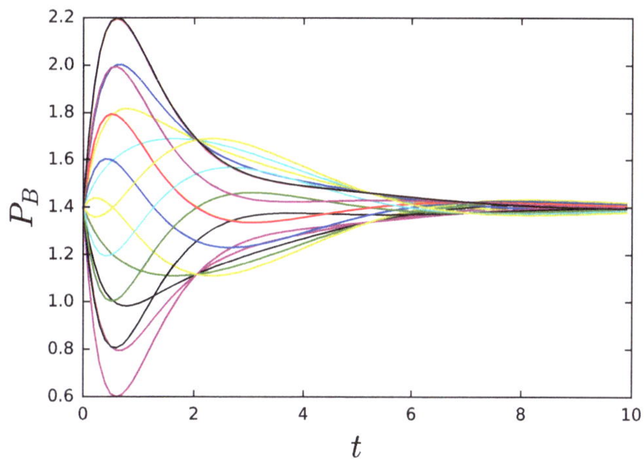

Fig. 16.3 Family of intracranial pressure solutions with respect to time

16.6 Quantitative Analysis

The equilibrium value of pressure is given by Eq. (16.6), when $\dot{u}$ is equal to the zero vector. When considering the parameter values of Table 16.1, we have

$$P_{eq} = \begin{bmatrix} P_{A,eq} \\ P_{B,eq} \end{bmatrix} = \begin{bmatrix} 7.041kPa \\ 2.255kPa \end{bmatrix} \tag{16.13}$$

The phase portraits presented in this section are ones that are created with the model parameter values set to those presented in Table 16.1. The behavior we observe in the phase space is indicative of two different time scales acting simultaneously. Figures 16.5 and 16.4 illustrate this behavior. In response to an initial condition with nonzero position yet zero velocity, the pressure P_B exponentially increases to a value on the order of microseconds (Fig. 16.4). Figure 16.5 is a zoomed-out version of Fig. 16.4. Here we see that P_B is observed to start returning to its equilibrium value on the order of seconds.

Figures 16.6 and 16.7 share initial conditions where at $t = 0$, $u_A = u_B = 0$ and where v_A and v_B are nonzero. The initial values of v_A and v_B for each solution path are shown on the border of Fig. 16.7. It is apparent that, at a very fast time scale, the dynamics quickly converge onto a manifold of lower dimension.

Table 16.1 Model parameter values

Parameter name	Symbol	Value [Reference]
Spring coefficient of arteries	k_A	10^5 N/m [10]
Spring coefficient of brain (KV)	k_B	1750 N/m [11]
Damping of arteries	η_A	10^2 N s/m [N/A]
Damping of brain (KV)	η_B	10^9 N s/m [5]
Formation rate of CSF	I	$6 \times 10^{-9}\ m^3/s$ [4]
Fluid resistance of arteries	R_a	800×10^6 Pa s/m^3 [3]
Fluid resistance of capillaries	R_c	800×10^6 Pa s/m^3 [3]
Fluid resistance to CSF formation	R_f	317×10^6 Pa s/m^3 [3]
Fluid resistance to CSF resorption	R_o	70×10^9 Pa s/m^3 [3]
Arterial boundary surface area	A_A	1×10^{-3} m^2 [N/A]
Ventricular boundary surface area	A_B	1×10^{-2} m^2 [N/A]

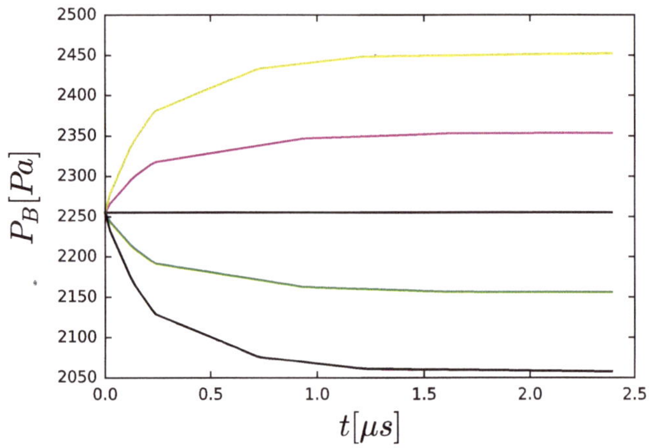

Fig. 16.4 Response of P_B to a nonzero initial displacement at time $t = 0$

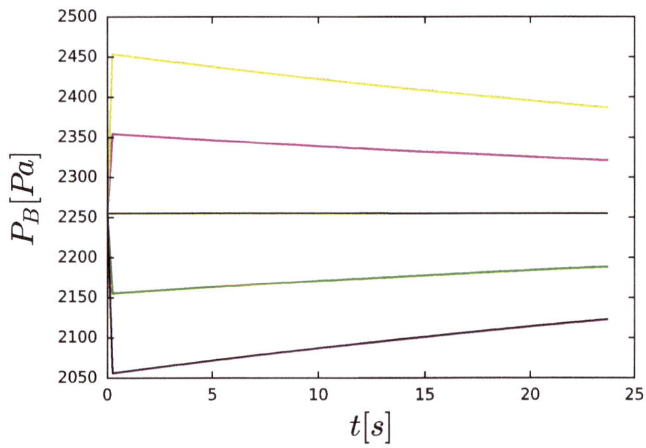

Fig. 16.5 Relaxation of P_B to steady-state as a response to a nonzero initial displacement at time $t = 0$

16.7 Stability Analysis

In order to analyze the stability boundaries of the system, first, we must rewrite the system (16.8) in first-order or state-variable form. Let us define

$$\dot{x} = v, \tag{16.14}$$

which allows (16.8) to be written in state-variable form as a first-order system as follows.

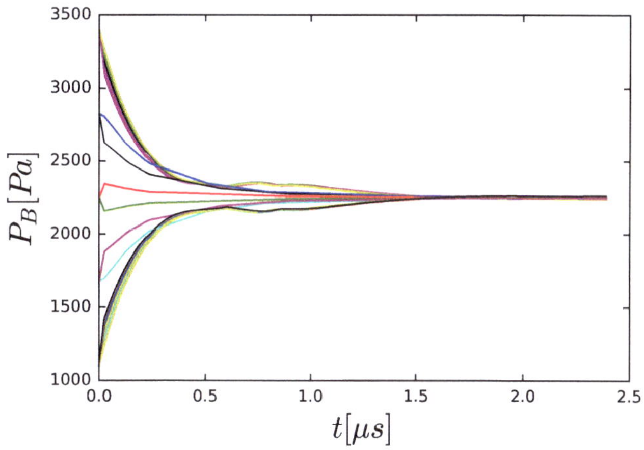

Fig. 16.6 Transient responses in P_B to nonzero initial velocity at time $t = 0$

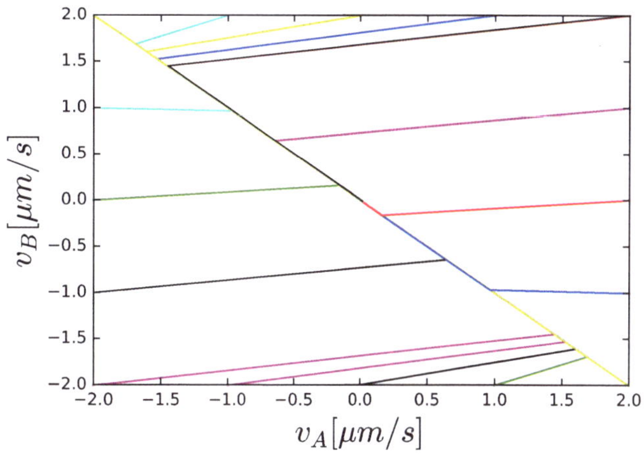

Fig. 16.7 Projection of the phase space onto the $v_A - v_B$ plane. Solution paths have initial conditions of $u_A = u_B = 0$, with v_A and v_B shown on the border of the plot

$$\begin{bmatrix} \dot{x} \\ \dot{v} \end{bmatrix} = \begin{bmatrix} 0 & I \\ -M^{-1}K & -M^{-1}(C - AE) \end{bmatrix} \begin{bmatrix} x \\ v \end{bmatrix}. \tag{16.15}$$

The elements of the matrix in Eq. (16.15) are 2×2 block matrices, where I is the 2×2 identity matrix. It is well-known that a system in this form has one and only one equilibrium point, which is located at the origin [12]. The steady-state solution is the solution that begins at the origin and remains at the origin for all time. To determine the stability of this equilibrium point, we can examine the eigenvalues of the 4×4 matrix in Eq. (16.15). If any of the eigenvalues have a positive-real part, then the steady-state solution is unstable [12].

In order to assess the stability of the steady-state solution, we first set the model parameters to those presented in Table 16.1, and calculated the eigenvalues of the 4×4 matrix in Eq. (16.15). Then, we selected one parameter and varied it over several orders of magnitude, while keeping the others fixed, numerically evaluating the eigenvalues at each point. This process was then repeated for all of the model parameters shown in Table 16.1. This method of exploring the parameter space found no points that yielded eigenvalues with positive-real part, and therefore, we found no parameter sets that cause the only equilibrium point to lose its stability. By expanding the allowable range of parameters to allow negative values, we were able to find points in the parameter space that cause a loss of stability of the system's equilibrium point. For example, by allowing the spring or damping coefficients to be negative, we can find parameter configurations that yield an unstable equilibrium point. However, the only configurations that we have found to cause such a loss of stability require some parameters to be negative, and therefore, physically unrealistic.

16.8 Conclusions and Future Work

In this paper we studied a linear mechano-hydraulic model of brain dynamics. We discussed its derivation, parameter values, and behavior. This simplified model allows us to study the separate effects of a hydraulic system, which was inspired by the Marmarou and Ursino-Lodi models of CSF dynamics, and a mechanical system, that predicts brain mechanics using a Kelvin-Voigt model. This allows us to directly incorporate recent research about the mechanics of the brain to further improve models of ICP dynamics. We also investigated the stability of the steady-state solution and found it to be stable for all of the physically-meaningful regions of the parameter space that were tested. The model suggests that we will need to consider two time scales to be able to study fast and slow behaviors.

In the future, we intend to study the necessary conditions for the equilibrium point to lose stability, as is the case when patients suffer from Lundberg A-Waves. This will involve expanding the model to include nonlinear effects and state-dependence of the model parameters.

References

1. Suarez, J.I.: Critical Care Neurology and Neurosurgery. Springer, New York (2004)
2. Wakeland, W., Goldstein, B.: A review of physiological simulation models of intracranial pressure dynamics. Comput. Biol. Med. **38**(9), 1024–1041 (2008)
3. Ursino, M., Lodi, C.A.: A simple mathematical model of the interaction between intracranial pressure and cerebral hemodynamics. J. Appl. Phycol. **82**(4), 1256–1269 (1997)
4. Marmarou, A., Shulman, K., Rosende, R.M.: A nonlinear analysis of the cerebrospinal fluid system and intracranial pressure dynamics. J. Neurosurg. **48**(3), 332–344 (1978)
5. Hasan, M.M., Drapaca, C.S.: A poroelastic-viscoelastic limit for modeling brain biomechanics. In: MRS Proceedings, vol. 1753, pp. mrsf14–1753. Cambridge University Press, Cambridge (2015)
6. Kyriacou, S.K., Mohamed, A., Miller, K., Neff, S.: Brain mechanics for neurosurgery: modeling issues. Biomech. Model. Mechanobiol. **1**(2), 151–164 (2002)
7. Goldsmith, W.: The state of head injury biomechanics: past, present, and future: Part 1 Crit. Rev. Biomed. Eng. **29**(5&6), 441–600 (2001)
8. Cutler, R.W.P., Page, L, Galicich, J., Watters, G.V.: Formation and absorption of cerebrospinal fluid in man. Brain **91**(4), 707–720 (1968)
9. Attwell, D., Buchan, A.M., Charpak, S., Lauritzen, M., MacVicar, B.A. Newman, E.A.: Glial and neuronal control of brain blood flow. Nature **468**(7321), 232–243 (2010)
10. Tanaka, M., Sugawara, M., Ogasawara, Y., Izumi, T., Niki, K., Kajiya, F.: Intermittent, moderate-intensity aerobic exercise for only eight weeks reduces arterial stiffness: evaluation by measurement of stiffness parameter and pressure–strain elastic modulus by use of ultrasonic echo tracking. J. Med. Ultrason. **40**(2), 119–124 (2013)
11. Kruse, S.A., Rose, G.H., Glaser, K.J., Manduca, A., Felmlee, J.P., Jack, C.R., Ehman, R.L.: Magnetic resonance elastography of the brain. Neuroimage **39**(1), 231–237 (2008)
12. Jordan, D.W., Smith, P.: Nonlinear Ordinary Differential Equations, 4th edn. Oxford University Press, Oxford (2007)

Chapter 17
Regional Variations in the Mechanical Strains of the Human Optic Nerve Head

Dan E. Midgett, Mary E. Pease, Harry A. Quigley, Mohak Patel, Christian Franck, and Thao D. Nguyen

Abstract The mechanical behavior of the lamina cribrosa (LC), the connective tissue that supports the retinal ganglion cells (RGCs), plays an important role in the development of glaucoma. We have developed an in-vitro inflation test to measure the three-dimensional displacement and strain fields in the posterior 250 μm of the LC in response to controlled pressurization. Human eyes were obtained from a tissue bank and the posterior sclera mounted in a pressure-controllable fixture. A laser-scanning microscope was used to acquire SHG images of the deforming tissue structure in response to pressure changes, and the fast-iterative digital volume correlation (FIDVC) algorithm developed by Bar Kochba et al. (Experimental Mechanics 55:261–274, 2015) was used to calculate 3D displacement fields from the SHG signal. Strains were calculated by fitting a high order polynomial function to the displacement field and taking the gradient. To compare for regional variation of the strains, the LC was divided about the central retinal artery and vein (CRAV) into the superior, nasal, inferior, and temporal quadrants and further divided into inner, middle, and outer radial regions centered on the CRAV and three equal regions through the thickness. These regions were statistically compared using a univariate repeated anova. We found that displacements and strains in the LC exhibited substantial variation in the plane but not through the thickness. The average shear strains were significantly smaller than the normal strains ($p = 0.0006$, 45 mmHg). For all eyes, the LC appeared as an elliptical plate. The normal strains were greater along the major axis compared to the minor axis of the LC for all eyes tested ($p = 0.03$, 45 mmHg). The maximum principal strains were also found to be greater in the middle inferior and temporal regions than in the superior and nasal regions ($p = 0.07$, 45 mmHg). Eyes exhibited two kinds of radial variation. Three out of six specimens exhibited the stiffest behavior around the CRAV with higher principal strains towards the periphery, while the remainder were more compliant around the CRAV with lower principal strain on the periphery. These regional variations in pressure-derived strains in the LC may help to explain the progression of axonal damage observed in glaucoma.

Keywords Digital volume correlation • Glaucoma • Lamina cribrosa • Biomechanics • Strains

17.1 Introduction

In glaucoma, the progressive loss of the retinal ganglion cells (RGCs) in the optic nerve head leads to permanent vision loss and eventual blindness. The intraocular pressure (IOP) is associated with the prevalence and severity of glaucoma damage, and the level of IOP is used to screen for the disease. Animal studies have shown that a chronic increase in IOP results in glaucoma damage, and lowering pressure slows the progression of damage in the majority of cases. However, there is a wide distribution of IOP among those with and without glaucoma. While glaucoma patients on average have higher IOP, nearly half of those with glaucoma damage have IOP in the normal range. Moreover, a significant number of those with higher IOP do not exhibit glaucoma damage [1–4]. These observations suggest that the susceptibility of the ONH to glaucoma damage may depend on the mechanical response of the tissues of the ONH and the surrounding sclera. Computational modeling studies of the posterior sclera have suggested that the elastic modulus of the LC and sclera strongly correlates with the strains experienced in the LC [5].

D.E. Midgett (✉) • T.D. Nguyen
Department of Mechanical Engineering, Johns Hopkins University, Baltimore, MD 21218, USA
e-mail: dmidget1@jhu.edu

M.E. Pease • H.A. Quigley
Wilmer Eye Institute, Johns Hopkins Medical Institute, 733 N Broadway, Baltimore, MD 21205, USA

M. Patel • C. Franck
School of Engineering, Brown University, Providence, RI 02912, USA

C.S. Korach et al. (eds.), *Mechanics of Biological Systems and Materials, Volume 6*,
Conference Proceedings of the Society for Experimental Mechanics Series, DOI 10.1007/978-3-319-41351-8_17

Both clinical and laboratory studies have measured differences in the mechanical behavior and structure of the posterior sclera, LC and ONH of normal and glaucomatous eyes. Structural changes in glaucoma include the remodeling of the anisotropic collagen fiber and elastin structure of the lamina cribrosa [4] and axial elongation and thinning of the sclera [6–8]. Hommer et al. [9] measured an increase in ocular rigidity with the progress of open-angle glaucoma. Downs et al. [10] found a larger relaxation modulus for scleral strips from monkey eyes with experimentally induced early glaucoma. Girard et al. measured a stiffening of the sclera [11]. Coudrillier et al. [12] reported a thickening of the peripapillary sclera, slower circumferential creep rates, and a stiffer behavior in the meridional direction of the peripapillary sclera in human eyes with glaucoma.

The current body of research indicates that glaucoma pathologically remodels the tissues of the ONH in a way that causes thinning of the nerve cross-section and excavation at the ONH, a phenomenon where the rim of the tissue making up the ONH narrows and deepens. Larger RGC's have been shown to have a greater susceptibility to glaucoma damage, indicating that deformation of the nerves by the surrounding connective tissues may be a cause of damage. The degree of axonal injury typically appears in the LC as an hourglass shape, with greater injury occuring at the superior and inferior poles [1]. Despite this, few studies have measured the mechanical behavior of the lamina cribrosa, which supports the retinal ganglion cells. In this study, we present a method to measure the three-dimensional displacements and strains experienced by the LC during inflation. The method uses laser-scanning second harmonic generation microscopy to image the collagen structures of the deforming volume of the lamina cribrosa and digital volume correlation for deformation mapping.

17.2 Methods

17.2.1 Tissue Preparation

Six Human donor eyes were received from the National Disease Research Interchange 24-h post-mortem. Eyes were stored at 4 °C and tested within 48 h. The sclera was cleaned of orbital tissues and the exposed nerve and episclera shaved off with a razor blade to expose the collagen beams of the lamina cribrosa. The eyes were then glued using Permabond 910 to a polycarbonate ring, centered on the ONH and placed 3 mm anterior of the equator. The LC and posterior sclera were kept hydrated with water for 20 min as the tissue adhesive dried. The cornea and lens along with 2 mm of the surrounding anterior sclera was then removed using a scalpel and the aqueous humor, retina, and choroid were removed. The exposed scleral edge was then scored using a razor blade and the edges glued flat to the back of the holder to provide a watertight seal. The assembly was allowed to dry for an additional 20 min (Table 17.1).

17.2.2 Imaging

The specimen was mounted on a custom inflation chamber and covered with PBS to ensure hydration. The specimen was aligned level to the imaging objective using the two tilt-correcting gears (Fig. 17.1). The pressure was controlled by the height of a water column connected to the holder reservoir and initially set to a baseline pressure of 5 mmHg. An inverted Zeiss LSM 710 NLO scope with a coherent Chameleon Ultra II laser was used to excite the exposed LC beams using backscattered Second Harmonic Generation (SHG) at 780 nm. The signal was collected using a short-pass 490 nm filter. The objective was centered on the central retinal artery and vein (CRAV). Images were acquired as a 2 × 2 tile using four 512 × 512 pixel images at 10× magnification. The 2 × 2 tiled z-stack was sized to cover most of the LC (2–3 mm in

Table 17.1 Human specimen ID and information

Eye ID	Gender	Age	Race	RE/LE
FC71	F	71	Caucasian	Both
FC49	F	49	Caucasian	RE
FC26	F	26	Caucasian	RE
MC57	M	57	Caucasian	RE
MH42	M	42	Hispanic	RE

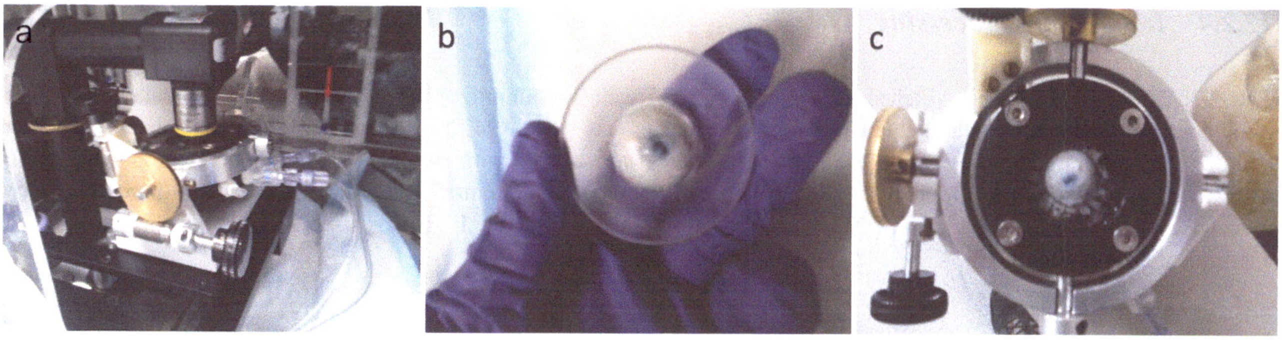

Fig. 17.1 Specimen preparation and imaging

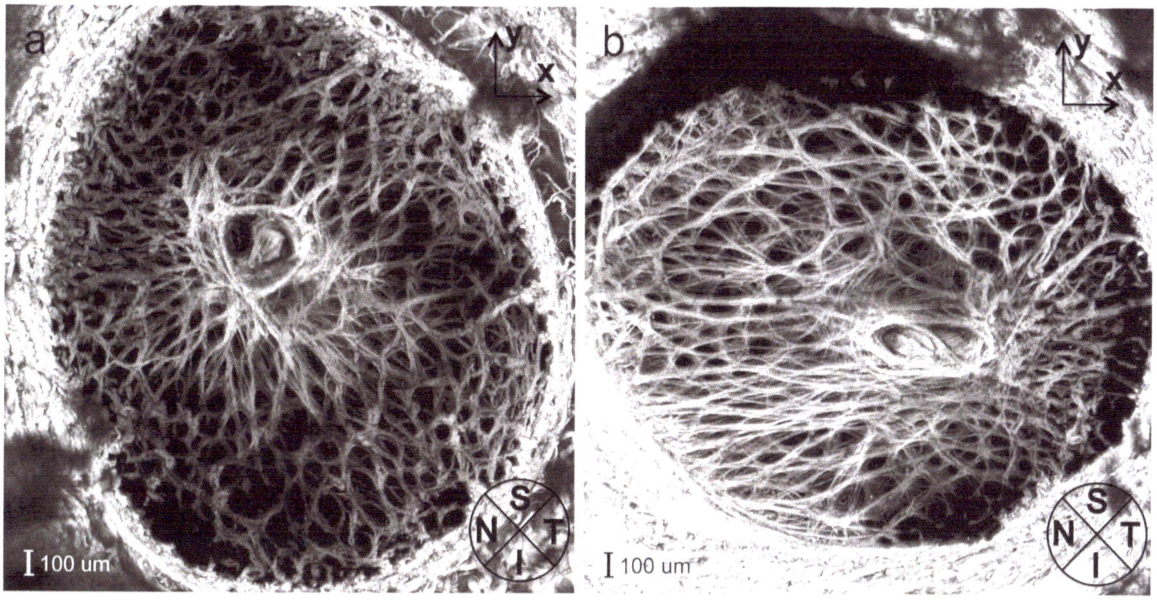

Fig. 17.2 Second harmonic generation images of the collagen beam structure of the LC

diameter). The zoom factor was varied between 0.6 and 0.8 to adjust the field of view and obtain the best full-field resolution of the visible LC signal (Fig. 17.2). The z-stacks were acquired with a short pixel dwell time (pdt) of 5 μs. Z-slices were spaced every 3 μm starting at the distance closest to the choroid at which features were first visible (approximately 200 μm in depth) and up to the outer surface of the LC. A duplicate set of 2×2 tiled z-stacks was acquired immediately afterwards.

17.2.3 Mechanical Testing

The specimens were allowed to equilibrate at the pressures of 5, 10, and 45 mmHg for 45 min after which a series of images were acquired as described in the imaging protocol. For each of the higher pressures the focus was readjusted to account for the average posterior displacement using a landmark within the tissue to minimize acquisition time. This adjustment was recorded and added back to the DVC displacement calculations. The imaging protocol was repeated at the new pressure.

17.2.4 Image Post-Processing

The 2 × 2 tiled images were stitched using the Zeiss software to minimize discontinuities, and the data exported as multistage tiff files for correlation. Final size of the images was 947 × 947 pixels after stitching and resolution varied from 2 to 2.5 μm based on the zoom factor. The SHG stacks were treated with Huygens Essential deconvolution algorithm using a theoretical point spread function (SVI, Hilversum, Netherlands) to remove noise and blur. Contrast-limited adaptive piecewise histogram equalization (CLAHE), applied through the program FIJI, was used to improve the contrast and subsequent correlation.

17.2.5 Digital Volume Correlation: (DVC)

Image sets were reconstructed into a three-dimensional matrix of image intensity in Matlab. X and Y dimensions corresponded to the horizontal and vertical axes of each tiled image, which were aligned to the nasal-temporal and inferior-superior directions of the eye. The Z dimension corresponded to the anterior position within the LC with 0,0,0 being defined as the upper-left corner of the image and the bottom of the stack (most anterior position). The Fast-Iterative DVC Algorithm developed by Bar-Kochba et al. [13] was used to correlate the image intensity of the reference and deformed volumes. A starting subset size of 128 × 128 × 64 pixels and a coarse calculation spacing of 32 × 32 × 16 pixels was selected for the first iteration with the spacing and size being refined in subsequent iterations. We used four iterations with a final calculation spacing of 4 × 4 × 2 pixels and subset size of 64 × 64 × 16. The resolution of our calculations was 2.0–2.5 μm/pixel in XY depending on the zoom factor used during imaging.

17.2.6 Error Measurements and Validations

To calculate the baseline displacement error and uncertainty, the duplicate 2 × 2 tiled z-stacks acquired back-to-back at each pressure were correlated. U_x and U_y are the DVC displacements in the nasal-temporal (NT) and inferior-superior (IS) directions respectively and U_z is the out-of-plane displacement. The DVC displacement field calculated between the two duplicates provides the baseline error under static conditions for these directions:

$$e_x = \frac{1}{M}\sqrt{\sum\nolimits_j^M U_j^2}, \tag{17.1}$$

$$e_y = \frac{1}{M}\sqrt{\sum\nolimits_j^M V_j^2}, \tag{17.2}$$

$$e_z = \frac{1}{M}\sqrt{\sum\nolimits_j^M W_j^2}, \tag{17.3}$$

where M the number of points in the volume, U_j, V_j, W_j are the displacements along the X, Y, and Z directions at the point j. These are root mean square errors. The errors for the displacements in the X and Y directions were considered separately due to the inherent bias in scan direction (x-direction).

In addition, we applied a displacement field representing a rigid body displacement and an equi-triaxial strain to one of the volumes using the matlab functions *affine3d* and *imwarp* in version R2015b and correlated the deformed volume to its duplicate. The error between the DVC calculation for the displacement and the applied displacement was calculated at each DVC grid point. A correlation threshold was selected to exclude points from regions of poor contrast with errors larger than 1 pixel.

17.2.7 Strain Calculation

Strains were calculated by fitting a high order polynomial function to the displacement field in each direction and taking the gradient. Gradients of the polynomial fit were used to calculate the deformation gradient and the six Green-Lagrange strain, components at each point:

$$\mathbf{F} = \begin{bmatrix} 1 + U_{x,x} & U_{x,y} & U_{x,z} \\ U_{y,x} & 1 + U_{y,y} & U_{y,z} \\ U_{z,x} & U_{z,y} & 1 + U_{z,z} \end{bmatrix}$$

$$\mathbf{E} = \frac{1}{2}\left(\mathbf{F}^{T}\mathbf{F} - \mathbf{I}\right), \quad (17.4)$$

The max principle strains and shear were calculated as follows, where ε_1 is the max principle in-plane strain, γ_{max} is the max in-plane shear strain, and $\mathbf{E}_{ij}$ are the Green-Lagrange strain components of $\mathbf{E}$.

$$\varepsilon_1 = \frac{E_{xx} + E_{yy}}{2} + \sqrt{\left(\frac{E_{xx} - E_{yy}}{2}\right)^2 + \left(E_{xy}\right)^2} \quad (17.5)$$

$$\gamma_{max} = \sqrt{\left(\frac{E_{xx} - E_{yy}}{2}\right)^2 + \left(E_{xy}\right)^2} \quad (17.6)$$

17.2.8 Statistics

The LC was divided about the central retinal artery and vein (CRAV) into sub volumes for statistical comparison. These regions were divided along 45-degree lines into the superior, nasal, inferior, and temporal quadrants, and further divided into outer, middle, and inner radial regions segmented at distances of 150 and 550 μm from the center of the CRAV. These were further divided into the topmost third, middle, and the bottom of the z-slices. A univariate repeated anova was used to look for statistically significant variations in strain between these 9 in-plane areas using the statistics toolbox matlab functions *fitrm*, for constructing the repeated measure model, and *anova* to run the comparison for the chosen independent variable.

17.3 Results

17.3.1 Imaging and Correlation Optimization

Since the collagen signal of the human lamina cribrosa is a natural contrast pattern the quality of the contrast varies greatly between samples. Factors such as the tilt and unevenness of the specimen surface play a role in the accuracy of correlation, along with the confounding effects of creep and noise in the imaging process. To ensure the accuracy of our results we used several quality tests that allow us to measure displacement error and uncertainty. The first setting that we optimized was the acquisition speed of our microscope. We looked at a broad spectrum of pixel dwell times in an early experiment to determine what the most accurate capture settings were. We found that the fastest setting had the lowest baseline error, indicating that DVC was more tolerant of noise and blur then it was of creep and other factors introduced with long acquisition times. The baseline errors measured for the pixel dwell times considered were: 3.5 μm for pixel dwell time of 80 μs, 2.8 μm for a pixel dwell time of 40 μs, and 1.6 μm for a pixel dwell time of 5 μs. With an average resolution of 2 μm/pixel this means we can expect a pixel dwell time of 5 μs to yield subpixel positional error. We similarly optimized the equilibration time by

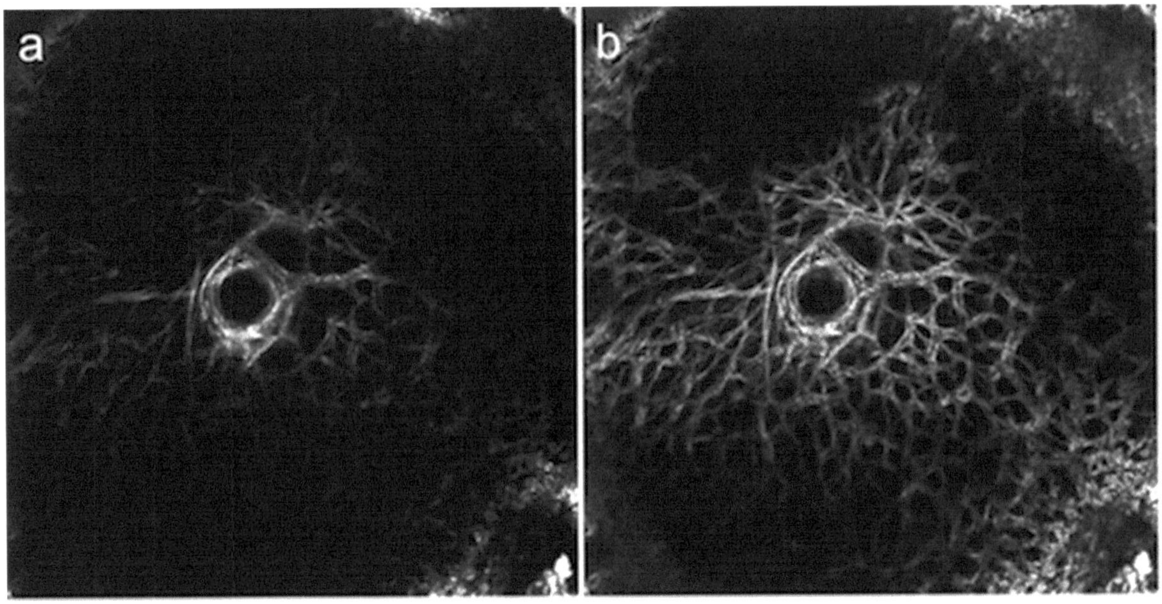

Fig. 17.3 Image enhancement from deconvolution and contrast limited adaptive histogram equalization

measuring the baseline error as a function of the time the specimen was held at a particular pressure. The error became subpixel for specimens after a hold-time of 30 min indicating this was the recommended time to equilibrate to a new pressure before imaging.

To reduce the correlation error we use two image post-processing methods to reduce blurring, improve contrast and reduce noise. The Huygens Essentials deconvolution package with theoretical psf for SHG to reduce the noise and blur, and the contrast-limited adaptive histogram equalization (CLAHE) from the program FIJI to equalize signal contrast. An example of the effect is shown in Fig. 17.3. Enhancement of the signal improved correlation in-plane and the effect was measured by applying a displacement field and measuring the correlation error to the known field. An example of the improvements gained by post-processing is shown in Fig. 17.4.

17.3.2 Displacements and Strains

All specimens examined exhibited displacements consistent with biaxial stretch with large in-plane normal strains but very little in-plane shear on average. Figure 17.5 shows example cross-section averaged displacement and strain results for FC26.

The LC were asymmetric in shape with either the nasal-temporal (NT) or inferior-superior (IS) direction being longer. The longest dimension of the LC experienced the greatest normal strain for each specimen. This was investigated by fitting ellipses to points marked on the LC-scleral interface. After the ellipse was fitted the projected axis-length in the NT and IS direction was measured. A repeated univariate anova was run to compare the strains in each direction based on the measured axial length of the LC in that direction. E_{mm} is the strain experienced in the longest direction (most aligned with the major axis) and E_{pp} the strain in the shortest (10 mmHg: $p = 0.08720$, 45 mmHg: $p = 0.02989$, Fig. 17.6a).

For all eyes, there appeared to be a significant difference in magnitude between the in-plane strains Exx and Eyy the IS and NT directions and the in-plane shear. Exx and Eyy were large though differing in magnitude, but in all specimens in-plane shear strains were very close to zero. To investigate the significance of this finding a repeated univariate anova comparing the strain components: Exy, Exx, and Eyy normal was run. The comparison showed a significant difference in magnitude between the three strain components (Fig. 17.6b). The largest difference was between the normal and shear strains but the normal strains along the major axis of the LC was significantly larger than that along the minor axis (10 mmHg: $p = 0.00175$, 45 mmHg: $p = 0.00062$). This difference was even more pronounced in the most accurately correlating regions in the middle and center of the LC surrounding the CRAV ($r > 550$ μm) (10 mmHg: $p = 0.00046$, 45 mmHg: $p = 0.00002$).

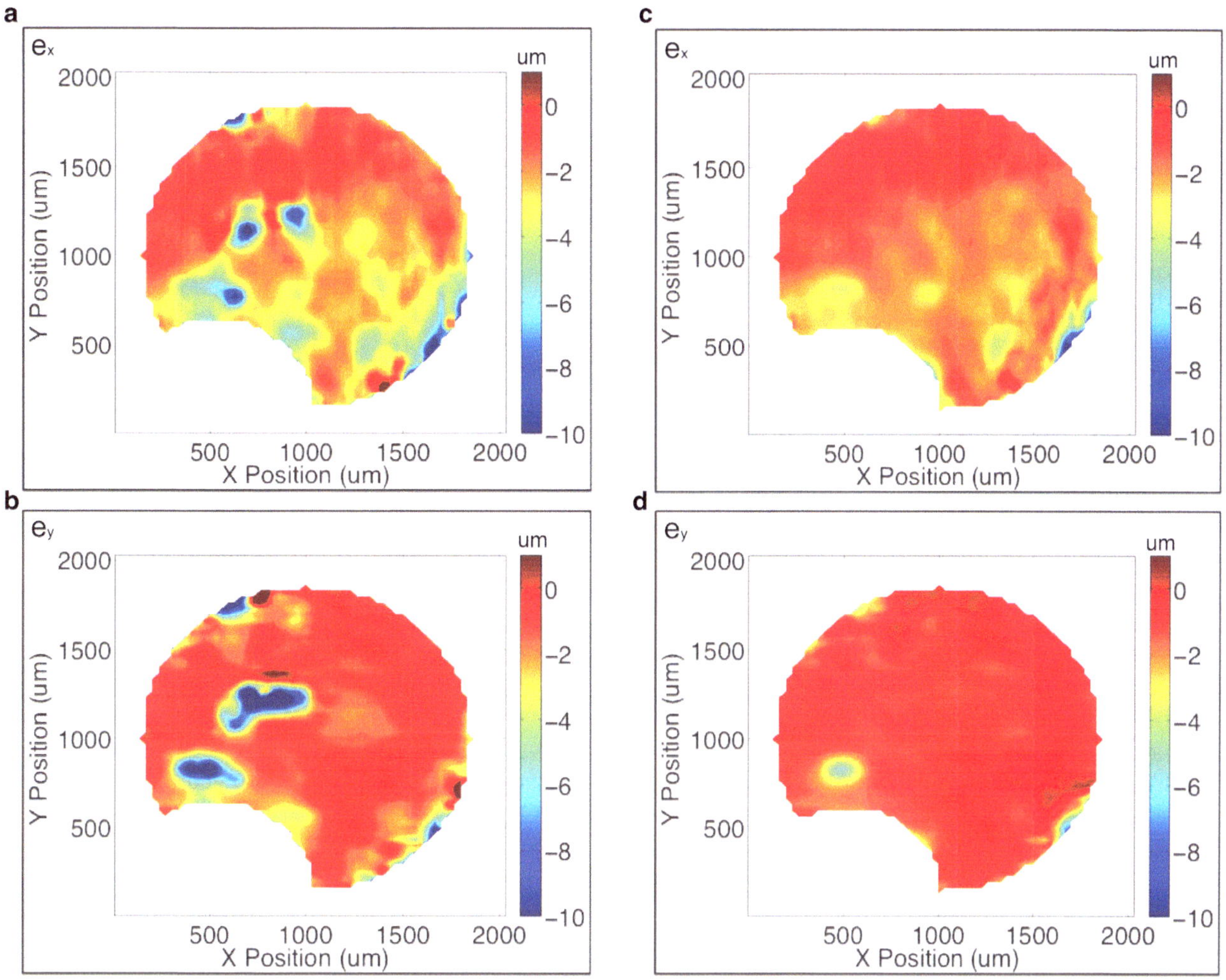

Fig. 17.4 Z-averaged displacement error fields from applied triaxial strain (**a**) X displacement error before contrast enhancement (**b**) Y displacement error before contrast enhancement (**c**) X displacement error after contrast enhancement (**d**) Y displacement error after contrast enhancement

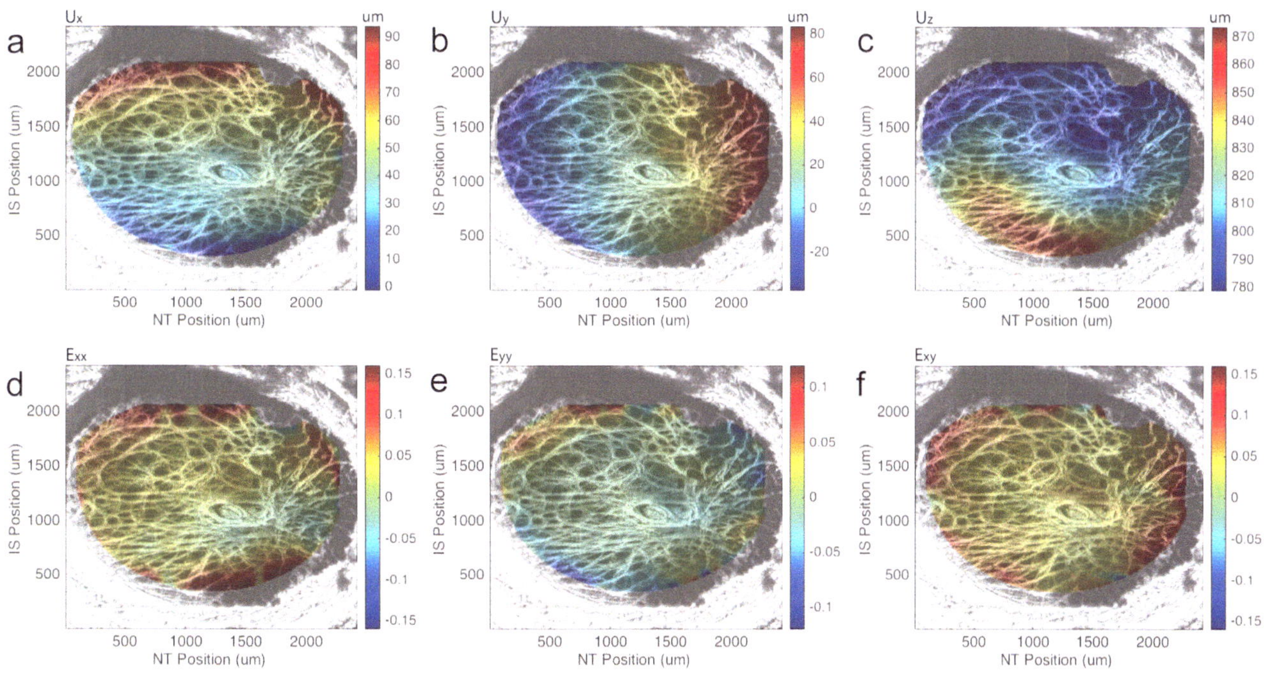

Fig. 17.5 Displacement and strain for FC26 (**a**) X Displacement (**b**) Y Displaement (**c**) Z displacement (**d**) E_{xx} normal strain (**e**) E_{yy} normal strain (**f**) E_{xy} shear strain

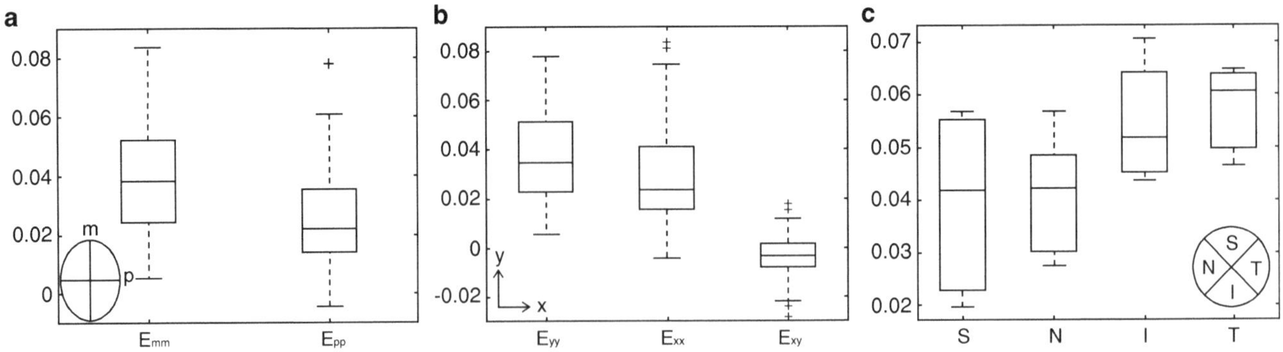

Fig. 17.6 Statistical Comparison of (**a**) strain along the long and short axes of the LC ellipse (**b**) the in-plane strain components (**c**) the maximum principle strain average of each LC quadrant

In the mid-center region surrounding the CRAV (550 μm radius) the maximum principal strains were shown to be larger in the inferior (I) and temporal (T) directions compared the superior (S) and nasal (N) directions (Fig. 17.6c) (10 mmHg: $p = 0.083123$, 45 mmHg: 0.071412). In paired t-tests between individual quadrants it was found that T > S (10 mmHg: $p = 0.08280$, 45 mmHg: $p = 0.06807$), I > N (10 mmHg: $p = 0.07785$, 45 mmHg: $p = 0.08225$) and T > N (10 mmHg: $p = 0.04964$, 45 mmHg: $p = 0.02581$).

Among the six eyes considered, we observed two different radial variations. Three out of the six were the stiffest around the CRAV with higher principal and shear strains towards the periphery (FC49, MH42, FC26) and the rest were more compliant around the CRAV and had less strain on the periphery (FC71, MC57).

17.4 Discussion

The main limitation of the study is insufficient resolution to calculate accurate out of plane strains and in some cases, lack of full-field data due to poor SHG contrast. We also did not measure the thickness of the lamina in the area we tested but we plan to do this in future experiments. In the future we will apply this method to study the effects of various treatments on the mechanical behavior of the posterior sclera and to study the mechanical changes of the LC as glaucoma progresses. A trial test with the agent ChaseABC has indicated that specimens can be measured via this method before and after treatment with only a slight degradation in SHG signal. This will allow us to measure, within the same eye, the in-vitro response of the LC before and after various treatments. We also plan to use this experiment in concert with an inverse finite element method to determine the material properties of the LC. This will help us to better test hypotheses and investigate material properties.

17.5 Conclusions

We developed a method to measure strains in the LC using inflation and DVC. The main findings of our work were:

1. Significant difference in magnitude between the in-plane strains Exy, Exx, and Eyy.
2. The six human lamina cribrosa studied were found to be approximately oval in cross-section. Normal strains in the direction most aligned with the major axis were found to experience the largest strain.
3. In the mid-central region it was found that the maximum principle strains in the inferior and temporal quadrants were greater than in the opposing superior and nasal quadrants.
4. Eyes exhibited two kinds of radial variation. Three out of the six were very stiff around the CRAV with higher principal and shear strains towards the periphery and the rest were more compliant around the CRAV and had less strain on the periphery.

References

1. Quigley, H.A.: Neuronal death in glaucoma. Prog. Retin. Eye Res. **18**, 39–57 (1997)
2. Burgoyne, C.F., Downs, J.C., Bellezza, A.J., Suh, J.K.F., Hart, R.T.: The optic nerve head as a biomechanical structure: a new paradigm for understanding the role of IOP-related stress and strain in the pathophysiology of glaucomatous optic nerve head damage. Prog. Retin. Eye Res. **24**, 39–73 (2005)
3. Morrison, J.C., Johnson, E.C., Cepurna, W., Jia, L.: Understanding mechanisms of pressure-induced optic nerve damage. Prog. Retin. Eye Res. **24**, 217–240 (2005)
4. Hernandez, M.R., Pena, J.D.O.: The optic nerve head in glaucomatous optic neuropathy. Arch. Ophthalmol. **115**, 389–395 (1997)
5. Sigal, I.A., Flanagan, J.G., Ethier, C.R.: Factors influencing optic nerve head biomechanics. Invest. Ophthalmol. Vis. Sci. **11**, 4189–4199 (2005)
6. Cone, F.E., Gelman, S.E., Son, J.L., Pease, M.E., Quigley, H.A.: Differential susceptibility to experimental glaucoma among 3 mouse strains using bead and viscoelastic injection. Exp. Eye Res. **91**, 415–424 (2010)
7. Nguyen, C., Cone, F.E., Nguyen, T.D., Coudrillier, B., Pease, M.E., Steinhart, M.R., Oglesby, E.N., Jefferys, J.L., Quigley, H.A.: Studies of scleral biomechanical behavior related to susceptibility for retinal ganglion cell loss in experimental mouse glaucoma. Invest. Ophthalmol. Vis. Sci. **54**, 1767–1780 (2013)
8. Cone-Kimball, E., Nguyen, C., Oglesby, E.N., Pease, M.E., Steinhart, M.R., Quigley, H.A.: Scleral structural alterations associated with chronic experimental intraocular pressure elevation in mice. Mol. Vis. **19**, 2023–2039 (2013)
9. Hommer, A., Fuchsjäger-Maryl, G., Resch, H., Vass, C., Garhofer, G., Schmetterer, L.: Estimation of the ocular rigidity based on measurement of pulse amplitude using pneumotonometry and fundus pulse using laser interferometry in glaucoma. Invest. Ophthalmol. Vis. Sci. **49**, 4046–4050 (2008)
10. Downs, J.C., Suh, J.K.F., Thomas, K.A., Bellezza, A.J., Hart, R.T., Burgoyne, C.F.: Viscoelastic material properties of the peripapillary sclera in normal and early-glaucoma monkey eyes. Invest. Ophthalmol. Vis. Sci. **46**, 540–546 (2005)
11. Girard, M.J.A., Suh, J.K.F., Bottlang, M., Burgoyne, C.F., Downs, J.C.: Biomechanical changes in the sclera of monkey eyes exposed to chronic IOP elevations. Invest. Ophthalmol. Vis. Sci. **52**, 5656–5669 (2011)
12. Coudrillier, B., Tian, J., Alexander, S., Myers, K.M., Quigley, H.A., Nguyen, T.D.: Biomechanics of the human posterior sclera: age- and glaucoma-related changes measured using inflation testing. Invest. Ophthalmol. Vis. Sci. **53**, 1714–1728 (2012)
13. Bar-Kochba, E., Toyjanova, J., Andrews, E., Kim, K., Franck, C.: A fast iterative digital volume correlation algorithm for large deformations. Exp. Mech. **55**, 261–274 (2015)

Chapter 18
Experimental Electromechanics of Red Blood Cells Using Dielectrophoresis-Based Microfluidics

Yuhao Qiang, Jia Liu, Michael Mian, and E. Du

Abstract This study focuses on electrically coupled mechanics of red blood cells (RBCs) using dielectrophoresis in a microfluidic chamber. When RBCs are exposed to non-uniform alternating current electric fields, they exhibit not only a net motion towards high electric field gradients but also morphological deformation at certain conditions. Through interdigitated electrode arrays, multiple cells can be trapped and stretched simultaneously in response to dielectrophoresis actuations. We utilize this approach to measure mechanical properties of individual RBCs. In addition, this dielectrophoresis platform provides a flexibility in different loading profiles, allowing us to measure both static and dynamic behavior of individual RBCs in response to cyclic stretching-relaxation loading. This approach can potentially lead to insights of the accumulative membrane failure of RBCs in blood and extracorporeal circulations.

Keywords Dielectrophoresis • Red blood cell • Cellular mechanics • Microfluidics • Stretching

18.1 Introduction

Biomechanics of healthy and diseased red blood cells (RBCs) has been a focus of cellular biomechanics [1–3]. RBCs experience marked deformation in blood circulation for several hundred thousand times during their 120-day normal lifespan [4]. RBC deformability is an important mechanical property in maintaining the homeostatic characteristics for proper function of human body [5]. RBC deformability can be significantly influenced by various pathophysiological processes such as malaria [6, 7], hypertension [8], and diabetes [9–11].

Various approaches have been presented to measure mechanical properties of single cells and cellular subcomponents, such as micropipette aspiration [11–14], optical tweezers [15–17], and microfluidics approaches [18]. Dielectrophoresis (DEP) in microfluidics has been demonstrated to be an effective method capable of high throughput measurements of single cells [19–21]. Generally, DEP refers to the induced movement of dielectric particles upon non-uniform electric field [22, 23]. In the case of deformable biological particles, such as RBCs, they exhibit not only movement but also morphological deformation under the electric-field induced forces [24].

In this paper, we present development and application of a DEP-based microfluidic device for mechanical measurements of single RBCs. Using programmable waveforms, different loading profiles can be generated to evaluate cellular deformation profiles in response to different levels of electrical frequencies and voltages. Additionally, besides static deformation, we studied dynamic behaviors of RBCs in response to an instantaneous stretch loading and a cyclic loading. DEP force was estimated using an ellipsoidal cell model [25].

18.2 Materials and Method

The microfluidic device consists of an interdigitated electrode array (IEA) chip (NanoSPR) and a 50 μm deep polydimethylsiloxane (PDMS) chamber (Fig. 18.1a). The gold IEA is 150 nm thick with 20 μm gap and 20 μm band width. DEP stretching of RBCs was achieved by feeding the IEA with an alternating current (AC) signal from a function generator (SIGLENT SDG830), and visualized through a GigE Camera (Imaging Source) mounted on a Nikon Eclipse TE2000-S inverted microscope (Fig. 18.1b). A sinewave was used to study the cellular response to different levels of voltage and frequency

Y. Qiang • J. Liu • M. Mian • E. Du (✉)
Department of Ocean and Mechanical Engineering, Florida Atlantic University, Boca Raton, FL 33431, USA
e-mail: edu@fau.edu

C.S. Korach et al. (eds.), *Mechanics of Biological Systems and Materials, Volume 6*,
Conference Proceedings of the Society for Experimental Mechanics Series, DOI 10.1007/978-3-319-41351-8_18

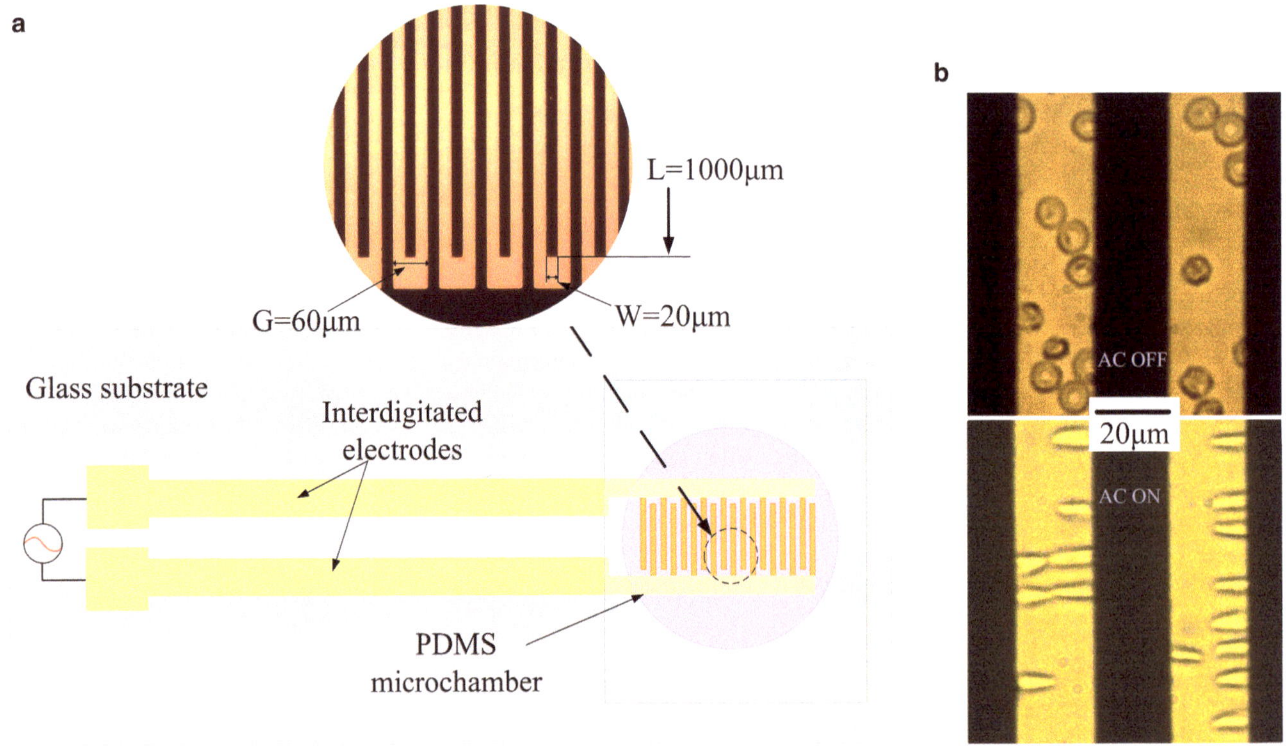

Fig. 18.1 DEP stretching of RBCs. (**a**) Schematic of the microfluidic device with an *inset* of IEA observed under the microscope. (**b**) *Upper panel*, RBCs are suspended in the medium and relaxed when the AC field is off. *Bottom panel*, RBCs are stretched in the electric field excited with a sinewave of 4 V_{rms} and 3 MHz

(0.5–7 V_{rms} and 0.1–30 MHz). An amplitude-shift keying of 500 mHz superposed on a sinusoid waveform (4 V_{rms} and 3 MHz) was created to study cellular dynamic behavior subjected to a cyclic stretching-relaxation loading.

The DEP working medium was prepared from an isotonic buffer (8.5 % sucrose, 0.3 % dextrose, and 0.1 % bovine serum albumin in deionized water) with an electrical conductivity of 55 mS m^{-1}, adjusted by adding phosphate-buffered saline (PBS, Lonza). Whole blood from healthy donors was obtained and washed with a centrifugation at 2000 rpm for 3 min at 21 °C. RBC pellet was aspirated and diluted into the DEP medium to a target 0.1 % hematocrit.

Behavior of RBCs in response to electric excitations was recorded by IC Capture (Imaging Source). Cellular size and geometry were analyzed using ImageJ (http://imagej.nih.gov/ij/). Configuration of deformed RBCs was fitted with ellipses, assuming symmetric geometry. Then cellular deformation was quantified by the ratio of major and minor axes of stretched cells. Gradient of electric field strength square, $\nabla \boldsymbol{E}^2{}_{rms}$ in the device was calculated using COMSOL Multipysics 5.0 (COMSOL, Inc). Correspondingly, values of $\nabla \boldsymbol{E}^2_{rms}$ at the free-end of individual stretched RBCs were determined.

To quantify the DEP force exerted on individual RBCs, we assume a stretched cell has an ellipsoidal structure. The time-averaged DEP force is estimated as [24, 25]

$$\langle \mathbf{F}_{DEP} \rangle = \frac{\pi abc}{4} \cdot \varepsilon_m \cdot Re(f_{CM}) \cdot \nabla \boldsymbol{E}^2_{rms}, \qquad (18.1)$$

where a and b are the major and minor axes of the RBCs, ε_m is the permittivity of the surrounding medium, and $\boldsymbol{E}_{rms}$ is the root-mean-square value of the electric field strength. Values for a and b for individual cells were determined from the experiments, c assumed to be a constant, 2 μm. $Re(f_{CM})$ is the real part of the Clausius–Mossotti factor (f_{CM}) based on the ellipsoid single-shell model for RBCs [26, 27],

Table 18.1 Typical values of parameters used to calculate f_{CM} of RBCs [21]

Cell	Membrane thickness (nm)	Relative permittivity, ε_r	Conductivity, σ (S/m)
RBCs	4.5	Membrane 4.44 ± 0.45	10^{-6}
		Interior 59 ± 6	0.31 ±0.03

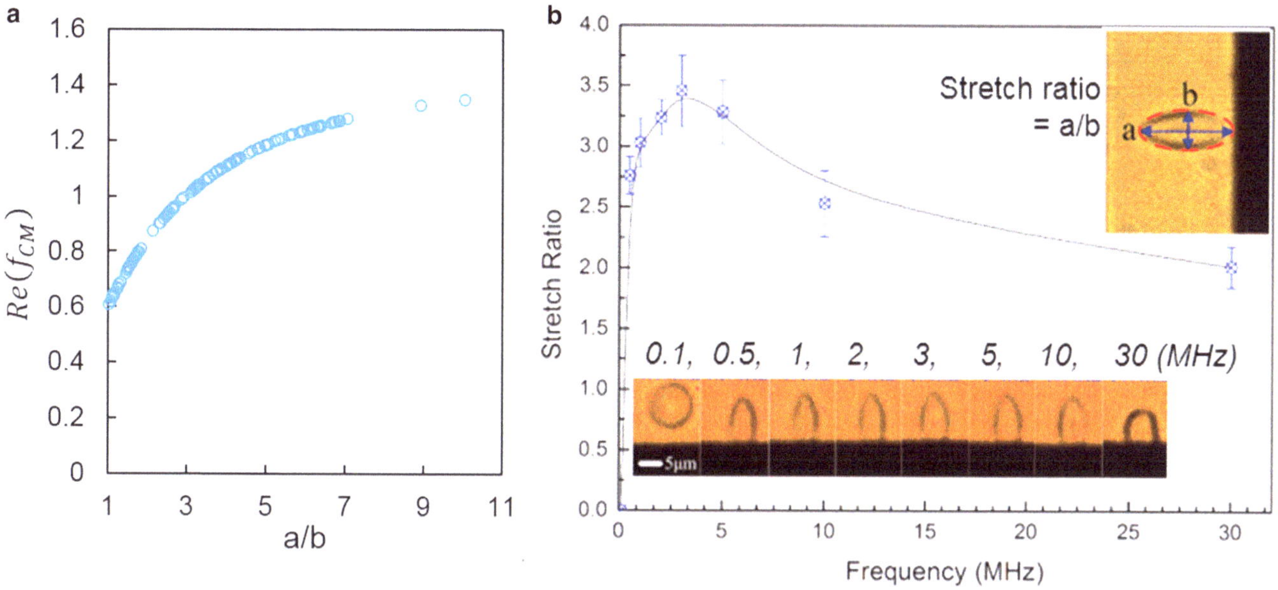

Fig. 18.2 (**a**) Frequency-dependent RBC deformation subjected to DEP stretching. RBC stretch ratio at a fixed electric voltage of 2 V_{rms}. *Inset* shows representative microscopic images of stretched RBCs. Data are presented in mean ± standard deviation (n = 10). (**b**) Value of f_{CM} as a function of stretch ratio of deformed RBCs

$$f_{CM} = \frac{1}{3}\frac{\left(\varepsilon^*_{mem}-\varepsilon^*_m\right)\left[\varepsilon^*_{mem}+A_1\left(\varepsilon^*_{cyto}-\varepsilon^*_{mem}\right)\right]+\rho\left(\varepsilon^*_{cyto}-\varepsilon^*_{mem}\right)\left[\varepsilon^*_{mem}-A_1\left(\varepsilon^*_{mem}-\varepsilon^*_m\right)\right]}{\left(\varepsilon^*_m+A_1\left(\varepsilon^*_{mem}-\varepsilon^*_m\right)\right)\left[\varepsilon^*_{mem}+A_1\left(\varepsilon^*_{cyto}-\varepsilon^*_{mem}\right)\right]+\rho A_2(1-A_2)\left(\varepsilon^*_{cyto}-\varepsilon^*_{mem}\right)\left(\varepsilon^*_{mem}-\varepsilon^*_m\right)} \tag{18.2}$$

where the subscripts *cyto*, *mem* and *m* stand for cytoplasm, membrane and medium, respectively. $\varepsilon^* = \varepsilon - i\sigma/\omega$ with ω being the angular field frequency, $i = \sqrt{-1}$, and ε and σ as the dielectric permittivity and conductivity, respectively. Values of these parameters for RBCs are listed in Table 18.1. $A_{i=1,2}$ is the depolarization factor, determined from

$$A_i = \frac{1-e_i^2}{2e_i^3}\left[log\left(\frac{1+e_i}{1-e_i}\right)-2e_i\right],\ \mathrm{i} = 1,\ 2 \tag{18.3}$$

where $e_1 = \sqrt{1-\left(\frac{b}{a}\right)^2}$, $e_2 = \sqrt{1-\left(\frac{b-2t}{a-2t}\right)^2}$, with t being the thickness of cell membrane. Values of $Re(f_{CM})$ for RBCs were calculated using a custom script in MATLAB R2016a (Mathworks).

18.3 Results and Discussion

From the ellipsoid model, Re(f_{CM}) is a function of stretch ratio of the particle. In our study, Values of Re(f_{CM}) of stretched RBCs ranged from 0.6 to 1.4 (Fig. 18.2a). As Re(f_{CM}) is a primary determinant of the magnitude of the DEP force exerted on cells, large deformation in RBCs was found to be within electrical frequencies of 1–10 MHz (Fig. 18.2b). Maximum stretch ratio was achieved at 3 MHz. A negative DEP was observed at 0.1 MHz where RBCs were repelled away from the electrode edges with highest gradient of electric field strength. These observations agree well with a previous report [21].

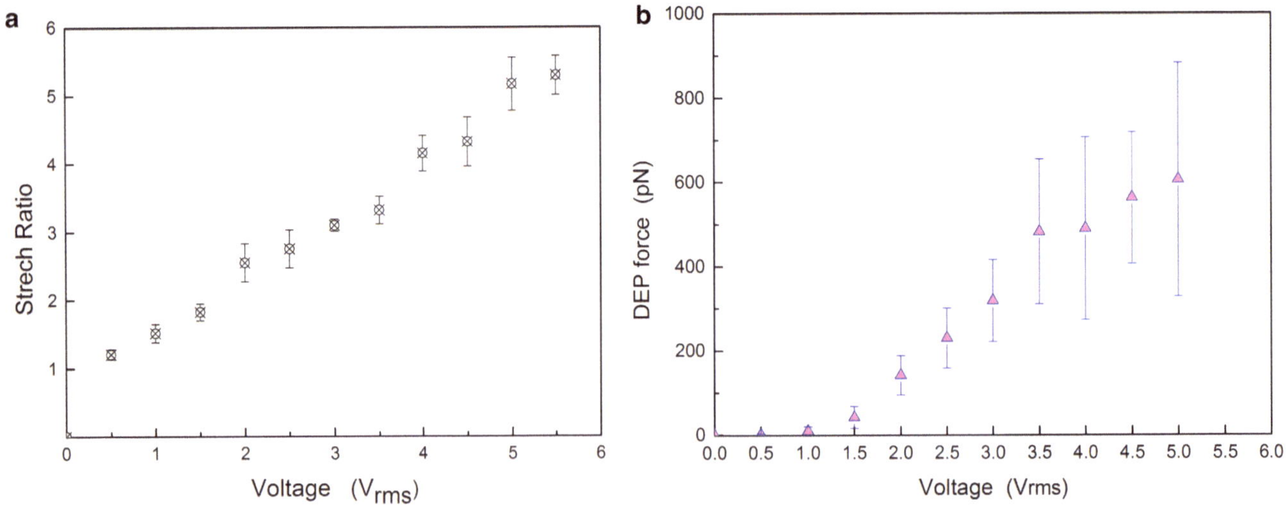

Fig. 18.3 RBC deformation in response to DEP stretching. (**a**) Stretch ratio of RBCs at different levels of electric voltages at 3 MHz (n = 10). (**b**) DEP forces as a function of the magnitude of the applied voltage. Data are presented in mean ± standard deviation (n = 10)

Cellular deformation increased with the applied voltage levels (Fig. 18.3a). DEP forces exerted on individual RBCs were determined from Eq. (18.1). The relationship between the DEP force and the applied voltage can be well fitted with a second-order polynomial (Fig. 18.3b), agreeing with the prediction by Eq. (18.1) where the force magnitude is proportional to $\nabla \boldsymbol{E}_{rms}^2$. These findings agree with previous reports [16, 21].

We further investigated the dynamic behavior of RBCs in response to a cyclic DEP loading. Each cycle started with a sudden application of a DEP actuation (4 V_{rms}, 3 MHz) and maintained for one second, followed by a sudden release of the DEP actuation and maintained for another second. Cellular response was quantified by a time sequence of the distance between the electrode edge and the free-end of the stretched cells. In the first cycle, each RBC deformed gradually from a fully relaxed state to a maximum deformed state and gradually recovered to the fully relaxed state (Fig. 18.4a), suggesting a typical viscoelastic behavior of RBCs. During the subsequent cycles, behavior of RBCs followed a similar but not identical trend (Fig. 18.4b); deformations of RBCs were found to be out of phase in later cycles (Figs. 18.4cd), suggesting a likely degradation in cellular mechanical properties.

18.4 Conclusion

We have developed a DEP-based microfluidic platform to measure the mechanical properties of individual RBCs. This electrically coupled mechanics method allowed probing multiple cells simultaneously. Importantly, it provided a flexibility in probing both quasi-static behavior of single cells and their dynamic behavior subjected to a cyclic loading. In blood circulation, RBCs experience repetitive mechanical stimulations in high shear-rate blood flows and when they travel through narrow capillaries and splenic sinusoids [28]. This method can be used to better understand the RBC membrane failure in blood circulation, from a perpective of cellular biomechanics.

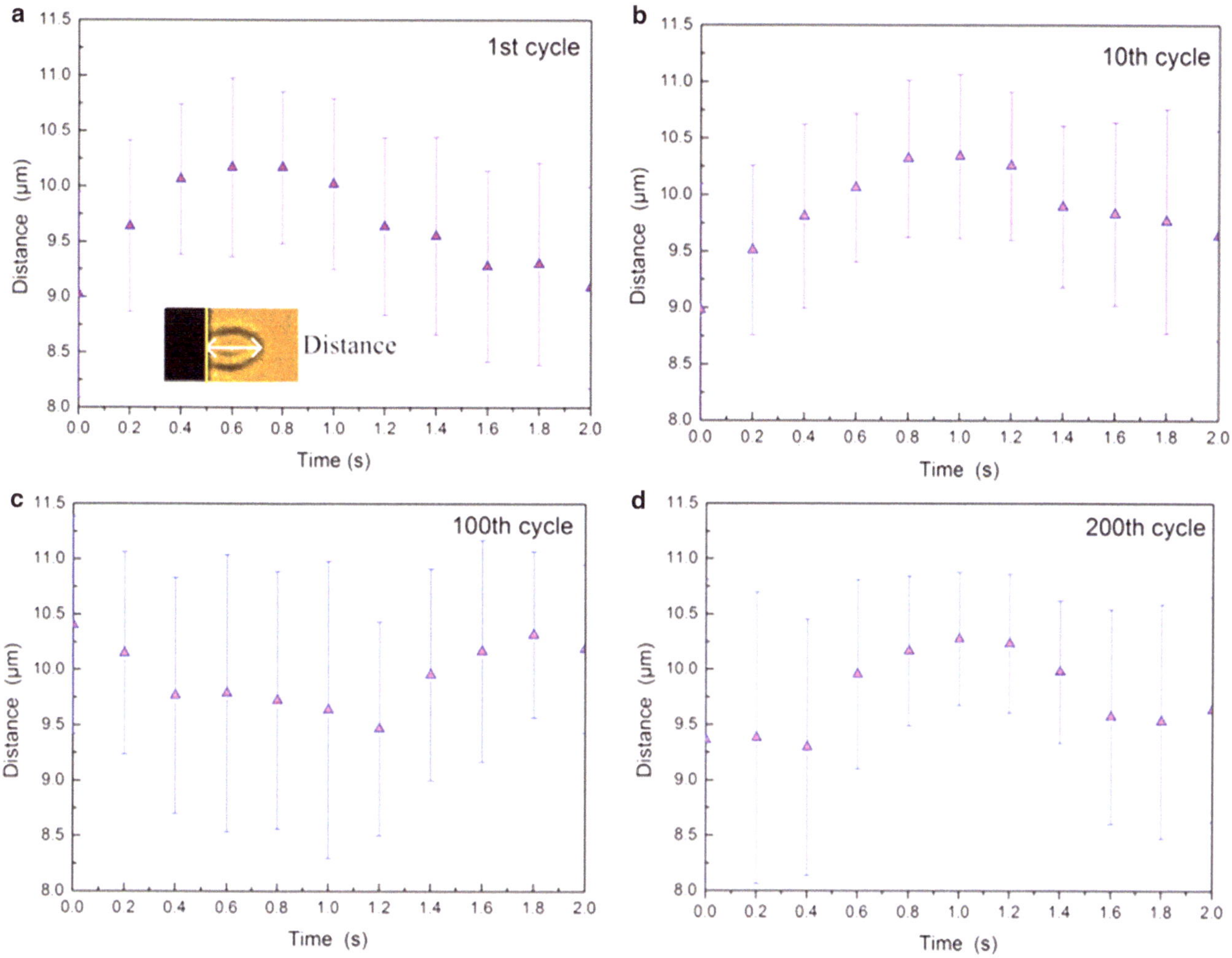

Fig. 18.4 Dynamic behavior of RBCs (n = 10) in response to cyclic stretching-relaxation loading in representative (**a**) cycle 1, (**b**) cycle 10, (**c**) cycle 100, and (**d**) cycle 200

References

1. Lee, G.Y., Lim, C.T.: Biomechanics approaches to studying human diseases. Trends Biotechnol. **25**, 111–118 (2007)
2. Diez-Silva, M., Dao, M., Han, J., et al.: Shape and biomechanical characteristics of human red blood.pdf. MRS Bull. **35**, 382–388 (2010)
3. Suresh, S.: Biomechanics and biophysics of cancer cells. Acta Biomater. **3**, 413–438 (2007)
4. Lenormand, G., Hénon, S., Richert, A., et al.: Direct measurement of the area expansion and shear moduli of the human red blood cell membrane skeleton. Biophys. J. **81**, 43–56 (2001)
5. Sakuma, S., et al.: Red blood cell fatigue evaluation based on the close-encountering point between extensibility and recoverability. Lab Chip **14**, 1135–1141 (2014)
6. Glenister, F.K.: Contribution of parasite proteins to altered mechanical properties of malaria-infected red blood cells. Blood **99**, 1060–1063 (2002)
7. Lim, C.T.: Single cell mechanics study of the human disease malaria. J. Biomed. Sci. Eng. **1**, 82–92 (2006)
8. Cicco, G., Pirrelli, A.: Red blood cell (RBC) deformability, RBC aggregability and tissue oxygenation in hypertension. Clin. Hemorheol. Microcirc. **21**, 169–177 (1999)
9. Brown, C.D., Ghali, H.S., Zhao, Z.H., et al.: Association of reduced red blood cell deformability and diabetic nephropathy-reply. Kidney Int. **67**, 2066–2067 (2005)
10. Tsukada, K., Sekizuka, E., Oshio, C., Minamitani, H.: Direct measurement of erythrocyte deformability in diabetes mellitus with a transparent microchannel capillary model and high-speed video camera system. Microvasc. Res. **61**, 231–239 (2001)
11. McMillan, D.E., Utterback, N.G., La Puma, J.: Reduced erythrocyte deformability in diabetes. Diabetes **27**, 895–901 (1978)
12. Aingaran, M., et al.: Host cell deformability is linked to transmission in the human malaria parasite *Plasmodium falciparum*. Cell. Microbiol. **14**, 983–993 (2012)
13. Hochmuth, R.M.: Micropipette aspiration of living cells. J. Biomech. **33**, 15–22 (2000)

14. Artmann, G.M., Sung, K.L., Horn, T., et al.: Micropipette aspiration of human erythrocytes induces echinocytes via membrane phospholipid translocation. Biophys. J. **72**, 1434–1441 (1997)
15. Dao, M., Lim, C.T., Suresh, S.: Mechanics of the human red blood cell deformed by optical tweezers. J. Mech. Phys. Solids **51**, 2259–2280 (2003)
16. Mills, J.P., Qie, L., Dao, M., et al.: Nonlinear elastic and viscoelastic deformation of the human red blood cell with optical tweezers. Mech. Chem. Biosyst. **1**, 169–180 (2004)
17. Haque, M.M., Moisescu, M.G., Valkai, S., Der, A., Savopol, T.: Stretching of red blood cells using an electro-optics trap. Biomed. Opt. Exp. **6**, 118–123 (2015)
18. Vanapalli, S.A., Duits, M.H.G., Mugele, F.: Microfluidics as a functional tool for cell mechanics. Biomicrofluidics **3**, 12006 (2009)
19. Engelhardt, H., Sackmann, E.: On the measurement of shear elastic moduli and viscosities of erythrocyte plasma membranes by transient deformation in high frequency electric fields. Biophys. J. **54**, 495–508 (1988)
20. Guido, I., Jaeger, M.S., Duschl, C.: Dielectrophoretic stretching of cells allows for characterization of their mechanical properties. Eur. Biophys. J. **40**, 281–288 (2011)
21. Du, E., Dao, M., Suresh, S.: Quantitative biomechanics of healthy and diseased human red blood cells using dielectrophoresis in a microfluidic system. Extreme Mech. Lett. **1**, 35–41 (2014)
22. Khoshmanesh, K., Nahavandi, S., Baratchi, S., et al.: Dielectrophoretic platforms for bio-microfluidic systems. Biosens. Bioelectron. **26**, 1800–1814 (2011)
23. Das, D., Biswas, K., Das, S.: A microfluidic device for continuous manipulation of biological cells using dielectrophoresis. Med. Eng. Phys. **36**, 726–731 (2014)
24. Morgan, H., Green, N.G.: AC Electrokinetics: Colloids and Nanoparticles. Research Studies Press, Philadelphia (2003)
25. Jones, T.B.: Electromechanics of Particles. Cambridge University Press, Cambridge, U.K. (2005)
26. Castellarnau, M., Errachid, A., Madrid, C., Juarez, A., Samitier, J.: Dielectrophoresis as a tool to characterize and differentiate isogenic mutants of *Escherichia coli*. Biophys. J. **91**, 3937–3945 (2006)
27. Zheng, L., Brody, J.P., Burke, P.J.: Electronic manipulation of DNA, proteins, and nanoparticles for potential circuit assembly. Biosens. Bioelectron. **20**, 606–619 (2004)
28. Mohandas, N., Gallagher, P.G.: Red cell membrane: past, present, and future. Blood **112**, 3939–3948 (2008)

Chapter 19
Microbuckling of Fibrous Matrices Enables Long Range Cell Mechanosensing

Brian Burkel, Ayelet Lesman, Phoebus Rosakis, David A. Tirrell, Guruswami Ravichandran, and Jacob Notbohm

Abstract When biological cells migrate, divide, and invade, they push and pull on individual fibers of the matrix surrounding them. The resulting fiber displacements are neither uniform nor smooth; rather, displacements localize to form dense fibrous bands that span from one cell to another. It is thought that these bands may be a mechanism by which cells can sense their neighbors, but this hypothesis remains untested, because the mechanism for band formation remains unknown. Using digital volume correlation, we measure the displacements induced by contractile cells embedded in a fibrous matrix. We find that cell-induced displacements propagate over a longer range than predicted by linear elasticity. To explain the long-range propagation of displacements, we consider the effect of buckling of individual matrix fibers, which generates a nonlinear stress-strain relationship. We show that fiber buckling is the mechanism that causes the displacements to propagate over a long range and the bands to form between nearby cells. The results thus show that buckling of individual fibers provides a mechanism by which cells may sense their distant neighbors mechanically.

Keywords Extracellular matrix • Mechanosensing • Fiber buckling • DVC • Traction force microscopy

19.1 Introduction

The extracellular matrix (ECM) provides a rich and complex milleu of information that cells use to modulate numerous processes and behaviors ranging from migration and division to tissue differentiation and cancer progression [1–5]. While the chemical cues and their signaling have been studied for some time, a potentially equally important and only recently appreciated facet of this signaling is the cue provided by the mechanical properties of the matrix. Properties like stiffness, pore size, or alignment elicit complex cellular responses which in turn generate localized traction forces on these fibrous matrices contributing to subsequent matrix remodeling [6]. Matrix fibers align due to cellular forces, and fiber displacements propagate over several cell diameters [7, 8]. While observations of these phenomena have now become common, our understanding of how these matrices deform due to localized cellular forces and what effect the remodeled structures have on cell behavior remain largely unknown.

Most previous studies investigating cell-matrix interactions have had two fundamental limitations. For one, they have a heavy reliance on linearly elastic gels like polyacrylamide. While these substrates are informative and can be tuned to have specific physical properties (e.g., stiffness), they exhibit linear elastic behavior which is incompatible with the behavior of natural substrates and tissues, which are fibrous in nature and exhibit nonlinear, inelastic, and inhomogeneous behaviors [9]. A second problem is that conventional experimental methods for measuring cell forces such as traction force microscopy (TFM) have been limited to cells plated on 2D substrates rather than embedded in 3D matrices. Numerous studies have highlighted the importance of studying cells in three dimensions as they have distinct physical structures, morphologies, and

B. Burkel • J. Notbohm (✉)
Department of Engineering Physics, University of Wisconsin-Madison, 1500 Engineering Drive, Madison, WI 53706, USA
e-mail: jknotbohm@wisc.edu

A. Lesman
School of Mechanical Engineering, Tel Aviv University, Tel Aviv 6997801, Israel

P. Rosakis
Department of Theoretical and Applied Mathematics, University of Crete, Heraklion 70013, Crete, Greece

D.A. Tirrell
Division of Chemistry and Chemical Engineering, California Institute of Technology, 1200 E. California Blvd, Pasadena, CA 91125, USA

G. Ravichandran
Division of Engineering and Applied Science, California Institute of Technology, 1200 E. California Blvd, Pasadena, CA 91125, USA

C.S. Korach et al. (eds.), *Mechanics of Biological Systems and Materials, Volume 6*,
Conference Proceedings of the Society for Experimental Mechanics Series, DOI 10.1007/978-3-319-41351-8_19

modes of locomotion compared to their 2D counterparts [10–13]. These shortcomings have highlighted a need for new methods and metrics for assessing cellular forces and the associated matrix displacements in three dimensions.

To quantify 3D displacements, we used a 3D image correlation algorithm known as Digital Volume Correlation (DVC), which accurately and quantitatively measures displacement fields [14]. Using this technique with fluorescently labelled ECM proteins and confocal microscopy, we ultimately demonstrated the importance of fiber buckling to the non-linear properties of fibrous matrices [7] as was predicted by previous modelling [8]. Buckling of fibers under compression was predicted to be a primary factor in both long range propagation of displacements and the formation of dense bundles of aligned fibers [7].

In this study through the use of modeling and experimentation, we continue to investigate the contributions of fiber buckling as a primary mechanism for long distance fiber displacements. Our current simulations incorporating fiber buckling more accurately match the properties of natural fibrous substrates than those predicted by linear elastic substrates like polyacrylamide. This has led to the hypothesis that the physical properties (such as propensity to buckle) of the extracellular microenvironment may provide a mechanism for cells to mechanically sense and migrate towards one another over long distances.

19.2 Methods

19.2.1 Labelled Fibrin

We embedded cells in fibrin matrices, which have been shown to promote cell growth, adhesion, and migration [15]. Fibrin gels were prepared by mixing various concentrations of fibrinogen (Omrix Biopharmaceuticals, Israel) ranging from 1 to 10 mg/mL with a 20 U/mL thrombin solution (Omrix) at a 1:1 ratio. To image the fibrin matrices with confocal microscopy, the fibrinogen was fluorescently labeled before mixing with thrombin. The fluorescent dye, 546 Alexa Fluor coupled to an amine-reactive succinimidyl ester (Invitrogen Life Technologies, Grand Island, NY), was first mixed with fibrinogen solution in a 5:1 dye-to-protein molar ratio for 1 h at room temperature and then filtered through a HiTrap desalting column packed with Sephadex G-25 resin (GE Healthcare, Milwaukee, WI) to separate the unreacted dye.

19.2.2 Labelled Collagen

In addition to fluorescently labelled fibrin, we also used a fluorescently labelled collagen. Collagen I is the most abundant protein of the ECM, and contributes significantly to the physical properties sensed by cells. Acidified type I collagen (BD Bioscience) was polymerized by neutralizing with a 2× HEPES buffer (pH 7.4) at room temperature [16]. Once polymerized, it was moved to 37 °C. To image the collagen gels with confocal microscopy, fluorescently labelled collagen was added 1:10 with unlabeled collagen before neutralization. To fluorescently label the collagen, 1 mg of Alexa Fluor 488 carboxylic acid, 2,3,5,6-tetrafluorophenyl ester (ThermoFisher) was added to 25 mg of acidified collagen. The reaction was allowed to proceed for 1 h at room temperature. The labelled collagen was dialyzed extensively with 10 K MWCO dialysis cassette (ThermoFisher) to remove unreacted dye.

19.2.3 Image Acquisition

To generate 3D volume stacks of fibrin matrices for DVC, a Sweptfield confocal scanner controlled by Nikon Elements Ar software was mounted on a Ti stand (Nikon Instruments, Melville, NY). Images were then collected with a 40× 1.15 numerical aperture water immersion objective with a QuantEM:512SC camera (Photometrics, Tucson, AZ). A pinhole setting of 60 μm and a collection time of 400 ms was used to increase the signal to noise while minimizing photobleaching. Volume stacks were collected with an out-of-plane (z) step size of 0.4 μm. The microscope was housed in a custom-built chamber heated to 37 °C with an Air-Therm ATX heater (World Precision Instruments, Sarasota, FL) with a precision of 0.1 °C. CO_2 conditions were maintained at 5 % by circulating 5 % CO_2 blood gas within the chamber. All experiments were performed >3 h after placing the specimens in the heated chamber to allow for thermal equilibration.

To generate 3D volume stacks of collagen matrices for DVC, a Yokogawa X1 Spinning Disk confocal scanner was mounted on a Ti stand (Nikon Instruments, Melville, NY). Images were then collected with a 20× 0.75 numerical aperture water immersion objective with a Zyla 3-Tap sCMOS camera. A 0.75 μm step size was used in the z direction.

19.2.4 Digital Volume Correlation

A custom MATLAB code for DVC was used based on the principles described by Franck et al. [14]. The simplest version of the correlation assumes the subvolumes undergo a mapping from the reference configuration x to the current configuration y that includes only a rigid body translation u:

$$y = x + u. \tag{19.1}$$

This simple mapping neglects rotation, stretching, and shearing of the subvolume, but it has the advantage that the DVC computation can be implemented efficiently with a cross correlation. Letting f represent the intensity of the high contrast pattern for a subvolume in the reference configuration and g the intensity in the current configuration, the cross correlation is written

$$c(u) = \int f(x) g(x + u) dx. \tag{19.2}$$

Equation (19.2) can be solved efficiently using the fast Fourier transform by finding the translation u such that the cross correlation c is a maximum:

$$c(u) = F^{-1}\left\{ F[f(x)]^{*} F[g(x)] \right\}. \tag{19.3}$$

In Eq. (19.3), F represents the Fourier transform, and the asterisk represents the complex conjugate. This DVC algorithm has been refined to account for stretches and rotations of the subvolume, leading to increased accuracy. See Bar-Kochba et al. for details [17]. The DVC approach is repeated for multiple subvolumes within the image stack, resulting in a full-field displacement measurement throughout the image volume. The displacements can be differentiated in space numerically to compute the strain tensor [14].

19.3 Results and Discussion

This study begins by examining how collagen and fibrin matrices deform due to localized forces and contractions. We used reconstituted, fluorescently labelled networks as a substrate representative of those seen in tissues and in wound healing. The fibrous networks were defined by brightly labelled protein polymers spreading in all three directions interspersed with pores of fairly uniform size (Fig. 19.1).

When cells are embedded in the fibrous matrix, the fibers displace due to the forces applied by the contractile cell. We measured those displacements in all three dimensions using DVC (Fig. 19.2a). To quantify how the displacements propagate through the fibrous matrix, we chose linear paths starting at the cell and propagating outward. We measured the displacement magnitudes u along those paths and plotted them versus distance r from the cell (Fig. 19.2b). A simple fit of the form $u \sim r^{-n}$ where n is a fitting parameter gave an average value of $n = 0.52$. In comparison, a contractile ellipsoidal inclusion in a linear elastic medium induces displacements that scale as $u \sim r^{-2}$ [7, 8]. Thus, cell-induced displacements in a fibrous matrix propagate over a longer range than predicted by linear elasticity.

The long range displacement propagation is thought to result from nonlinearity due to the fibrous structure. As shown in Fig. 19.1, typically only two fibers intersect at each node, and some of those fibers have free ends in space. This is consistent with previous studies of collagen networks that have shown an average coordination number (number of fibers meeting at a node) to be approximately 3.4 [18, 19]. This is below the coordination number of 6 required for mechanical stability of a 3D network of fibers supporting only axial tension and compression. Thus some other deformation mechanism such as bending of fibers must play a role in the mechanical response.

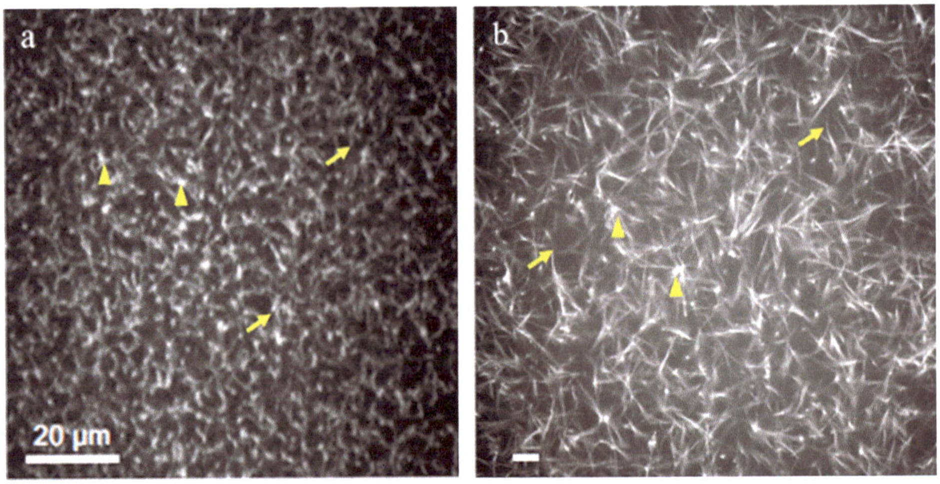

Fig. 19.1 Typical image of fluorescently labelled fibrin (**a**) and collagen (**b**) matrices. *Yellow arrows* show places where fibers dangle and *arrowheads* show where multiple fibers interconnect

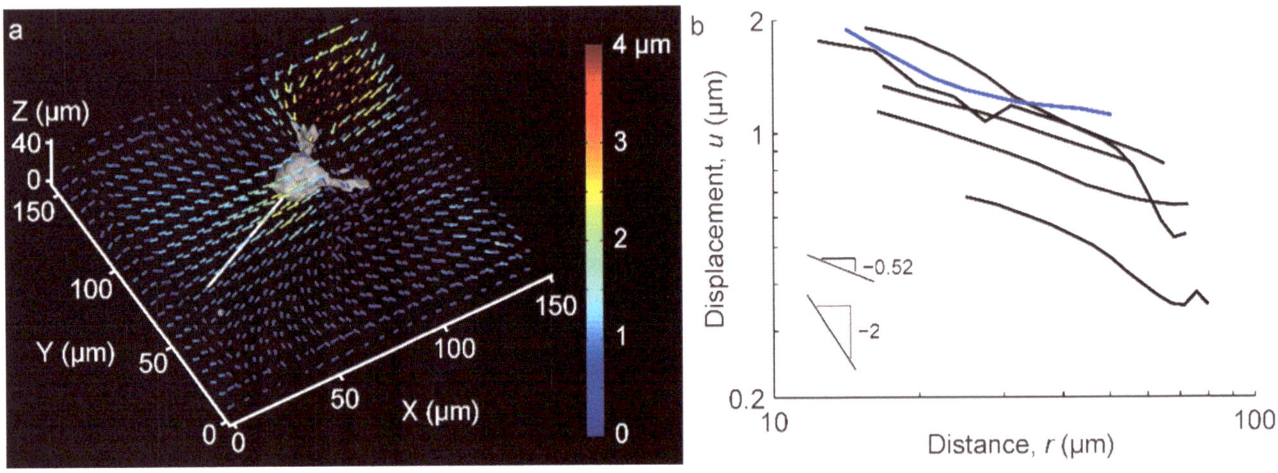

Fig. 19.2 (**a**) A contracting cell (*gray*) in a fibrous matrix. The colored quivers show 3D matrix displacements induced by the cell, as measured with DVC. (**b**) Scaling of displacements with distance from the cell. The displacement magnitude is calculated, and linear paths (similar to the white line in panel **a**) are drawn proceeding outward from the cell body. Displacement magnitude u is quantified along those paths and plotted versus distance from the cell body. Each line shows an average of multiple paths for a single cell. The *blue line* shows the displacements in panel **a**. On average, the lines have a slope of -0.52, indicating displacements scale as $u \sim r^{-0.52}$. Adapted from Notbohm et al. [7]

We suggest here that fibers bend, and therefore fibers under compression buckle. We argue here that buckling may provide a mechanism for long range propagation of displacements. Consider a spherical cell contracting in a 3D matrix. Assume that fibers under compression buckle easily and therefore the matrix supports no compressive stress. The stress field due to the cell will therefore consist of only positive radial stresses σ_{rr}, with angular stresses equal to zero. The radial equilibrium equation will then become

$$\frac{d\sigma_{rr}}{dr} + 2\frac{\sigma_{rr}}{r} = 0. \qquad (19.4)$$

The solution to this equation shows stresses scale as $\sigma_{rr} \sim r^{-2}$ [7] and displacements scale as $u \sim r^{-1}$ [8], which indicates displacements that propagate over longer range than the $u \sim r^{-2}$ scaling predicted by linear elasticity. Thus, compression weakening due to buckling could be the mechanism for long range propagation of displacements.

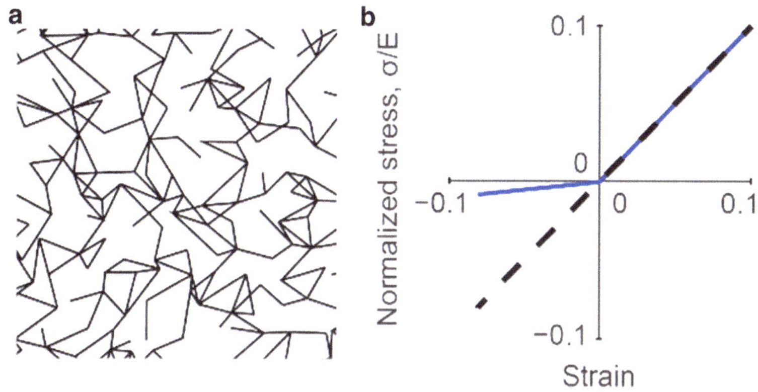

Fig. 19.3 Fiber network details. (**a**) Fiber network with coordination number $C = 3$. (**b**) Stress–strain curve for a linear fiber ($\rho = 1$, *dashed line*) and a fiber that weakens in compression due to buckling ($\rho = 0.1$, *solid line*). Adapted from Notbohm et al. [7]

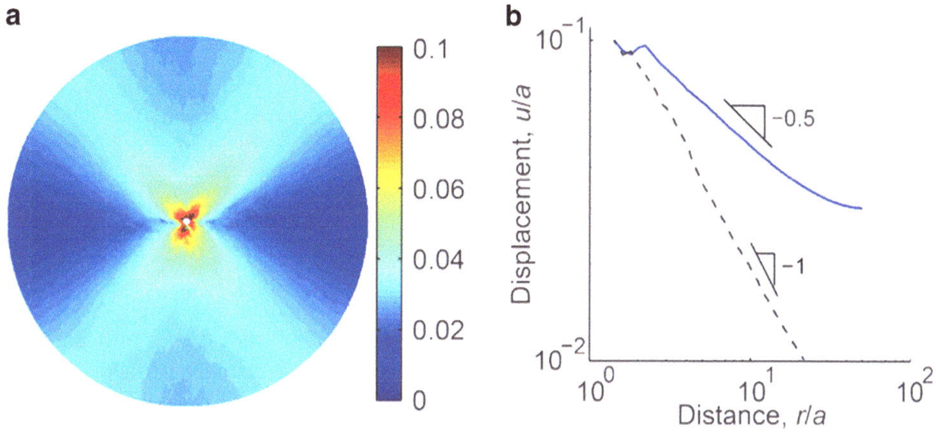

Fig. 19.4 Buckling causes displacements to propagate over a long range. (**a**) A contractile cell is modeled by a contracting circle of radius a. Normalized displacement u/a is shown for a model that simulates buckling ($\rho = 0.1$). (**b**) Displacements are averaged around a circle of radius r about the center of the model for a simulation with a linear fiber ($\rho = 1$, *dotted line*) and a buckling fiber ($\rho = 0.1$, *solid line*). Plots shown here are for a coordination number of $C = 3$. Adapted from Notbohm et al. [7]

To test the hypothesis that fiber buckling facilitates long range propagation of cell-induced displacements, we developed a model for a fibrous network (Fig. 19.3) that simulated buckling as a loss of stiffness in compression. To do so, we considered a piecewise stress–strain curve having a ratio of stiffness in compression to stiffness in tension given by the constant ρ. For $\rho < 1$ (e.g., $\rho = 0.1$ in Fig. 19.3b, solid line), the fibers lose resistance to compression, hence mimicking buckling. We compare this model to one with $\rho = 1$ (Fig. 19.3b, dashed line), thus giving a comparison between buckling and no buckling.

In preliminary simulations, we model the contractile cell as a circle contracting in fibrous matrix (Fig. 19.4a). When fiber buckling is present ($\rho = 0.1$), displacements u propagate over a long range with a fit to distance from the circle r matching $u \sim r^{-0.35}$. In contrast, when fibers do not buckle ($\rho = 1$), displacements decay as $u \sim r^{-0.80}$, which more closely matches the solution from linear elasticity $u \sim r^{-1}$. Therefore, we conclude that buckling is sufficient to induce displacements that propagate over a long range.

Since fiber buckling appears to be sufficient for long distance alignment and propagation of fiber displacements, it may allow the cell to mechanically probe its distant environment. Mechanical guidance cues are not undocumented, and cells have previously been shown to migrate toward increasing stiffness [4]. To ascertain if cells can leverage this basic property of fiber buckling to sense their environments, 3D stacks of fluorescently labeled cells and fibrin matrix were imaged simultaneously. As expected, cell-cell mechanical interactions were observed through matrix fiber remodeling between pairs cells (Fig. 19.5). Dense bundles of aligned fibers stretched between the cell bodies, and appear to interact over a range of 100 μm. This range is approximately 5 times the diameter of a cell body, indicating that cells may sense one another through long range interactions facilitated by the nonlinear properties of the matrix.

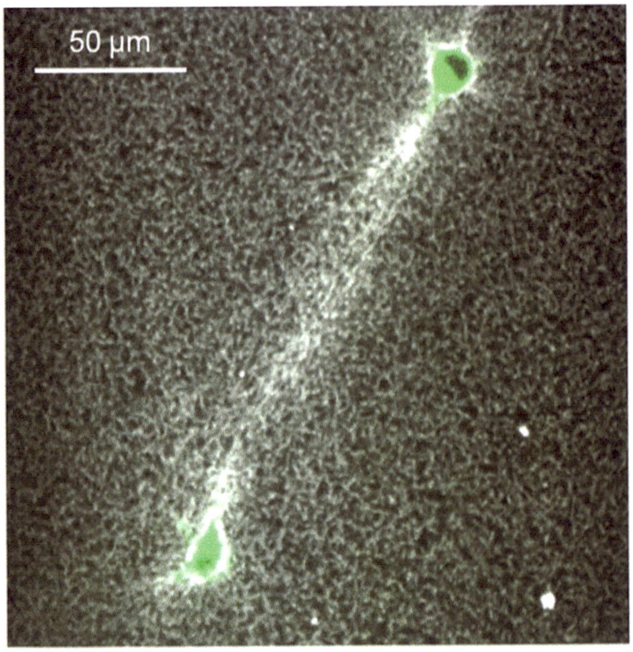

Fig. 19.5 A pair of 3T3 fibroblasts (*green*) in a 3D fibrin matrix (*gray*) interact mechanically by applying force to the matrix, resulting in local matrix deformations (*white*). The bands of matrix fibers may be a mechanism for long range cell mechanosensing

19.4 Conclusions

In this study, we utilized simulations to identify fiber buckling as a primary contributor to the nonlinearities that are observed in native fibrous matrices and tissues. To accurately measure and quantify these displacements due to localized contractile events, we also described a powerful correlation technique, DVC. Together these tools provide additional insight into how cells might read, interpret, and navigate in their 3 dimensional environment. Certainly chemical cues continue to be important in determining how cells sense their surroundings, but this work suggests a rich response to the mechanical environment may exist as well. These tools provide an enticing means to explore this new frontier.

Acknowledgements This work was funded by grants from the National Science Foundation (Division of Materials Research 0520565 and 1206121) and from the California Institute for Regenerative Medicine (RB5-07398). J.N. was supported by the National Science Foundation Graduate Research Fellowship under Grant No. DGE-1144469. A.L. was supported in part by a Rothschild Foundation fellowship.

References

1. Riching, K.M., et al.: 3D collagen alignment limits protrusions to enhance breast cancer cell persistence. Biophys. J. **107**(11), 2546–2558 (2014)
2. Lesman, A., et al.: Contractile forces regulate cell division in three-dimensional environments. J. Cell Biol. **205**(2), 155–162 (2014)
3. Provenzano, P.P., et al.: Collagen reorganization at the tumor-stromal interface facilitates local invasion. BMC Med. **4**(1), 38 (2006)
4. Lo, C.M., et al.: Cell movement is guided by the rigidity of the substrate. Biophys. J. **79**(1), 144–152 (2000)
5. Engler, A.J., et al.: Matrix elasticity directs stem cell lineage specification. Cell **126**(4), 677–689 (2006)
6. Provenzano, P.P., et al.: Contact guidance mediated three-dimensional cell migration is regulated by Rho/ROCK-dependent matrix reorganization. Biophys. J. **95**(11), 5374–5384 (2008)
7. Notbohm, J., et al.: Microbuckling of fibrin provides a mechanism for cell mechanosensing. J. R. Soc. Interface **12**(108), 20150320 (2015)
8. Rosakis, P., Notbohm, J., Ravichandran, G.: A model for compression-weakening materials and the elastic fields due to contractile cells. J. Mech. Phys. Solids **85**, 17 (2015)
9. Notbohm, J., et al.: Quantifying cell-induced matrix deformation in three dimensions based on imaging matrix fibers. Integr. Biol. (Camb) **7**(10), 1186–1195 (2015)
10. Petrie, R.J., et al.: Nonpolarized signaling reveals two distinct modes of 3D cell migration. J. Cell Biol. **197**(3), 439–455 (2012)
11. Petrie, R.J., Koo, H., Yamada, K.M.: Generation of compartmentalized pressure by a nuclear piston governs cell motility in a 3D matrix. Science **345**(6200), 1062–1065 (2014)
12. Beningo, K.A., Dembo, M., Wang, Y.L.: Responses of fibroblasts to anchorage of dorsal extracellular matrix receptors. Proc. Natl. Acad. Sci. U. S. A. **101**(52), 18024–18029 (2004)

13. Cukierman, E., et al.: Taking cell-matrix adhesions to the third dimension. Science **294**(5547), 1708–1712 (2001)
14. Franck, C., et al.: Three-dimensional full-field measurements of large deformations in soft materials using confocal microscopy and digital volume correlation. Exp. Mech. **47**(3), 12 (2007)
15. Lesman, A., et al.: Engineering vessel-like networks within multicellular fibrin-based constructs. Biomaterials **32**(31), 7856–7869 (2011)
16. Wozniak, M.A., et al.: ROCK-generated contractility regulates breast epithelial cell differentiation in response to the physical properties of a three-dimensional collagen matrix. J. Cell Biol. **163**(3), 583–595 (2003)
17. Bar-Kochba, E., Toyjanova, J., Andrews, E., Kim, K.-S., Franck, C.: A fast iterative digital volume correlation algorithm for large deformations. Exp. Mech. **55**(1), 13 (2015)
18. Licup, A.J., et al.: Stress controls the mechanics of collagen networks. Proc. Natl. Acad. Sci. U. S. A. **112**(31), 9573–9578 (2015)
19. Lindstrom, S.B., et al.: Biopolymer network geometries: characterization, regeneration, and elastic properties. Phys. Rev. E Stat. Nonlin. Soft Matter Phys. **82**(5 Pt 1), 051905 (2010)

Chapter 20
The Growth and Mechanical Properties of Abalone Nacre Mesolayer

Anqi Zhang, Yan Chen, MariAnne Sullivan, and Barton C. Prorok

Abstract Abalone nacre has long been interesting to material engineers due to its fascinating structure and outstanding mechanical behavior. It is a self-organized biogenic mineral composite which has a hierarchical architecture composite. The main component of this structure is "block and mortar" like tablets. Between the tablets structures is a layered structure which consist of both organic and inorganic parts called mesolayer. This research focus on the growth behavior of the mesolayer structure and its mechanical influence on nacre. Fresh-grown nacre is harvested by the "flat pearl" technique under different temperature schemes. SEM observation of those nacre reveals temperature scheme will inhibits the growth of tablets structure, and form other layered structure. The mechanical property of harvested "flat pearl" is tested by nano-indentation. In this work, we aim to show the mechanic significance of mesolayer in nacre structure, and hope to establish its strengthening and growth mechanism.

Keywords Biomimetic • Nanoindentation • Biomineralization • Multilayer composites

20.1 Introduction

Abalone nacre is a nature composite material composed by rather brittle calcium carbonate and organic materials [1]. It has a surprisingly high strength and fracture toughness, which believed contributed by the special self-assembled "brick and mortar" structure, giving its name "column nacre" [2, 3]. This structure can be shown by SEM (Scanning Electronic Microscope) images clearly (Fig. 20.2.).

Another feature of the shell can be found through the observation on the cross-section of abalone shell, especially the wild abalone shell, some black lines could be seen in the nacre (Fig. 20.1). This structure is called mesolayer [4]. The existence of such structure shows the abalone nacre having a structure of hierarchy.

Under SEM, those "black lines" showing like some layers with complex structure, which are named "mesolayer". Backscattered electron image shows that the composition of one part in the mesolayer is not calcium carbonate revealed by the different phase contrast (the darker area shown in Fig. 20.2). Both of nacre tablets and inorganic part of Mesolayer were shown in the same color in the back scattered image were aragonite [2]. And, the image also suggested that darker part in the image organic composition. EDS (Energy Dispersive Spectrometry) mapping shows the element distribution of such structure (Fig. 20.3). In the dark area of the backscattered SEM image, a high carbon level exists. This also suggest the organic property of such layer.

Although the mechanical behavior of nacre was extensively studied, nacre shows a significantly better mechanical property compare to the aragonite along [5]. The influence of Mesolayer structure to the mechanical behavior of the whole shell is still not entirely unveiled [6, 7]. This can be attributed to the fact that most research use the farm raised abalones, which do not have many mesoleyers in their body.

A. Zhang • Y. Chen • M. Sullivan • B.C. Prorok (✉)
Department of Mechanical Engineering, Auburn University, 275 Wilmore Laboratories, Auburn, AL 36849, USA
e-mail: prorok@auburn.edu

C.S. Korach et al. (eds.), *Mechanics of Biological Systems and Materials, Volume 6*,
Conference Proceedings of the Society for Experimental Mechanics Series, DOI 10.1007/978-3-319-41351-8_20

Fig. 20.1 Optical microscope image of abalone shell

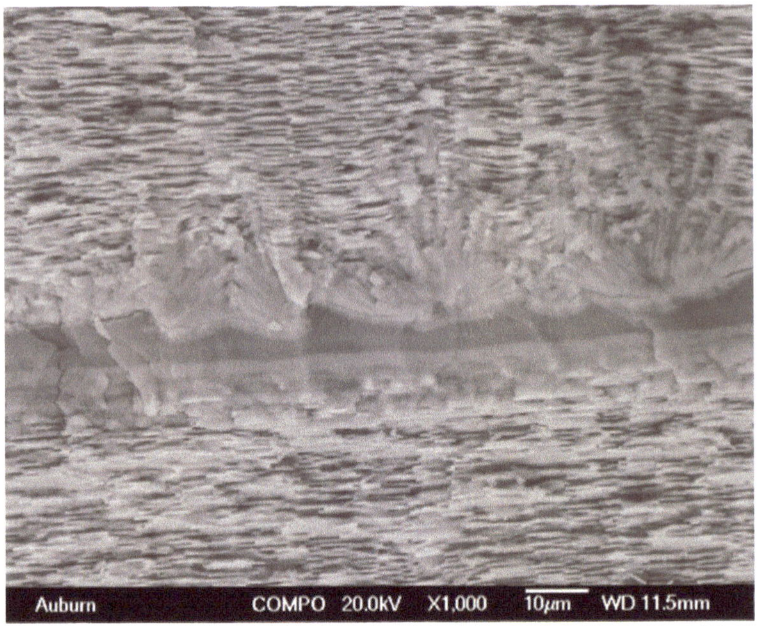

Fig. 20.2 Backscattered electron SEM image of an abalone nacre with mesolayer in it

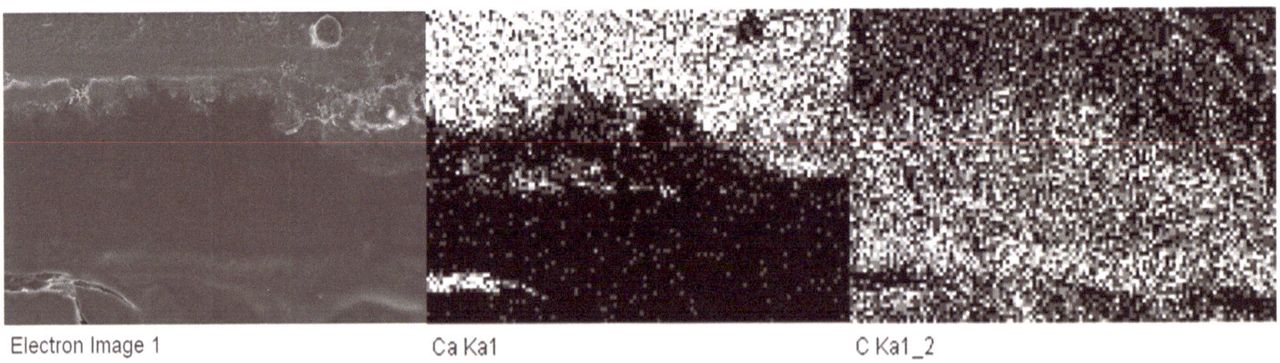

Fig. 20.3 EDS mapping analysis for the mesolayer in abalone nacre

20.2 Experiment Procedure

Abalones was acquired from California and feed in our aquarium, which comprised by 10 different aquarium tanks. The growth of abalone nacre and Mesolayer was on a "Flat Pearl". A technic pioneered by researchers in UC Santa Barbara [8]. Thin glass slides were inserted between abalone mantle and shell (Fig. 20.4) and those abalones were kept in different tanks

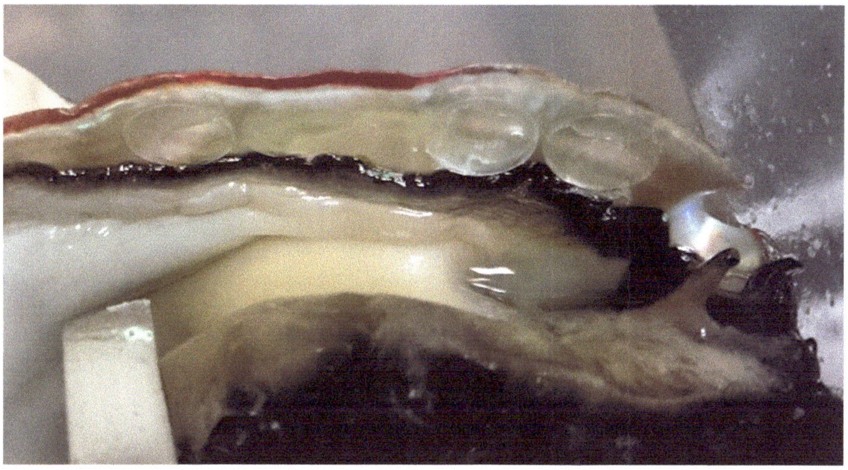

Fig. 20.4 Thin glass slides were stuck on the inner side of abalone shell and wait for mantle to cover them

with different temperature settings for 10 weeks. Some of those tanks were set with constant temperature while some other had their temperature changed during the 10 weeks. The formed flat pearl were taken out from the animal. Microstructure observation used scanning microscope, on the top and cross section of flat pearl. Mechanical property was tested by MTS Nanoindenter XP with continuous measurements.

20.3 Results

20.3.1 Nacre and Mesolayer Growth

For those samples grown under the variation temperature during the 10 weeks, some interruption layers were did produced. However, when backscattered SEM image were taken, it could be obviously told that those interruption structures are not mesolayer (Fig. 20.5). Neither organic composition nor prismatic structure were observed in the interruption layer grown through "Flat Pearl". It shows that the mesolayer is not a result of temperature variation.

From those samples, some growing interruption structure were also observed (Fig. 20.6). But none of them could be proved to be mesolayer.

An assumption about the mesolayer formation could be made here: in some periods, abalone might produce very small amount of Ca^{2+} due to some complicate reasons, which leads to a fact that the calcium carbonate is unable to nucleate and, then, the organic secretion accumulate layer by layer and finally form a whole thick layer of organic compostion. When the Ca^{2+} secretion recover to the usual level, calcium carbonate will nucleate again. As at this time calcium carbonate will nucleate and grow on the soft organic layer, the prismatic structure can be formed first and then convert to nacre structure gradually. Based on this, we might try to introduce some organic composition into the Flat Pearl to mimic the mesolayer structure.

20.3.2 Mechanical Properties Tested by Nano-Indenter

The value of Nano-indenter tested elastic modulus are quite similar in farm raised abalone nacre, wild abalone nacre and lab raised nacre through Flat Pearl (Fig. 20.7). For those samples, nanoindentation elastic modulus is 47 ($\pm$4.10) GPa for wild abalone, 48 ($\pm$3.07) GPa for farm raised abalone, and 49 ($\pm$2.11) GPa for bio fabricated sample obtained by Flat Pearl in lab. No matter where the nacre come from, no matter how it is obtained, the elastic modulus is quite stable for this structure. At the meantime, with the assistance of SEM image, it is sure that complete nacre structure could be produced through Flat Pearl.

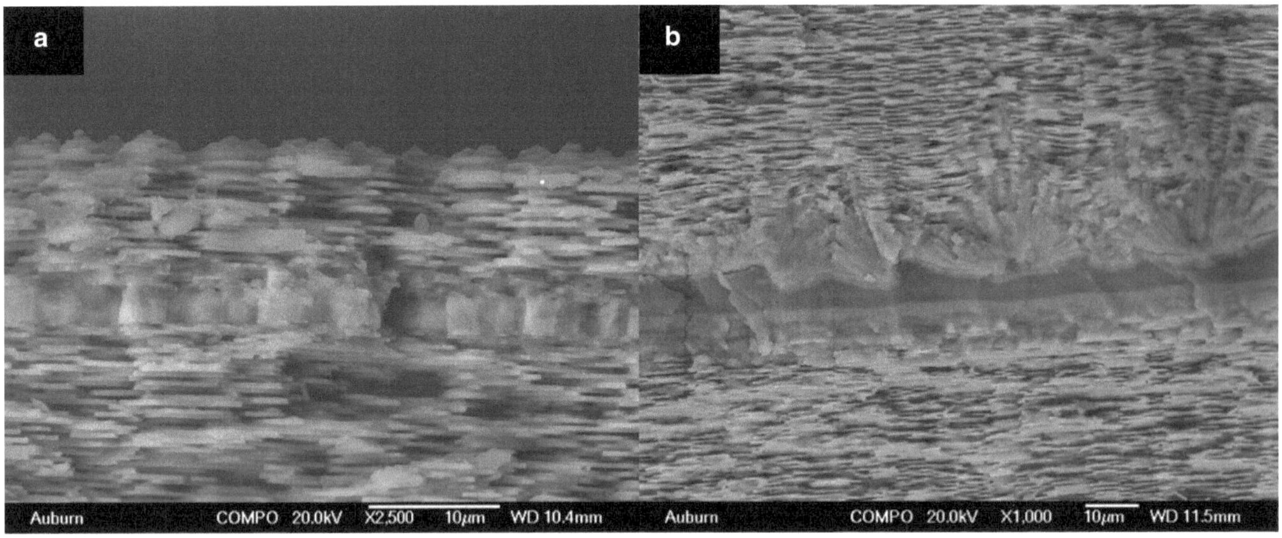

Fig. 20.5 Comparison of interruption obtained in the lab and mesolayer in wild abalone shell. (**a**) 60 °F, 10th week, Tank 6, Experiment 4, Backscattered electron image; (**b**) Mesolayer in wild abalone shell, Backscattered electron image

Fig. 20.6 Some interruption layer were covering nacre tablets

For the nacre with mesolayer structure, the nanoindentation elastic modulus was 43 (±2.74) GPa for wild abalone and 36 (±6.04) GPa for farm raised abalone. Those values are a little lower than the value of nacre without mesolayer. However, the lowering of apparent elastic modulus can also be attributed to existence of compliant structure underneath.

Thus, the mechanism of how this composite structure, i.e. mesolayer, influence the fracture property of nacre is still under investigation.

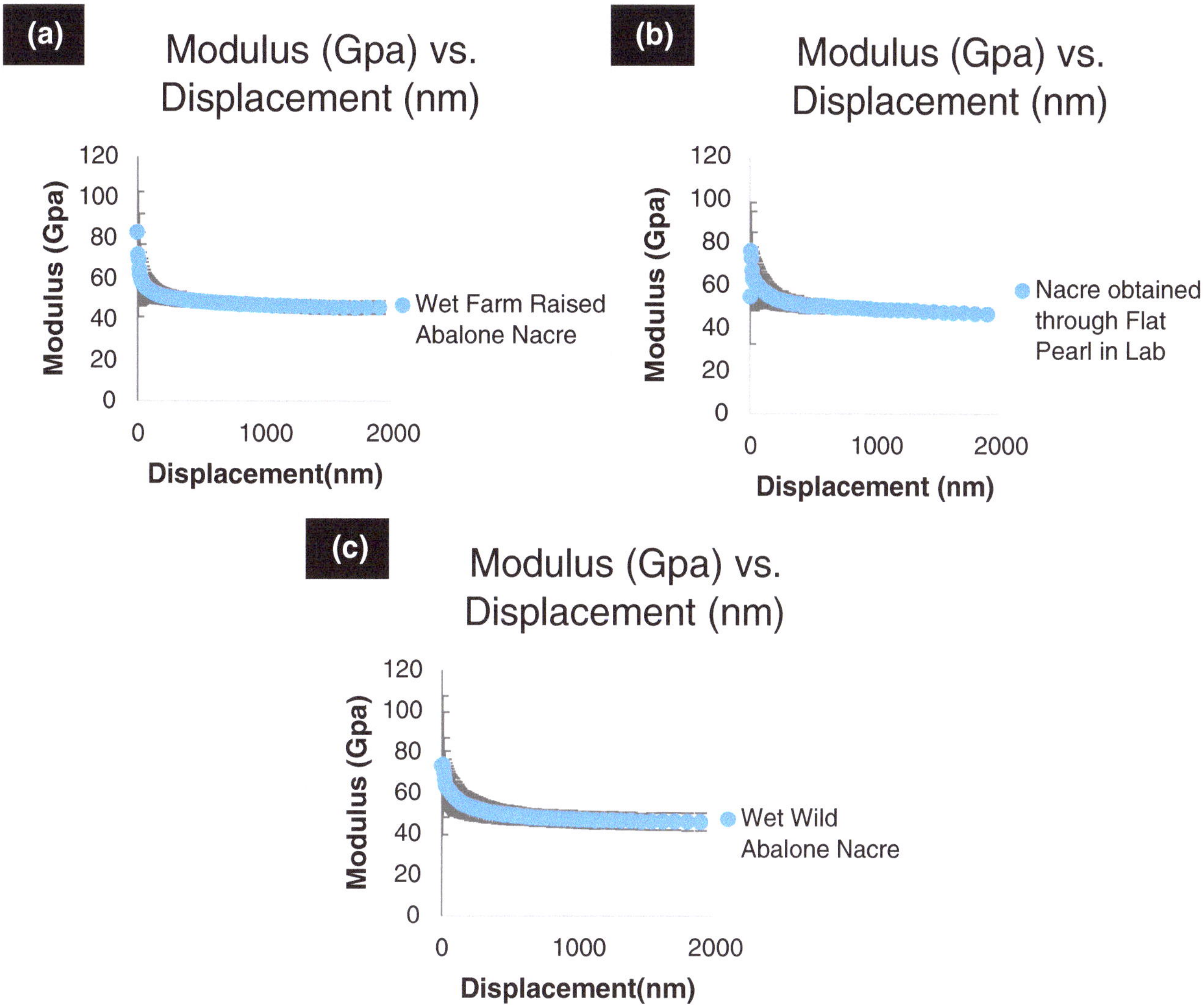

Fig. 20.7 Modulus vs. displacement tested by nanoindentation for nacre without mesolayer. (**a**) Modulus of farm raised abalone nacre (combined by data of five samples); (**b**) modulus of lab raised (by *Flat Pearl*, 60 °F, 10th week, Experiment 4, Tank 10) abalone nacre; (**c**) modulus of wild abalone nacre (combined by data of three samples)

20.4 Conclusion

For the growth mechanism of mesolayer, many environment factors may lead to its growth. And, by now, the details are unknown. However, it is possible that its structure could be mimicked through Flat Pearl method. And by observing the grown nacre in lab, the details about nacre and mesolayer growth process might be learned.

For the mechanical properties effected by mesolayer, although a lower elastic modulus tested by Nano-indenter is shown in the nacre structure with mesolayer in it than in the nacre structure without mesolayer, it is still not sure that it caused by the mesolayer. The mechanical behavior of mesolayer in nacre is still under investigation.

References

1. Heinemann, F., Launspach, M., Gries, K., Fritz, M.: Gastropod nacre: structure, properties and growth — biological, chemical and physical basics. Biophys. Chem. **153**, 126–153 (2011)
2. Zhang, X., Fan, Z., Lu, Q., Huang, Y., Kaplan, D.L., Zhu, H.: Hierarchical biomineralization of calcium carbonate regulated by silk microspheres. Acta Biomater. **9**, 6974–6980 (2013)
3. Barthelat, F., Li, C.-M., Comi, C., Espinosa, H.D.: Mechanical properties of nacre constituents and their impact on mechanical performance. J. Mater. Res. **21**, 1977–1986 (2006)
4. Zaremba, C.M., et al.: Critical transitions in the biofabrication of abalone shells and flat pearls. Chem. Mater. **8**, 679–690 (1996)
5. Zhang, J., Zhang, Y., Xu, K., Ji, V.: Young's modulus surface and Poisson's ratio curve for orthorhombic crystals. J. Chem. Crystallogr. **38**(10), 733–742 (2008)
6. Sullivan, M.A., Prorok, B.C.: Controlling abalone shell architecture with temperature. In: Mechanics of Biological Systems and Materials, vol. 7, pp. 27–31 (2015)
7. Zhou, B., Prorok, B.C.: A new paradigm in thin film indentation. J. Mater. Res. **25**(09), 1671–1678 (2010)
8. Fritz, M., Belcher, A.M., Radmacher, M., Walters, D.A., Hansma, P.K., Stucky, G.D., Morse, D.E., Mann, S.: Flat pearls from biofabrication of organized composites on inorganic substrates. Nature **371**(6492), 49–51 (1994)

Chapter 21
Evaluation of Precise Optimal Cyclic Strain for Tenogenic Differentiation of MSCs

Yasuyuki Morita, Toshihiro Sato, Sachi Watanabe, and Yang Ju

Abstract Although there are a number of papers relating to tenogenic differentiation of human bone marrow mesenchymal stem cells (hBMSCs) using uniaxial cyclic stretching stimulation with homogeneous strain field, it has been pretty hard to figure out the optimal normal strain in the stretch direction for the differentiation. In the present study, our group has developed a non-uniform strain field system to elucidate the optimal normal strain in one-time experiment in principle. A relationship between the normal strain of membrane and expression levels of the differentiation marker proteins, type I collagen (Col I) and tenascin-C (Tnc), derived from stretched cells was obtained. Finally, the rigorous optimal normal strains were clarified 7.9 and 8.5 % for Col I and Tnc, respectively. Additionally, we found that a dependence of protein expression levels with the normal strain of membrane was different in each protein, which would be crucial in the field of embryology and regenerative medicine.

Keywords Digital image correlation (DIC) • Optimal strain • Mechanical stimulation • Tenogenic differentiation • Mesenchymal stem cell (MSC)

21.1 Introduction

Tendon tissue engineering in vitro could be the best medical treatment since tendon tissues have poor self-curing ability. The main problems are, however, the low number of cells obtained from explanted tendon tissue since tendons are relatively acellular and low cell density of tenocytes [1]. Hence, bone marrow mesenchymal stem cells (BMSCs) are commonly used in this field of study due to their high proliferative capacity and pluripotency [2, 3]. Promoting differentiation of BMSCs into tenocytes effectively in vitro is one of the vital issues. Mechanical stimulation [4] would have advantages from cost and simplicity perspectives although there are some techniques, e.g. biochemical stimulation [5], micro-/nano-structure [6], co-culture [7], etc. for BMSC-to-teñocyte differentiation. Thus, a number of research groups have investigated the differentiation by using cyclic uniaxial stretch stimulation [8]. Our group has been also involved in this research field [9, 10]. In general, this kind of experiment has been carried out by means of PDMS elastic chamber like shown in Fig. 21.1. Cells are seeded onto the bottom of the chamber which is coated with cell-adhesion factor, and then mechanical loading is imposed to the cells by stretching the chamber uniaxially. Song et al. indicated that 1.0 Hz of cyclic frequency works effectively for proliferation of human BMSCs (hBMSCs) in vitro [11]. Although it has been found that the optimal strain rate of cyclic uniaxial stretch for enhancing tenogenic differentiation of hBMSCs was roughly 8–10 % in the stretch direction [9, 10, 12], the rigorous strain rate of the uniaxial stretching which is crucial for tendon tissue engineering has not been elucidated yet. The main reason must be that the numerous experimental conditions of the strain rate need to be performed though those experiments are impractical from perspectives of cost and efficiency.

Recently, our group has developed a new 2D culture system using a non-uniform strain field [13]. Then by using the system, the axial strain threshold of cells was determined with high reliability, and the preferential axial strain of cells was suggested firstly as a new characteristic of cells [13]. Exploring cellular functions and activities using non-uniform deformation fields has been getting more important and paid attention nowadays [13–15]. In the present report, we determined the optimal strain rate of cyclic stretching for hBMSC-to-tenocyte differentiation described above by means of the 2D culture system of a non-uniform strain field. Digital image correlation (DIC) method [16] was applied to quantify the non-uniform strain distribution of the membrane. Expression levels of marker proteins for tenogenic differentiation, type

Y. Morita (✉) • T. Sato • S. Watanabe • Y. Ju
Department of Mechanical Science and Engineering, Graduate School of Engineering, Nagoya University,
Furo-cho, Chikusa-ku, Nagoya 464-8603, Japan
e-mail: morita@mech.nagoya-u.ac.jp

C.S. Korach et al. (eds.), *Mechanics of Biological Systems and Materials, Volume 6*,
Conference Proceedings of the Society for Experimental Mechanics Series, DOI 10.1007/978-3-319-41351-8_21

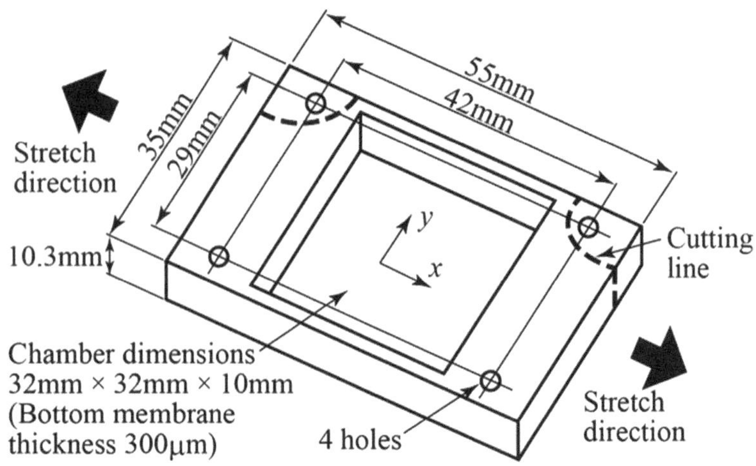

Fig. 21.1 Geometry of a PDMS elastic chamber for cell culture with cyclic mechanical stretching

I collagen (Col I) and tenascin-C (Tnc), were evaluated by immunofluorescent staining method. These measurements made it possible to obtain a relationship between strain rate of cyclic stretching and expression levels of these proteins, which correspond to tenogenic differentiation level. The rigorous optimal strain rate for the differentiation could be determined from the relationship. No previous reports give such rigorous optimal strain rate of cyclic mechanical stimulation for tenogenic differentiation, and it must make a significant contribution to manipulation of tenogenic differentiation in tissue engineering in vitro.

21.2 Materials and Methods

21.2.1 Cell Preparation

Human bone marrow mesenchymal stem cells (hBMSCs; UE6E7T-3; Health Science Research Resources Bank, Tokyo, Japan) were maintained in low-glucose Dulbecco's modified Eagle's medium containing 10 % fetal calf serum, 0.5 % GlutaMax, and 0.05 % gentamicin (Invitrogen, Carlsbad, CA) at 37 °C under an atmosphere of 5 % CO_2 in a humidified incubator. The hBMSCs were cultured in 25 cm^2 culture flasks (BD Biosciences, Franklin Lakes, NJ) at an initial density of 1.0×10^4 cells/cm^2 for expansion without differentiation. The medium was replaced with fresh medium every 3 days. At near-confluence, which occurred every 5–7 days, cells were detached from culture flasks with 0.25 % trypsin (Wako Pure Chemical Industries, Ltd., Osaka, Japan)/1 mM EDTA (B3232; Takara Bio, Shiga, Japan) and seeded into new culture flasks.

21.2.2 Application of Cyclic Stretch with Inhomogeneous Strain Field

The commercially-available PDMS-elastic chambers (STB-CH-10; Strex, Osaka, Japan: Fig. 21.1) were used in the present study. As in ordinary use, the chambers are put in a simple uniaxial stretching device (STB-140; Strex), which is used in researches regarding mechanical cyclic stretching stimulation with homogeneous strain distribution [9, 10, 17]. The chamber is fixed by grappling hooks at the four corners and mechanically deformed by the stretching device. In this study, the upper side holes of the chamber were removed (see the cutting lines in Fig. 21.1) and the upper side of the chamber was not held by the hooks to induce non-uniform deformation (i.e. inhomogeneous strain distribution). The bottom of the chamber, however, was undulated by the applied inhomogeneous strain field as the bottom membrane was thin ($t = 100$ μm). Therefore, we cut way the bottom membrane and adhered a stiffened thicker PDMS ($t = 300$ μm) as a new bottom with adhesive (SILPOT 184; Dow Corning Tray, Tokyo, Japan).

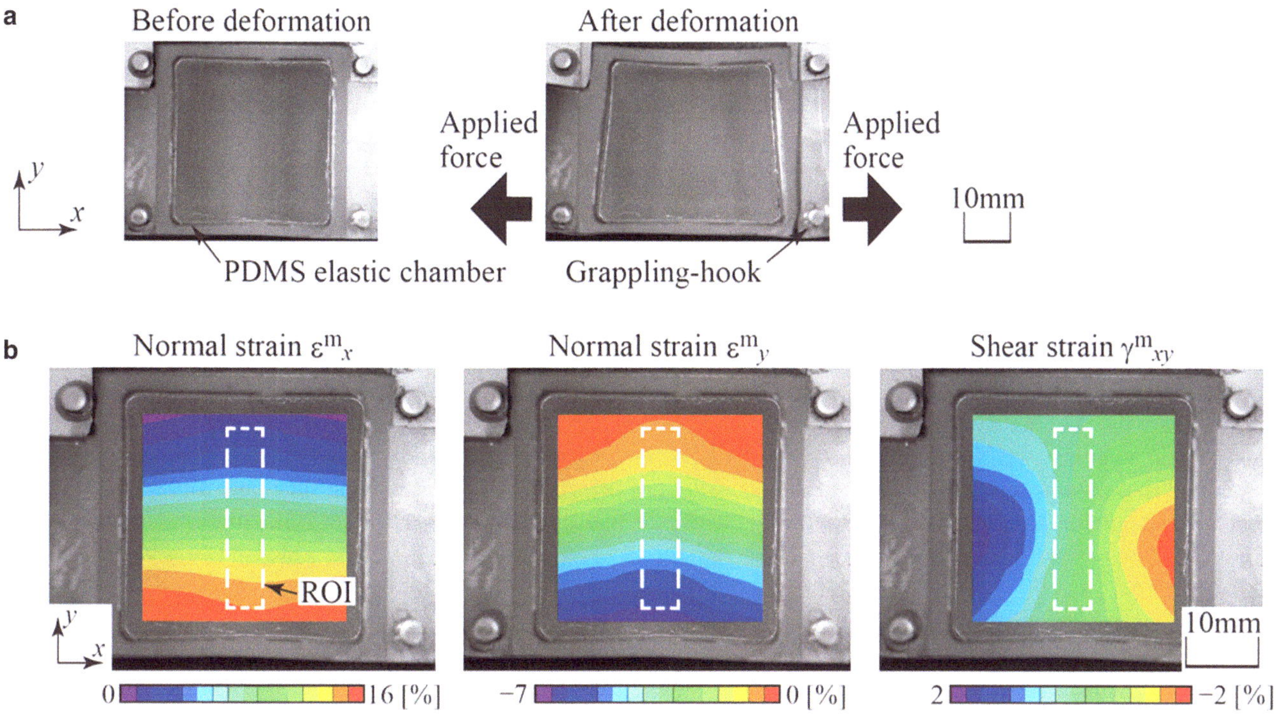

Fig. 21.2 Deformation of a PDMS elastic chamber for cyclic non-uniform stretching. (**a**) Photographs before deformation (*left*) and after deformation (*right*) with random speckle patterns. (**b**) Strain distribution of the membrane. *Left*: normal strain ε^{m}_{x} in the *x*-axis. *Middle*: normal strain ε^{m}_{y} in the *y*-axis. *Right*: shear strain γ^{m}_{xy} in the *x*-*y* plane

Then, the cultured hBMSCs were plated onto the bottom of the chambers at a density of 1.0×10^4 cells/mL and cultured for 2 days without cyclic stretching. The chemical treatment of the chamber would like to refer to Ref [13]. Finally, 1-Hz cyclic stretching stimulation of the non-uniform strain field was applied to the hBMSCs in the culture environment for 48 h.

The actual non-uniform strain distribution of the membrane put the cells was evaluated by the DIC method. This analysis was performed using commercially available software (Vic-2D; Correlated Solutions, Columbia, SC). The displacement sensitivity of the software was 0.01 pixels at minimum. Thus, the displacement sensitivity in the present study corresponded to 0.2 μm. Although the hBMSCs were seeded throughout the whole area of the membrane, the region of interest (ROI; 5×26 mm^2) was set in the central area of the membrane (see the dotted rectangle in Fig. 21.2b) as the normal strain ε^{m}_{x} of the membrane in the direction of stretch (*x*-axis) changed smoothly along the orthogonal (*y*) direction. In addition, the center of the membrane was not susceptible to the edge effect, which means that the edge area exhibited undesirable drastic strain changes and high/low cell density. As a result, the inhomogeneous strain in the ROI of the membrane had normal strain ε^{m}_{x} values of nearly 2.0–15.0 %.

21.2.3 Determination of Expression Levels of Differentiation Marker Proteins

Col I molecules self-assemble into highly organized fibrils that form collagen firers [18]. Cross-linking of these fibers in the extracellular matrix gives them a high tensile strength and provides mechanical strength for tendon tissue [19]. Tnc is thought to be involved in ECM formation, contributing to the mechanical stability of tendon tissue through its interactions with collagen fibrils and decorin, a proteoglycan [20]. Immunofluorescent staining method was employed to evaluate the expression levels of these marker proteins derived from the stretched cells. The staining procedure was taken after Apte et al. [21]. The stretched cells after testing were fixed in methanol (Wako Pure Chemical Industries, Osaka, Japan) at −20 °C for 10 min. After removing the methanol, blocking treatment was conductied by PBS containing 5 % fetal calf serum. The cells were then incubated with the primary antibody overnight at 4 °C. The primary antibody was employed Goat Anti-Type I Collagen-UNLB (Southern Biotechnolgy, Birmingham, AL) for Col I diluted 1 in 150 in 5 % fetal calf serum, Tenascin-C Rabbit Monoclonal Antibody (Epitomics, Burlingame, CA) for Tnc diluted in 1 in 200 in 5 % fetal calf serum. After washing in PBS, the cells were incubated for 30 min at room temperature with secondary antibody, which was Alexa Fluor 488 Donkey Anti-Goat IgG (H+L) (A-21447; Invitrogen, Carlsbad, CA) for Col I, Rhod (TRITC) AffiniPure Donkey Anti Rabbit IgG (H+L) (711-025-152; Jackson ImmunoResearch Laboratories, West Grove, PA) in 1 in 200 in 5 % fetal calf serum both. In this study, the cellular nuclei were also stained by DAPI (268298-10MG; Calbiochem, Darmstadt, Germany) to figure out the position of each cell. After washing in PBS again, the cells with the chamber were encapsulated by Fluoromount-G (Southern Biotechnolgy). Then, the sample was observed by a confocal fluorescence microscopy (A1Rsi-N; Nikon, Tokyo, Japan).

Qualitative evaluation of the expression levels of the differentiation maker proteins was conducted by measuring the fluorescent intensity as follows. Firstly, the region (6.3×10^{-2} mm^2) which had the same value of the normal strain $\varepsilon^{m}{}_{x}$, in the ROI was trimmed by Photoshop (Adobe, San Jose, CA) software. The mean number of cells was 25.6 ± 0.2 (mean ± SEM) in one region. Secondly, high-frequency noise was removed by reference to the DAPI fluorescence image which could recognize the cell position. The mean fluorescent intensity per cell $\bar{I}$ was then determined by dividing the sum of the fluorescent intensity by the number of cells contained in the region. The respective $\bar{I}$ were obtained in each $\varepsilon^{m}{}_{x}$ from 2.0 to 15.0 % every 0.6 %. Finally, normalizing process was performed by dividing the respective $\bar{I}$ in each $\varepsilon^{m}{}_{x}$ by the maximum value of $\bar{I}$ in a certain $\varepsilon^{m}{}_{x}$. Nine and six data for Col I and Tnc, respectively, were acquired under the same experimental condition. These two assessments, the determination of $\varepsilon^{m}{}_{x}$ of the membrane and the fluorescent intensity of the stretched cells, made it possible to obtain the correlation with the tenogenic differentiation levels of hBMSCs to the mechanical cyclic stretching stimulation.

21.3 Results and Discussion

Figure 21.2a presents the PDMS elastic chamber before and after deformation with speckle patterns. Then, the displacement distribution of the membrane obtained by the DIC analysis. Figure 21.2b exhibits the strain distribution of the membrane. The $\varepsilon^{m}{}_{x}$, $\varepsilon^{m}{}_{y}$, and $\gamma^{m}{}_{xy}$ values represent the normal strain in the x-axis, the normal strain in the y-axis, and the shear strain in the x-y plane of the membrane, respectively. The normal strain $\varepsilon^{m}{}_{x}$ in the direction of stretch was increased continuously from 2.10 ± 0.20 % (mean ± SD) in the upper side (free end) to 15.90 ± 0.14 % in the lower side (fixed end) in the ROI of the membrane. The normal strain $\varepsilon^{m}{}_{y}$ in the orthogonal direction was changed from −1.21 ± 0.01 % in the upper side to −6.92 ± 0.02 % in the lower side in the ROI (corresponding to nearly $-\nu\varepsilon^{m}{}_{x}$. ν: Poisson's ratio of the membrane). The distribution of the shear strain $\gamma^{m}{}_{xy}$ was almost uniform at 0.10 ± 0.27 %. The strain value in the direction of mechanical cyclic stretch stimulation, $\varepsilon^{m}{}_{x}$, was important in this study. Then, the data (x, y, $\varepsilon^{m}{}_{x}$) in the ROI could be obtained by this experimental process, where x, y indicate the position of the ROI in the membrane.

As an example, Fig. 21.3 shows fluorescence micrograph of the differentiation marker protein Col I derived from the stretched cells in the ROI of the membrane. The image size corresponds to the ROI of the membrane, and the normal strain $\varepsilon^{m}{}_{x}$ increased with the lower side of the image. If the fluorescent intensities of both marker proteins were high, we could recognize that the hBMSC-to-tenocyte differentiation was accrued actively [10]. Overall, you can see that there was a certain strain region of the ROI at which the fluorescent intensity grew higher, despite the intensities were fluctuated in the fluorescence micrograph of the protein shown in Fig. 21.3. Finally, the data ($\varepsilon^{m}{}_{x}$, $\overline{I_{ColI}}$) and ($\varepsilon^{m}{}_{x}$, $\overline{I_{Tnc}}$) could be obtained through this experimental process.

Figure 21.4 shows the relationship between the normalized $\bar{I}$ which corresponds to the expression levels of the differentiation marker proteins, Col I (Fig. 21.4a) and Tnc (Fig. 21.4b), and the normal strain $\varepsilon^{m}{}_{x}$ in the ROI. It can be seen that the expression level of each protein had a peak with respect to the normal strain $\varepsilon^{m}{}_{x}$. Namely, it suggests that an optimal normal strain $(\varepsilon^{m}{}_{x})_{opt}$ in the membrane definitely exists for effective hBMSC-to-tenocyte differentiation with mechanical cyclic stretch stimulation. To determine the $(\varepsilon^{m}{}_{x})_{opt}$, the data were fitted by polynomial approximation curve (solid line in Fig. 21.4) using the lowest order (4th) curve which could trace them congruously. Consequently, the optimal normal strains $\varepsilon^{m}{}_{x}$ in the stretch direction of the membrane for the expression levels of the proteins were 7.9 and 8.5 % for Col I and Tnc, respectively. Chen et al. investigated mRNA expression levels regarding the tenogenic differentiation of hBMSCs under 3 % or 10 % uniaxial stretch with 1 Hz and 48 h, and they concluded that the 10 % stimulation was better [12]. More precisely, our group studied mRNA and protein expression levels about the tenogenic differentiation of hBMSCs under 5 %, 10 %, or 15 % stretching stimulation with 1 Hz and 48 h, and we also obtained that the 10 % elongation promoted the differentiation [9]. The result shown in the present study is consistent with these previous reports [9, 12], although they conducted the experiments with coarse condition about the strains (3 % or 10 %, and 5 %, 10 % or 15 %). Furthermore, our group indicated that uniaxial cyclic strain of 8 % with 1 Hz and 48 h encouraged the protein expression

Fig. 21.3 Fluorescence micrographs of the differentiation marker proteins Col I derived from the stretched cells in the ROI of the membrane. *Inset*: enlarged picture in a certain area of the ROI

levels for the tenogenic differentiation of hBMSCs rather than 10 % stretch stimulation [10]. The result in this study, which the optimal normal strains were 7.9 and 8.5 % for Col I and Tnc, respectively, also corresponds with our previous report [10] using uniaxial stretching stimulation. This indicates that the non-uniform strain field suggested in this study is quite valid for determination of the optimal strain of the tenogenic differentiation with mechanical cyclic stimulation. In this manner, not a few experiments has to be conducted to figure out the optimal cyclic strain for the tenogenic differentiation if you use a uniaxial strain chamber. However, if you employ the non-uniform strain chamber, you can clarify the optimal strain for the tenogenic differentiation in one-time experiment in principle.

Finally, we would like to touch the thing which there is a discrepancy in the shape of the approximation curves between Col I and Tnc as shown in Fig. 21.4. Namely, the approximation curve of the Col I has a wide skirt with respect to the normal strains $\varepsilon^{m}{}_{x}$ of the membrane, while the one of the Tnc has a peaked shape relatively. The difference of the sensitivity of each

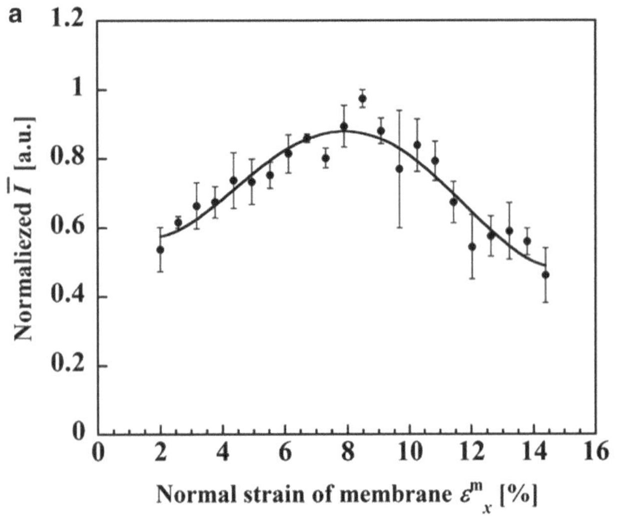

Fig. 21.4 Relationship between the normalized $\bar{I}$ which corresponds to the expression levels of the differentiation marker proteins, (**a**) Col I and (**b**) Tnc, and the normal strain ε^{m}_{x} in the ROI. Data are normalized to the maximum value of the mean fluorescent intensity per cell $\bar{I}$ (defined as 1). Data shown represent mean ± SEM

protein expression level to the normal strains ε^{m}_{x} of the membrane is attracting our interest. Col I protein is one of the most essential components in most connective tissues and the connective tissues are subjected to a wide variety of intensity of mechanical stimulations. Therefore, cells who synthesize Col I must permit a wide range of intensity of mechanical stimulation. On the other hand, Tnc protein is a specific component contributing to connective tissues like tendon. Then, cells who express Tnc could tolerate only certain narrow intensity of mechanical stimulations. We consider that the sensitivity difference of the protein expression level to mechanical stimulation would be crucial in embryology and regenerative medicine. Our group would like to focus on this issue in the future.

21.4 Conclusions

In tenogenic differentiation of hBMSCs using cyclic uniaxial stretch stimulation, its optimal strain value has not been clarified for a long time. However, a cyclic stretch stimulation with non-uniform strain field we proposed elucidated the precise value. The value was 7.9 and 8.5 % for type I collagen and tenascin-C, respectively. These values were roughly consistent with experiments using uniaxial cyclic stretch stimulation which were conducted under rough conditions regarding strains. Additionally, this study found out that there was a discrepancy of expression levels of each protein with respect to the normal strains of the membrane. This experimental technique using the non-uniform strain field must be best suited for determination of the optimal strain not only in tenogenic differentiation of hBMSCs, but also in various cellular differentiations.

References

1. Bullough, R., Finnigan, T., Kay, A., et al.: Tendon repair through stem cell intervention: cellular and molecular approaches. Disabil. Rehabil. **30**, 1746–1751 (2008)
2. Pittenger, M.F., Mackay, A.M., Beck, S.C., et al.: Multilineage potential of adult human mesenchymal stem cells. Science **284**, 143–147 (1999)
3. Awad, H., Butler, D.L., Boivin, G.P., et al.: Autologous mesenchymal stem cell-mediated repair of tendon. Tissue Eng. **5**, 267–277 (1999)
4. Kuo, C.K., Tuan, R.S.: Mechanoactive tenogenic differentiation of human mesenchymal stem cells. Tissue Eng. A **14**, 1615–1627 (2008)
5. Haddad-Weber, M., Prager, P., Kunz, M., et al.: BMP12 and BMP13 gene transfer induce ligamentogenic differentiation in mesenchymal progenitor and anterior cruciate ligament cells. Cytotherapy **12**, 505–513 (2010)
6. Yin, Z., Chen, X., Chen, J.L., et al.: The regulation of tendon stem cell differentiation by the alignment of nanofibers. Biomaterials **31**, 2163–2175 (2010)

7. Luo, Q., Song, G., Song, Y., et al.: Indirect co-culture with tenocytes promotes proliferation and mRNA expression of tendon/ligament related genes in rat bone marrow mesenchymal stem cells. Cytotechnology **61**, 1–10 (2009)
8. Lui, P.P.Y., Rui, Y.F., Ni, M., Chan, K.M.: Tenogenic differentiation of stem cells for tendon repair - what is the current evidence? J. Tissue Eng. Regen. Med. **5**, e144–e163 (2011)
9. Morita, Y., Watanabe, S., Ju, Y., Xu, B.: Determination of optimal cyclic uniaxial stretches for stem cell-to-tenocyte differentiation under a wide range of mechanical stretch conditions by evaluating gene expression and protein synthesis levels. Acta Bioeng. Biomech. **15**, 71–79 (2013)
10. Morita, Y., Suzuki, S., Ju, Y., Kawase, N.: Differences between protein expression and extracellular matrix state on uniaxial stretching for tenogenic differentiation. J. Mech. Med. Biol. **14**, 1450025 (2014)
11. Song, G., Ju, Y., Soyama, H., et al.: Regulation of cyclic longitudinal mechanical stretch on proliferation of human bone marrow mesenchymal stem cells. Mol. Cell Biomech. **4**, 201–210 (2007)
12. Chen, Y.J., Huang, C.H., Lee, I.C., et al.: Effects of cyclic mechanical stretching on the mRNA expression of tendon/ligament-related and osteoblast-specific genes in human mesenchymal stem cells. Connect. Tissue Res. **49**, 7–14 (2008)
13. Morita, Y., Watanabe, S., Ju, Y., Yamamoto, S.: In vitro experimental study for the determination of cellular axial strain threshold and preferential axial strain from cell orientation behavior in a non-uniform deformation field. Cell Biochem. Biophys. **67**, 1249–1259 (2013)
14. Balestrini, J.L., Skorinko, J.K., Hera, A., et al.: Applying controlled non-uniform deformation for in-vitro studies of cell mechanobiology. Biomech. Model. Mechanobiol. **9**, 329–344 (2010)
15. Richardson, W.J., Metz, R.P., Moreno, M.R., et al.: A device to study the effects of stretch gradients on cell behavior. J. Biomech. Eng. **133**, 101008 (2011)
16. Sutton, M.A.: Digital image correlation for shape and deformation measurements. In: Sharpe, W.N. (ed.) Springer Handbook of Experimental Solid Mechanics, pp. 565–600. Springer, New York (2008)
17. Inoh, H., Ishiguro, N., Sawazaki, S.I., et al.: Uni-axial cyclic stretch induces the activation of transcription factor nuclear factor B in human fibroblast cells. FASEB J **16**, 405 (2002)
18. Eyre, D.R., Paz, M.A., Gallop, P.M.: Cross-linking in collagen and elastin. Annu. Rev. Biochem. **53**, 717–748 (1984)
19. Lejard, V., Brideau, G., Blasis, F., et al.: Scleraxis and NFATc regulate the expression of the pro-alpha1(I) collagen gene in tendon fibroblasts. J. Biol. Chem. **282**, 17665–17675 (2007)
20. Elefteriou, F., Exposito, J.Y., Garrone, R., Lethias, C.: Binding of tenascin-X to decorin. FEBS Lett. **495**, 44–47 (2001)
21. Apte, M.V., Haber, P.S., Darby, S.J., et al.: Pancreatic stellate cells are activated by proinflammatory cytokines: implications for pancreatic fibrogenesis. Gut **44**, 534–541 (1999)

Chapter 22
Effect of Fiber Architecture on the Cell Functions of Electrospun Fiber Membranes

F. Sultana, M. Vaughan, and M. Khandaker

Abstract The objective of this research was to determine the effect of fiber architecture on the cell viability of the electrospun nanofiber (ENF) membrane. Three different ENF membranes (consists of single layer, four layers and six layers of cross-direction ENF layer) were developed and deposited on one side of the acrylic mold. The fibrous side of the mold was placed on the top of a well plate. Mouse osteoblast cell adhesion and proliferation on each of the different kinds of membranes were observed using a fluorescent microscope technique. Results shows that the cell adhered and proliferated on each kind of ENF fiber membranes. Tests shows that the designed ENF membrane has no adverse effect on osteoblast cells, fiber attached to the fiber, and strongly dependent on the fiber layer numbers in the membrane.

Keywords Polycaprolactone • Electrospun nanofiber • Membrane • Tissue engineering • Cell culture disc

22.1 Background

Electrospinning is a process by which fibers with micro to nanometer diameters can be obtained from an electrostatically driven jet of polymer solution [1]. These fibers have a high surface area to volume ratio, which can have numerous industrial applications as barrier fabrics, wipes, medical, and pharmaceutical uses [2, 3]. An innovative method has been developed in the authors research and patent application "Method and apparatus for the controlled deposition of branched electrospun fiber on biomedical implants and material" has been awarded [4].

A single layer of aligned unidirectional electrospun nanofiber (ENF) can be produced between two parallel collectors using the electrospinning unit. These fibers can be harvested at different angles to produce an ENF membrane on a substrate. The biological functions of an ENF membrane depends on the fiber architecture in the ENF membrane [2]. When the thickness of ENF membrane is larger than micron, the biological functions of the membrane decreased significantly due to loss of porosity. The loss of porosity was higher when fibers are organized in membranes randomly in place of aligned fashion. Due to these facts, most of the ENF membrane currently available in market has limited application in producing 3D scaffold (requires thickness more than millimeter) for tissue engineering applications. The effect of fiber architecture, consists of single layer, four layers and six layers of cross-direction ENF layers, of the ENF membrane on the osteoblast cell functions is not known and was studied in this research.

22.2 Materials and Methods

Polycaprolactone (PCL) aligned cross-direction ENF membrane was used to produce PCL ENF membrane. PCL solution will be prepared by ultrasonic (Sonics & Materials, Inc., model # Vibra-cell VCX 130) mixing of 7.69 wt% of PCL pellets (pellet size ~3 mm, average Mn 80,000) with acetone (laboratory reagent $\geq$99.5 %). The sonication process will be carried out at approximately 80 °C for an hour. The solution will be poured into a glass syringe in an infusion pump (Harvard

F. Sultana
Department of Math and Computer Science, University of Central Oklahoma, Edmond, OK 73034, USA

M. Vaughan
Department of Biology, University of Central Oklahoma, Edmond, OK 73034, USA

M. Khandaker (✉)
Department of Engineering and Physics, University of Central Oklahoma, Edmond, OK 73034, USA
e-mail: mkhandaker@uco.edu

C.S. Korach et al. (eds.), *Mechanics of Biological Systems and Materials, Volume 6*,
Conference Proceedings of the Society for Experimental Mechanics Series, DOI 10.1007/978-3-319-41351-8_22

Apparatus, mode # PHD ULTRA) for fiber production. PCL fibers will be ejected from the glass syringe via charged needle (23G blunt needle, aluminum hub, 1″ length, model # BX 25). The needle will be charged by high voltage power source (Gamma High Voltage Research, Inc., model # ES 30 series). Four layers of aligned fibers will be deposited on a grounded custom made silicon mold chamber, where the direction of fiber will be 90° between two adjacent layers.

Cell viability test on a multilayers of ENM membrane was conducted on an acrylic mold (Fig. 22.1) to independently test the fiber architecture and cell functions. The fibrous side of the mold was placed on the top of the well plate. This study adapts standard cell sub-culturing techniques as well as adaptations of standard protocols. Mouse osteoblast cells (ATCC cell line # MT3T3E1) were seeded at a density of 70,000 cells/ml on each well of fiber scaffold samples. The cells were then fixed with methanol or neutral buffer formalin. DAPI stain is used in observation of the cell adhesion and a click-Edu stain for proliferation. Each of these staining procedures are fluorescent. Each experiment is run for 48 h to allow the osteoblast to settle in the media during the experiment. The resulting stain was viewed with a fluorescent microscope. The qualitatively measurement of cell adhesion and proliferation on the stained fiber were conducted from the captured images of 1.46 mm^2 window area.

22.3 Results

An electrospin setup was used to produce multilayers of ENM membrane. The first step to fabricate an ENM membrane is to build a uni-direction ENF layer comprised of a single layer of aligned ENF (Fig. 22.2a). The second step is to stack structure comprised of multiple layers of ENF to produce an ENF membrane, where the fibers in each layer are aligned cross-directionally (~90°) to the fibers in adjacent layers. ENM membrane consists of four and six layers of cross-direction ENF membrane was successfully created (Fig. 22.2b) and viewed under microscope for evaluating the architecture of ENF membrane.

The standard fixing method of methanol has shown to damage the PCL fibers and even dissolve the matrices. Neutral buffer formalin was able to fix the cells to preserve the PCL fiber membrane.

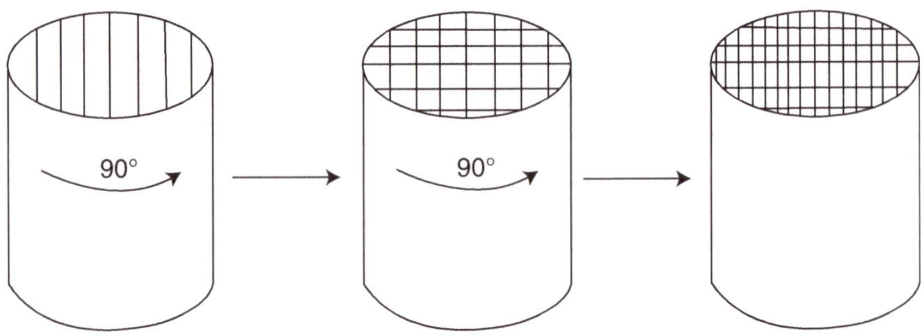

Fig. 22.1 Schematic representation of production of ENF membrane composed of layers of uni-direction fiber layers stacked in ~90° orientation

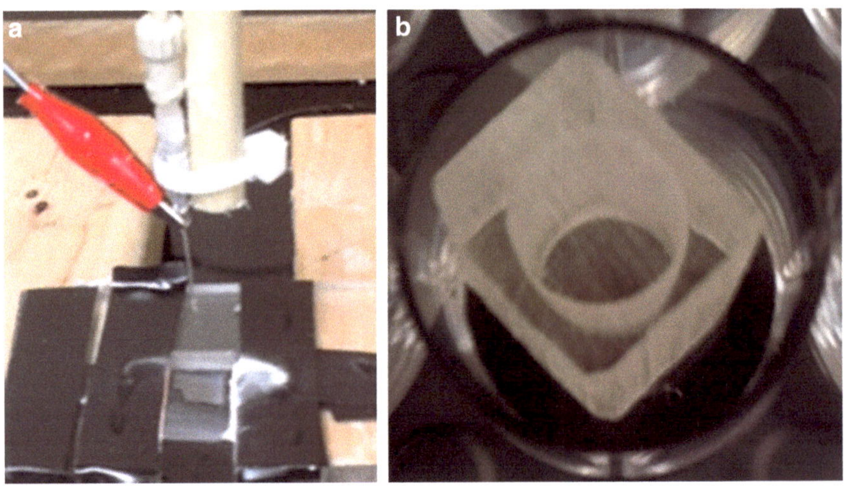

Fig. 22.2 (**a**) A single layer of aligned unidirectional electrospun nanofiber (ENF) and (**b**) ENF membrane composed of layers of ~90°

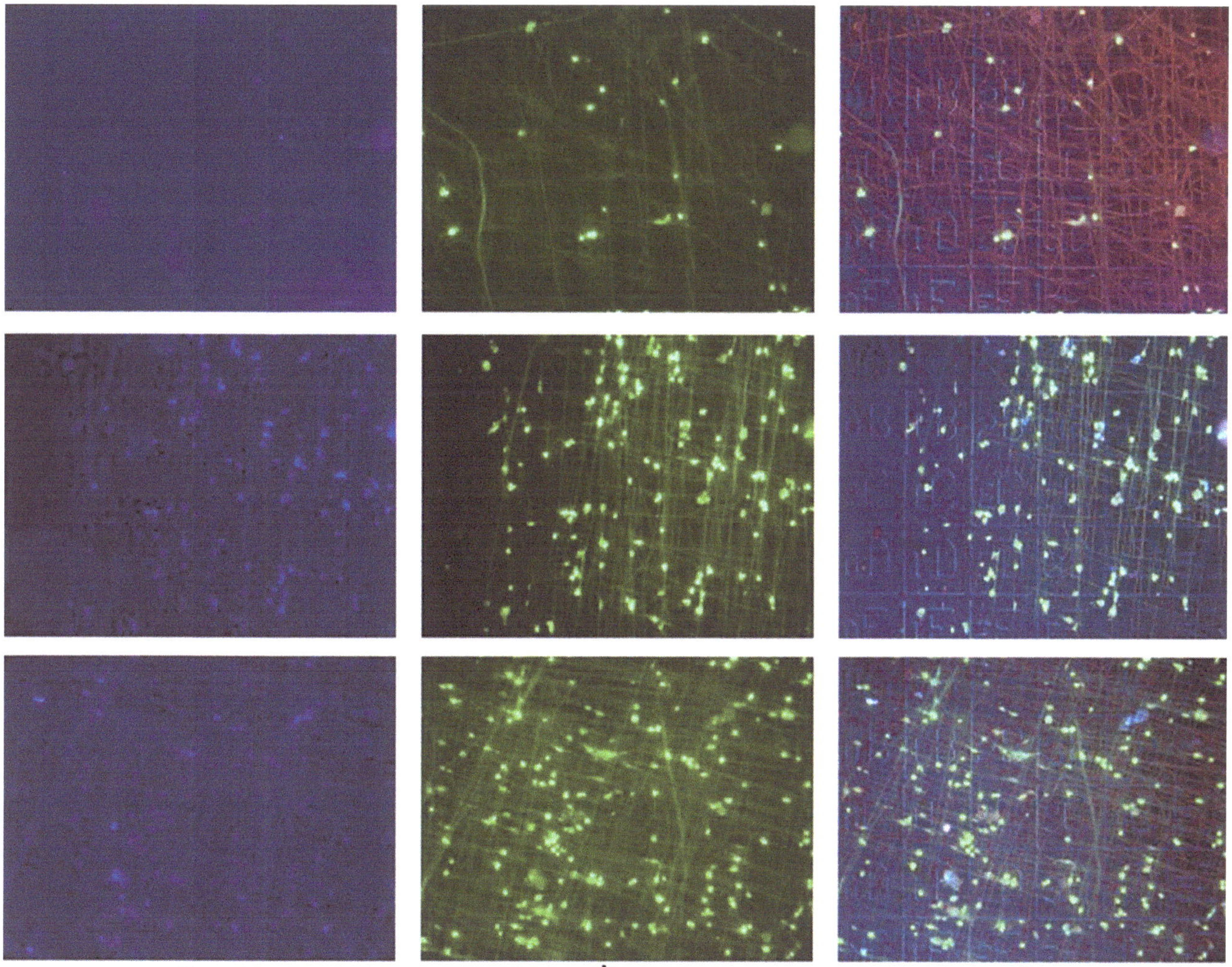

Fig. 22.3 Cell viability tests on ENF membranes consists of single layer (*top row*), four layers (*middle row*) and six layers (*bottom row*) of cross-directions fibers. Figure shows (**a**) cell adhesion, (**b**) cell proliferation and (**c**) combined adhesion and proliferation stained image after 48 h of cell culture

The results of a single, four and six layers of PCL fibers showed adhesion and proliferation of osteoblasts under standard microscopy light. Higher layers of PCL fibers in the membrane obtained higher adhesion and proliferation of osteoblast cells. The ENF membrane consists of six layers of cross-direction ENF resulted in an randomly distributed ENF membrane structure. Higher adhesion and proliferation of osteoblast cells was observed in the higher layers of layers of PCL fiber sample than the sample made with lower layers of fibers.

The experiments involving the PCL fibers showed no adverse effects on the adhesion (Fig. 22.3a) and proliferation (Fig. 22.3b) of the osteoblast cells. Figure 22.3c shows the combined cell adhesion and proliferation image for the ENF membranes. Tests show the designed ENF membrane has no adverse effect on osteoblast cells and fiber attached to the fiber. Due to electrostatic charge, it was impossible to produce aligned cross partnered fiber mat more than four layers. Fiber mat pattern become random after depositing three layers of aligned fiber on top of each other.

22.4 Conclusions

This research was to evaluate the effect of architecture on the ossteointergration of an electrospun fiber membrane made with Polycaprolactone (PCL) fibers. PCL fibers was produced using electrospin process. A technique was developed to harvest fiber on a substrate. Cell viablity on the PCL substrate was evaluated. The test showed that the cells adhered to the fiber thread within 48 h of cell culture and able to proliferate along the fiber strand. Osteoblast cell adhered more at the junction of cross-linked fiber strand in the ENF membrane. The results indicates that ENF membrane consists of four to six layers of fibers are suitable for further biological evaluations such as differentiation, protein adsorption.

Acknowledgements This research was made possible by grant 5P20GM103447 from the US National Institutes of Health and an on-campus faculty grant program from the University of Central Oklahoma Office of Research and Grants.

References

1. Deitzel, J., et al.: Controlled deposition of electrospun poly(ethylene oxide) fibers. Polymer **42**, 8163–8170 (2001)
2. Bosworth, L.A., Turner, L.-A., Cartmell, S.H.: State of the art composites comprising electrospun fibres coupled with hydrogels: a review. Nanomed. Nanotechnol. Biol. Med. **9**(3), 322–335 (2013)
3. Mizuno, H., et al.: Biomechanical and biochemical characterization of composite tissue-engineered intervertebral discs. Biomaterials **27**(3), 362–370 (2006)
4. Khandaker, M., Snow, P.: Method and apparatus for the controlled deposition of branched electrospun fiber on biomedical implants and material (U.S. Patent Application No. 14/734147, International Patent Application No. PCT/US15/45183, and U.S. Patent 9,359,694)

Chapter 23
Controlling hESC-CM Cell Morphology on Patterned Substrates Over a Range of Stiffness

Brett N. Napiwocki, Max R. Salick, Randolph S. Ashton, and Wendy C. Crone

Abstract There is a consensus in the cardiac stem cell biology field that human embryonic stem cell derived–cardiomyocytes (hESC-CMs) are immature and do not resemble human adult cardiomyocytes, either phenotypically or transcriptionally. One striking difference between hESC-CMs and mature adult cardiomyocytes is their morphology. hESC-CMs grown in vitro are pleomorphic in shape and have no clear sarcomere organization; conversely, adult cardiomyocytes are rod-shaped with an average length-to-width (aspect) ratio of 7:1 and display a highly organized internal cytoskeletal structure. By combining multiple cues, i.e. substrate stiffness and topographical features, it may be possible to create a more physiologically-relevant model that better recapitulates the architecture of the native human heart which will aid in regenerative medicine therapies, disease modeling, drug testing, developmental and cardiotoxicity studies. Prior work in our lab used microcontact printing on glass slides to control the cell shape to improve the maturation of hESC-CMs. Since then, new work has focused on patterning methods on more compliant substrates using both microcontact printing, as well as a sacrificial polyvinyl alcohol (PVA) film. In this proceeding, the advantages and disadvantages of the above methods will be discussed in relationship to hESC-CM maturation.

Keywords Cardiomyocyte • Stem cell • Cardiac bioengineering • Microcontact printing • PDMS

23.1 Introduction

Cardiovascular disease (CVD), which includes myocardial infarctions, strokes and hypertension, is a life-threatening condition worldwide. One of the main reasons CVD is so lethal is because primary cardiomyocytes in the postnatal human heart are essentially postmitotic and are unable to regenerate a damaged heart [1]. Therefore, an improved model of cardiogenesis is needed to develop methods that could potentially help elevate or restore cardiac function back to normal levels. One potential and exciting cell source that can be used in these models are cardiomyocytes derived from human pluripotent stem cells (hPSC-CMs), including both embryonic (hESC-CMs) and induced pluripotent (hiPSC-CMs); however, the immature phenotype of these CMs has thus far limited their clinical application.

Current hPSC-CMs resemble fetal cardiomyocytes in terms of gene expression, metabolism, calcium handling properties and spontaneous contraction in culture [2]. Another important distinction between mature and immature CMs is cell morphology. It has long been observed that adult CMs in vivo are rod-shaped with myofibrils aligned along the longitudinal axis of the cell, whereas hPSC-CMs in vitro are pleomorphic in shape and display less myofibril organization. Disruption in

B.N. Napiwocki • R.S. Ashton
Department of Biomedical Engineering, University of Wisconsin-Madison, Engineering Research Building, 1500 Engineering Drive, Madison, WI 53706, USA

M.R. Salick
Materials Science Program, University of Wisconsin-Madison, Engineering Research Building, 1500 Engineering Drive, Madison, WI 53706, USA

W.C. Crone (✉)
Department of Biomedical Engineering, University of Wisconsin-Madison, Engineering Research Building, 1500 Engineering Drive, Madison, WI 53706, USA

Materials Science Program, University of Wisconsin-Madison, Engineering Research Building, 1500 Engineering Drive, Madison, WI 53706, USA

Department of Engineering Physics, University of Wisconsin-Madison, Engineering Research Building, 1500 Engineering Drive, Madison, WI 53706, USA
e-mail: crone@engr.wisc.edu

C.S. Korach et al. (eds.), *Mechanics of Biological Systems and Materials, Volume 6*,
Conference Proceedings of the Society for Experimental Mechanics Series, DOI 10.1007/978-3-319-41351-8_23

the organization of myofibrils alters the contractile activity of heart muscle cells which can lead to heart disease. To help promote maturation, several studies have used microfabrication techniques to control CM alignment. One method that has thus far yielded promising results is microcontact printing of the substrate on which cells are seeded to limit their attachment to certain regions.

Microcontact printing is the process of forming patterns of self-assembled monolayers on a substrate through conformal contact. By patterning lanes of extracellular matrix (ECM) proteins on glass substrates, our lab has been able to control the cytoskeletal organization within a pure population of hPSC-CMs. Our research has shown that lane widths between 30 and 80 μm promote highly aligned CMs with an organized myofilament structure, while CMs in widths greater than 100 μm do not develop aligned myofibrils [3]. While the sarcomere organization of these hESC-CMs resembles that of adult CMs, other characteristics such as circumferential gap junction expression and low calcium propagation speeds were still representative of an immature phenotype.

To further improve cardiomyocyte maturation, our current research focuses on patterning lanes of proteins onto more compliant, physiologically-relevant substrates. In the native human heart, the extracellular matrix is believed to be 5–15 kPa [4]. On the other hand, diseased hearts are accompanied by an accumulation of collagenous extracellular matrix proteins and can have an increased stiffness of 35–70 kPa which changes the electrical conductivity of the heart and ultimately impairs cardiac function leading to arrhythmias [5]. For comparison, the substrate used in the above micropattern experiment was glass which is about hundred 1000 times stiffer (GPa) than both normal and fibrotic cardiac tissue (10–100 kPa). In this proceeding, the advantages and disadvantages of microcontact printing on glass vs. PDMS will be discussed in relationship to hESC-CM maturation.

23.2 Methods

23.2.1 hESC Culture

Cardiomyocytes were derived from an engineered H9 hESC line containing a cardiac troponin T GFP (cTnT-GFP) promoter. This cTnT-GFP promoter line exhibits resistance to the antibiotic zeocin along with cardiac troponin T expression indicated by the presence of GFP, allowing for the purification of cTnT-expressing cells. hESCs were maintained on vitronectin-coated polystyrene plates and E8 media (replaced daily) prior to differentiation [6]. E8 media was produced with DMEM/F-12 (Life Technologies), L-ascorbic acid 2-phosphate sesquimagnesium salt hydrate, sodium selenite, sodium bicarbonate, holotransferrin, insulin, βFGF (WiCell), and TGFβ (Peprotech). Unless otherwise noted, all reagents were purchased from Sigma-Aldrich. hESCs were passaged at a 1:12 ratio every 4 days using versene (Life Technologies) and were seeded with ROCK inhibitor (Tocris).

23.2.2 hESC Cardiomyocyte Differentiation

To produce a high percentage of cardiomyocytes derived from the cTnT-promoting H9 hESC line ("hESC-CMs"), a modified version of the small molecule Wnt-agonist method of differentiation was used [7]. Briefly, on day-3 the cTnT-GFP hESCs were seeded onto Matrigel-coated (BD Biosciences) 12-well plates at a density of 400,000 cells per well and fed with E8 medium supplemented with 5 μm ROCK inhibitor (Tocris). On day-1, the cells were again fed with E8 medium. On day 0 (exactly 72 h after initial seed), cells were fed with RPMI (Life Technologies) supplemented with B27-insulin (Life Technologies), 10.5 μm of the Wnt agonist CHIR 99021 (Tocris) and 1 μg/ml of insulin (Sigma). On day 1, cells were fed with RPMI supplemented with B27-insulin. On day 3, cells were treated with RPMI supplemented with B27-insulin and 2.5 μm of the Wnt inhibitor IWP4 (StemGent). On day 5, cells were fed with PRMI containing B27-insulin and then treated with RPMI containing B27-complete on days 7, 8, 9, and every day thereafter. Upon observation of contractions (typically between days 9–12), zeocin treatments were applied in which 100 μg/mL/day was added to the RPMI-B27-complete medium for three consecutive days to remove non-cTnT expressing cell types. The resulting cells were confirmed by flow cytometry to be ~98 % cardiomyocytes.

23.2.3 Fabrication of PDMS with Varying Elastic Moduli

As previously described, two types of commercially available polydimethylsiloxane (PDMS), Sylgard 184 elastomer and Sylgard 527 gel (Dow Corning), were blended together to create PDMS substrates with tunable mechanical properties [8].

Briefly, Sylgard 184 was made per manufacturer's directions by mixing ten parts base to one part curing agent. Sylgard 527 was prepared per manufacturer's directions by mixing equal weights of part A and part B. Both Sylgard 184 and 527 were mixed for 5 min with a glass stir rod. After each blend was properly mixed, they were then combined in different mass ratios of Sylgard 184:527 and again mixed for 5 min. The mass ratio used in this study was 1:10, in addition to pure Sylgard 184 and 527. Once mixed, the PDMS was poured into 100 mm diameter petri dishes and cured overnight at 60 °C. The following day the samples were cut into the correct shape with a razor blade and UV sterilized prior to coating with Matrigel. For mechanical testing, each PDMS blend was poured into a tension sample-shaped Teflon mold and cured under similar conditions.

23.2.4 *Mechanical Characterization*

The tension test samples were fabricated with a total length of 35 mm, a gauge length of 8.0 mm, a gauge width of 1.5 mm and a thickness of 2 mm using a PDMS stencil [9]. After fabrication, each sample was sprinkled with glass beads with diameters of 30–50 μm (Polysciences, Inc.) to allow for optical strain measurement during the test. Uniaxial tensile testing was performed on the samples using an Instron 5548 MicroTester mechanical testing machine with a 10 N load cell. Time-lapse microscopy was used to observe the displacement of the beads during the tensile test. During testing, the neck region of the sample was imaged as quickly as possible using QCapture Pro and a Q-Imaging MicroPublisher 5.0 RTV Camera. Using an in house Matlab code, the strains were calculated by measuring the vertical (axial) distance between pairs of beads during the test [10]. Samples were stretched at a rate of 2.00 mm/min until failure. At least six measurements from two different sample preparations were analyzed per condition. The elastic modulus for each sample is the slope of the linear regression of the stress-strain curves.

23.2.5 *Micropattern Design and Stamp Production*

The technique in this study utilized microcontact printing on gold-coated glass substrates and PDMS substrates. In order to use this technique, several steps were required to make a PDMS mold with the correct design. First a template with the feature designs was created in AutoCad (autodesk) and sent to the Stanford Microfluidics Foundry for fabrication of a photomask and a 4-in. patterned Si wafer (Stanford University, Santa Clara County, CA). For this experiment, the features consisted of lane widths ranging from 20 to 160 μm. Using soft photolithography techniques, the Si wafer was spin coated with an SU-8 negative photoresist. The photomask was placed above the Si wafer and subsequentially exposed to UV light which imprinted the intended designs on the Si wafer. The silicon master was rendered inert by overnight exposure in vapors of (tridecafluoro-1, 1, 2, 2-tetrahydrooctyl) trichlorosilane. PDMS (Sylgard 184) was then poured over the patterned Si wafer and cured over night at 60 °C, creating a reverse replica of the patterned Si wafer. Fisherbrand microscope cover glass (24 × 50 #1) slides were cleaned by sequential washes in toluene and methanol followed by sonication for 1 min in acetone.

23.2.6 *μCP of ω-Mercaptoundecyl Bromoisobutyrate on Au-Coated Slides*

Once the glass slides were dried, 35 Å of titanium (Ti) followed by 180 Å of gold (Au) were deposited onto the surface using a CHA-600 Telemark focused electron-beam evaporation system (Wisconsin Center for Applied Microelectronics). Afterwards, the gold-coated slides were rinsed with absolute ethanol prior to use. To make micropatterned substrates, PDMS stamps were coated with ω-mercaptoundecyl bromoisobutyrate (2 mM in absolute ethanol), dried under nitrogen, and then brought in conformal contact with Ti/Au-coated glass slides. Lastly, the micropatterned slides were washed in absolute ethanol for 2 h and then dried under nitrogen.

23.2.7 Surface-Initiated Activator Generated by Electron Transfer Atom Transfer Radical Polymerization (SI-AGET ATRP) of OEGMEMA

The macromonomer OEGMEMA (8 g, 16.7 mmol), water (7.5 mL), methanol (7.5 mL), copper(II) bromide (0.08 mmol, 17.9 mg) and 2 ′2-Bipyridine (0.24 mmol, 37.5 mg) were added to a Schlenk flask, sealed, and stirred for 30 min. Then, the flask was degassed with three freeze-vacuum-thaw cycles, and backfilled with nitrogen. Next, the mixture was transferred by syringe to a reaction flask containing micropatterned glass slides under vacuum. To start the reaction, L-ascorbic acid (0.8 mmol, 140.9 mg) in deionized water was injected into the flask. The reaction continued for 16 h to generate micropatterned thick PEGMEMA brushes. Polymerization was stopped by adding air, and the slides were rinsed with ethanol, water, and ethanol sequentially before drying under nitrogen. Prior to cell seeding, the slides were washed with ethanol for 2 h.

23.2.8 Adsorption of Adhesion Ligands

Once the glass substrates were fabricated, they were coated with DMEM (Life Technologies) containing 83 μg/mL Matrigel (WiCell) overnight at 37 °C. Since PEGMEMA is bioinert, the Matrigel could only adhere to the gold-coated feature regions on the glass slides. For PDMS substrates, a PDMS mold with the intended features was coated overnight at 37 °C with DMEM containing 83 μg/mL Matrigel. Afterwards, the Matrigel-coated PDMS mold was dried with N_2 and placed on top of a sacrificial polyvinyl alcohol (PVA) film [11]. After 1 h, the mold was removed and the PVA film was placed on top of the PDMS substrates in such a way that the Matrigel was sandwiched between the PVA film and the PDMS substrate. The PVA film plus PDMS substrate was incubated at 37 °C for 20 min and then washed with PBS to dissolve the PVA film but still leave the Matrigel lanes on the PDMS substrates. The remaining areas between Matrigel lanes on the PDMS substrates were backfilled with Pluronic F127 (Sigma-Aldrich) for an additional hour to prevent cell outgrowth from the lanes. Both the glass slides and PDMS substrates were rinsed with PBS prior to cell seeding.

23.2.9 HESC-CM Seeding and Culture

After zeocin purification, the pure population of hESC-CMs was washed with PBS and dissociated from their original culture dish via 5 min exposure to 0.25 % Trypsin (Life Technologies). Trypsin was used to ensure break up of the significant amount of cell–cell binding and extracellular matrix deposits that formed in these cultures during the differentiation process. These cells, referred to as pre-seed, were then seeded into new Matrigel-coated wells using 20 % serum-containing media. After 10 days the pre-seed hESC-CMs were again exposed to 0.25 % Trypsin for 5 min and then seeded onto the micropatterned substrates using 20 % serum-containing media. Prior experimentation had resulted in better isolation of single cells using the pre-seed method, as opposed to seeding the hESC-CMs directly from the differentiation culture plates which resulted in more CM aggregates than single cells. The hESC-CMs were seeded at a density of 30,000 cells/cm^2 which was enough cells to cover the substrate while avoiding excess "piling up" of seeded cells in multilayers. The following day, the media was replaced with B27-supplemented RPMI media containing 8 μg/mL Matrigel and changed every 3 days thereafter.

23.2.10 Immunofluorescence

After 5 days of culture the samples were stained and imaged. The cells were washed once with PBS and then exposed to 4 % paraformaldehyde (PFA) (Electron Microscopy Sciences) for 15-min at room temperature. The cells were washed again with PBS and then treated with 0.1 % Triton (Sigma) for 6 min at room temperature to permeabilize the cell membrane. Again, the cells were washed with PBS and treated for 30 min with a blocking solution consisting of PBS, 2 % FBS, 0.1 % Triton, 11.2 mg/mL glycine, and 50 mg/mL BSA. Afterwards, phalloidin conjugated to tetramethylrhodamine B isothiocyanate (TRITC) (Sigma) was applied at a 50 mg/mL concentration to label actin filaments and DAPI was applied at a 1:1000 dilution to label nuclei. Cells were washed with PBS and transferred to coverslips, where they were mounted using ProLong

Gold Antifade (Life Technologies). To visualize the transfer of Matrigel from the patterned PDMS mold to the flat PDMS stamp, the primary antibody rabbit anti-laminin (1:500, Sigma-Aldrich) was incubated with the flat PDMS stamp overnight at 4 °C. The next day the flat PDMS stamp was washed once with PBS for 5 min and then treated for 1 h with blocking solution containing the secondary antibody, Cy3-conjugated mouse anti-rabbit (1:1000 dilution, Abcam). Samples were imaged on a Nikon A1RSi Confocal Microscope with an attached Photometrics CoolSNAP HQ2 camera.

23.2.11 Nuclear Alignment Measurement

As described in our prior publication, nuclear alignment was used as a measure of cardiomyocyte alignment on the different micropatterned lane widths [3]. Using the Nikon Elements—AR software, ROIs were automatically determined in the DAPI (nuclear) channel of the confocal images. Overlapping nuclei were manually separated to keep them from being quantified as a single nucleus. The major axis of the nuclei was then determined automatically, and the angle of this axis was used to determine cell alignment. Cells were then quantified as "aligned" if their major axis fell within 20° of the feature axis on which the cell was adhered.

23.3 Results and Discussion

23.3.1 Mechanical Properties of Different PDMS Blends

To obtain the Young's modulus for each PDMS blend, tensile testing was performed. The stiffness for each blend are as follows: Sylgard 184 = 1749.23 ± 33.57 MPa, 1:10 = 55.16 ± 6.56 kPa and Sylgard 527 = 2.7 ± 0.24 kPa. These moduli are in close agreement with previously reported values [8]. As already mentioned, the stiffness of the native human heart is 5–15 kPa while fibrotic hearts can range from 35 to 70 kPa [4, 5]. Using Sylgard 527 and the 1:10 PDMS blend is useful for studying how substrate stiffness impacts the maturation of hESC-CMs when the cells are grown on substrates that mimic the stiffness of the native human heart and diseased heart, respectively. In contrast, hESC-CMs grown on glass (GPa) substrates are not representative of in vivo conditions. It has been reported that CMs grown on rigid culture substrates (GPa) show a progressive loss of rhythmic beating while those grown on physiologically-relevant substrates develop optimal contractile work and have well-developed striations [12]. Additionally, it is now widely known that matrix elasticity impacts the lineage commitment of stem cells; however, the impact of matrix stiffness on hESC-CM maturation is less defined and will be the subject of future work.

23.3.2 Immunofluorescence of Matrigel Transfer

As described in the methods section, microcontact printing on glass slides is completed through an ATRP reaction that generates micropatterned thick PEGMEMA brushes on the substrate surface. Since PEGMEMA is bioinert, the whole glass slide can be coated in Matrigel solution and only the bare gold-coated glass regions will bind to the Matrigel. This method is highly reproducible and Matrigel covers the regions uniformly and consistently (Fig. 23.1a). In contrast, microcontact printing can't be performed directly on more compliant substrates because the tackiness of the substrate causes the stamp to adhere to the substrate and distorts the pattern once removed. Instead, microcontact printing was performed on a sacrificial layer of PVA which was then placed on the more compliant PDMS substrates and dissolved, resulting in patterned regions of Matrigel which were then backfilled with Pluronics F127 to prevent cellular outgrowth (Fig. 23.1b).

Although some regions of the pattern transfer over nicely, others remain un-patterned and don't include Matrigel due to air bubbles between the PVA film and substrate surface. Additionally, any wrinkles in the PVA film will distort the underlying pattern transfer. These infidelities in pattern transfer affect hESC-CM morphology as the cell assumes the shape of the underlying Matrigel pattern. If un-patterned regions of Matrigel are present, then hESC-CMs will not adhere to that region which causes discontinuities in cell-cell contact and calcium signaling. Regions in which the Matrigel pattern transferred over correctly with no defects resulted in hESC-CM attachment on the PDMS substrates that was indistinguishable from those adhered to glass substrates when viewing phase contrast images.

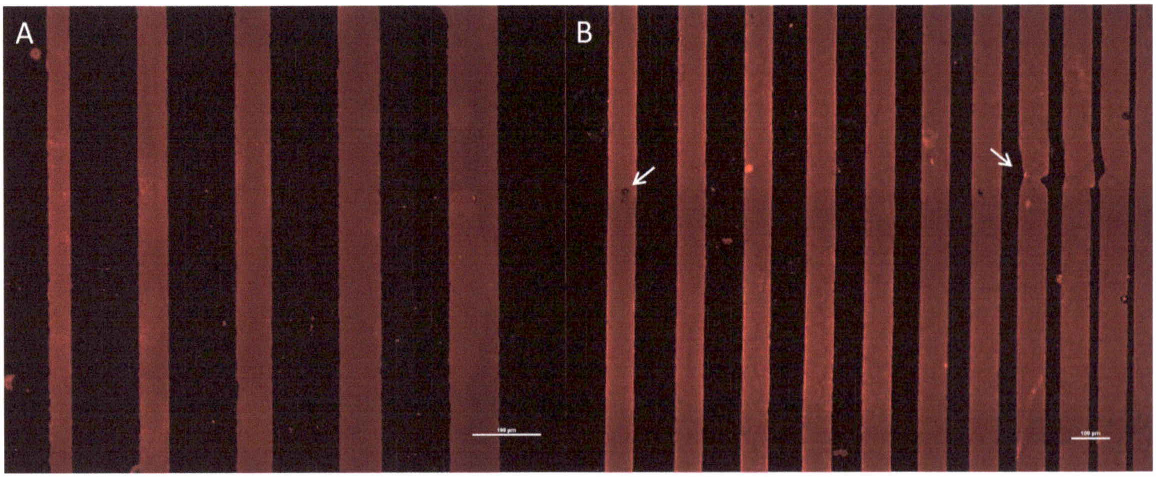

Fig. 23.1 Immunofluorescence images of patterned Matrigel (*red*), visualized with an anti-laminin antibody, on (**a**) glass and (**b**) 5 kPa substrates. *White arrows* indicate un-patterned Matrigel regions and pattern distortions due to a wrinkled PVA film. Scale bar = 100 μm

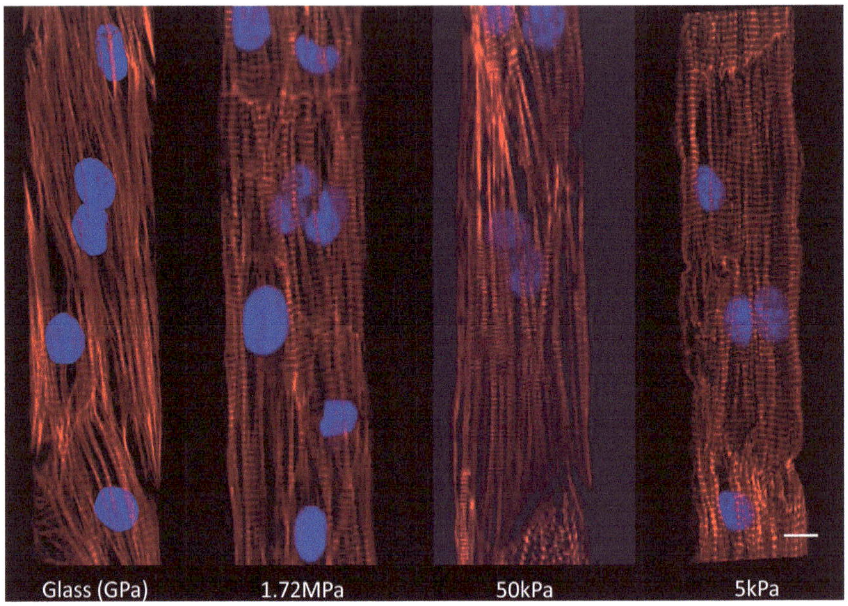

Fig. 23.2 hESC-CMs grown on different substrate stiffnesses all with a lane width of 36 μm. *Red* = F-actin, *blue* = DAPI. Scale bar = 10 μm

23.3.3 Sarcomere Organization of hESC-CMs

One indication of CM maturation is cell shape. Adult CMs are rod-shaped with well-aligned myofibrils while immature, un-patterned hESC-CMs are pleomorphic and have no clear myofibril organization. Sarcomeres, the basic mechanical unit of striated muscle, when aligned in series make up a myofibril. To assess myofibril organization qualitatively, hESC-CMs were seeded on different lane widths of Matrigel on both glass and PDMS substrates and stained with phalloidin which labels F-actin. In agreement with previous results on glass substrates [3], hESC-CMs seeded on PDMS substrates on lane widths under 100 μm developed aligned, organized myofibrils that ran parallel to the micropattern direction. In comparison, those grown on lane widths greater than 100 μm had aligned sarcomeres only along the edges of the micropattern while the middle of the feature had myofibrils that were less aligned. hESC-CMs grown on a lane width of 36 μm, irrespective of substrate stiffness, were able to span the entire lane width and formed well-aligned myofibrils (Fig. 23.2). Future work aims to further examine hESC-CM myofibril organization and measure force production as a function of substrate stiffness.

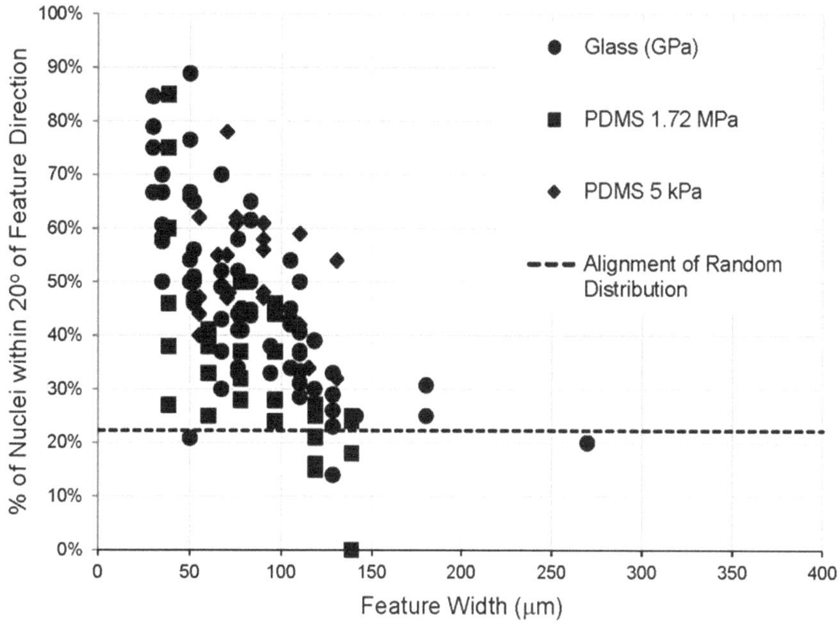

Fig. 23.3 Quantification of hESC-CM nuclear alignment depending on substrate stiffness and micropattern lane width. PDMS results described in this paper (*square and diamond*) match the glass results (*circle*) obtained by Salick et al. [3]

23.3.4 *Nuclear Alignment of hESC-CMs*

One method used to determine cell alignment is to look at nuclear alignment which was done using Nikon Elements—AR software. As previously described, cells were categorized as "aligned" if their major axis was within 20° of the feature axis to which they were adhered. In cells that have no directionality the nuclei would be expected to be aligned 22 % of the time. In our previously published results, hESC-CMs seeded on glass substrates had nuclei that were aligned in lane widths of 40–100 μm. Larger lane widths generated less aligned nuclei which followed the same trend as the sarcomere organization. The same nuclear alignment trend was observed in hESC-CMs seeded on 1.72 MPa and 5 kPa PDMS substrates (Fig. 23.3). Therefore, we conclude that hESC-CMs will develop aligned nuclei when seeded on lane widths under 100 μm, regardless of the underlying matrix elasticity.

23.4 Conclusion

In conclusion, although hPSC-CMs hold enormous therapeutic potential, their immaturity and likeness to fetal human CMs rather than mature, adult CMs is hampering their usefulness. As described above, microcontact printing was used on a wide range of substrate stiffnesses to control the alignment of hESC-CMs. Microcontact printing on more compliant PDMS substrates required an additional intermediate step involving a sacrificial PVA film to accurately transfer Matrigel to the PDMS substrates. Sarcomere organization and nuclear alignment followed the same trends on all substrate stiffnesses, namely hESC-CMs were aligned if patterned on lane widths under 100 μm. In addition to having more physiologic relevance, lower stiffness substrates have several other advantages in their use. Since PDMS substrates can be embedded with fluorescent beads, traction force microscopy can be performed to measure the force generated by the hESC-CMs, a measure of maturation. Furthermore, unlike glass substrates, PDMS substrates allow for the addition of applied loads, both static and cyclic stretch, which have been shown to improve CM maturation. In order for hPSC-CMs to reach the level of maturity seen in their adult counterparts, it is likely that multiple signals will be needed that mimic the in vivo myocardial milieu. Topographical cues combined with a physiologically-relevant substrate stiffness open up the possibility of adding additional cues which could increase the level of maturation currently seen in hPSC-CMs and make them more therapeutically useful.

Acknowledgments The authors would like to thank the Graduate School and the Office of the Vice Chancellor for Research and Graduate Education at the University of Wisconsin-Madison for funding this work. Additional thanks are given to Dr. Timothy Kamp of the University of Wisconsin-Madison for providing the cTnT H9 hESC line used in the experiments.

References

1. Bergmann, O., et al.: Evidence for cardiomyocyte renewal in humans. Science **324**, 98–102 (2009)
2. Feric, N.T., Radisic, M.: Maturing human pluripotent stem cell-derived cardiomyocytes in human engineered cardiac tissues. Adv. Drug Deliv. Rev. **96**, 110–134 (2016)
3. Salick, M.R., et al.: Micropattern width dependent sarcomere development in human ESC-derived cardiomyocytes. Biomaterials **35**(15), 4454–4464 (2014)
4. Engler, A.J., et al.: Matrix elasticity directs stem cell lineage specification. Cell **126**(4), 677–689 (2006)
5. Berry, M.F., et al.: Mesenchymal stem cell injection after myocardial infarction improves myocardial compliance. Am. J. Phys. Heart Circ. Phys. **290**(6), H2196–H2203 (2006)
6. Chen, A., et al.: Shrink-film configurable multiscale wrinkles for functional alignment of human embryonic stem cells and their cardiac derivatives. Adv. Mater. 23.48, 5785–5791 (2011)
7. Lian, X., et al.: Robust cardiomyocyte differentiation from human pluripotent stem cells via temporal modulation of canonical Wnt signaling. Proc. Natl. Acad. Sci. **109**(27), E1848–E1857 (2012)
8. Palchesko, R.N., et al.: Development of polydimethylsiloxane substrates with tunable elastic modulus to study cell mechanobiology in muscle and nerve. PLoS One **7**(12), e51499 (2012)
9. Johnson, B.D., Bauer, J.M., Niedermaier, D.J., Crone, W.C., Beebe, D.J.: Experimental techniques for mechanical characterization of hydrogels at the microscale. Exp. Mech. **44**(1), 21–28 (2004)
10. Salick, M.R.: Engineered microenvironments for the maturation and observation of human embryonic stem cell derived cardiomyocytes. University of Wisconsin-Madison, Madison, WI (2015)
11. Yu, H., et al.: A novel and simple microcontact printing technique for tacky, soft substrates and/or complex surfaces in soft tissue engineering. Acta Biomater. **8**(3), 1267–1272 (2012)
12. Engler, A.J., et al.: Embryonic cardiomyocytes beat best on a matrix with heart-like elasticity: scar-like rigidity inhibits beating. J. Cell Sci. **121** (22), 3794–3802 (2008)

Chapter 24
Cytoskeletal Perturbing Drugs and Their Effect on Cell Elasticity

Martha E. Grady, Russell J. Composto, and David M. Eckmann

Abstract The cytoskeleton is primarily responsible for providing structural support, localization and transport of organelles, and intracellular trafficking. The structural support is supplied by actin filaments, microtubules, and intermediate filaments, which contribute to overall cell elasticity to varying degrees. We evaluate cell elasticity in five different cell types with drug-induced cytoskeletal derangements to probe how actin filaments and microtubules contribute to cell elasticity and whether it is conserved across cell type. Specifically, we measure elastic stiffness in chondrocytes, fibroblasts, endothelial cells, hepatocellular carcinoma, and fibrosarcoma using atomic force microscopy. We subject all five cell lines to two cytoskeletal destabilizers: cytochalasin D and nocodazole, which disrupt actin and microtubule polymerization, respectively. Non-cancer cells treated with cytochalasin D show a decrease of 60–80 % in moduli values compared to untreated cells of the same origin, whereas the nocodazole-treated cells show no change. Alternatively, cancer cells exhibit increased stiffness as well as stiffness variability when subjected to nocodazole. Overall, we demonstrate actin filaments contribute more to elastic stiffness than microtubules but this result is cell type dependent. Lastly, we show that disruption of microtubule dynamics affects cancer cell elasticity, suggesting therapeutic drugs targeting microtubules be monitored for significant elastic changes.

Keywords Atomic force microscopy • Cell mechanics • Elasticity • Cytoskeleton • Cancer

24.1 Introduction

The interdependence of cell elasticity and cytoskeletal components is a critical step toward understanding the mechanics of living tissue. Cellular responses and their microarchitecture react and adapt to their environment and disease state [1, 2]. Changes in cell elasticity have been implicated in the pathogenesis of many human diseases including vascular disorders, malaria, sickle cell anemia, arthritis, asthma and cancer [3]. Therefore, there is a practical need to measure cell mechanics quantitatively to understand how diseased cells differ from healthy ones. In particular, investigating the mechanical properties of cancer cells may help to better understand the physical mechanisms responsible for cancer metastasis.

Common techniques to measure the mechanics of cells, include the use of magnetic beads, optical tweezers, micropipette aspiration, and atomic force microscopy (AFM) [2]. Since the application of AFM to living cells, it has readily been adapted to characterize cell topography as well as mechanical properties. The AFM has been further adapted for microrheology [4, 5], frequency modulation [6, 7], and creep [8–10] experiments to study the viscoelastic properties of various cell lines. Of interest to this work, is the evaluation of elasticity of different cell lines [11]. Based on elastic modulus measurements, many groups show that cancerous cells are softer than their non-transformed counterparts [12, 13]. This stands to reason since

M.E. Grady
Department of Materials Science and Engineering, School of Engineering and Applied Science, University of Pennsylvania, 3231 Walnut Street, Philadelphia, PA 19104, USA

Department of Anesthesiology and Critical Care, School of Medicine, University of Pennsylvania, 3620 Hamilton Walk, Philadelphia, PA 19104, USA

R.J. Composto
Department of Materials Science and Engineering, School of Engineering and Applied Science, University of Pennsylvania, 3231 Walnut Street, Philadelphia, PA 19104, USA

D.M. Eckmann (✉)
Department of Anesthesiology and Critical Care, School of Medicine, University of Pennsylvania, 3620 Hamilton Walk, Philadelphia, PA 19104, USA
e-mail: david.eckmann@uphs.upenn.edu

C.S. Korach et al. (eds.), *Mechanics of Biological Systems and Materials, Volume 6*,
Conference Proceedings of the Society for Experimental Mechanics Series, DOI 10.1007/978-3-319-41351-8_24

increased deformability of cancer cells allows them to metastasize and infiltrate tissues [14]. Such behavior is attributed to modification in the cytoskeletal organization of the cells [15, 16].

The influence of cytoskeletal drugs on cell elasticity has allowed the decoupling of cytoskeletal components and their contribution to overall cell elasticity. While many AFM studies examine drug responses of one cell type, or one healthy and one cancerous cell type, our goal is to present a comprehensive AFM study across multiple cell lines, healthy and cancerous, that examines the effect of cytoskeletal destabilizers on cell elasticity. We find destabilizing the microtubule network produces no effect in healthy cell lines, while producing a significant increase in elastic moduli in the fibrosarcoma line. We aim to elucidate the functional relationship between cellular mechanics and cytoskeletal stability and whether that relationship is conserved across different cell types.

24.2 Methods

24.2.1 *Cell Culture*

Five cell lines were cultivated for this study: human dermal fibroblasts, bovine chondrocytes, HUVECs, HT 1080 (fibrosarcoma), and HUH-7 (hepatocellular carcinoma). Fibroblasts and HUVECs were cultured in Fibrolife (Lifeline Cell Technologies) and Vasculife (Lifeline Cell Technologies), respectively. HT 1080 fibrosarcoma cells, HUH-7 hepatocarcinoma cells, and chondrocytes were cultured in DMEM L-glutamine supplemented with 10 % fetal bovine serum and 1 % penicillin streptomycin. All cells were incubated at 37 °C in a humidified atmosphere with 5 % CO_2. Cells were grown to confluence and harvested by trypsinization. Coverslips were coated for 30–40 min with 5 μg/mL fibronectin (BD Biosciences) dissolved in PBS prior to plating. Cells were initially plated at 50–75 k density on glass coverslips (22 × 40 mm) for all studies, and were incubated for 48 h prior to experiments. Three conditions were studied: control (recording buffer only), 2.5 μM cytochalasin D (Sigma Aldrich) in recording buffer for 30 min, and 10 μM nocodazole (Sigma Aldrich) in recording buffer for 30 min. Recording buffer solution was made as follows: 10 % 10× HBSS (Gibco), 1 % Heparin, 1 % Glutamax, and 1 % FBS in DI water. Solution was then pH-balanced to 7.4 using NaOH.

24.2.2 *Tubulin and Actin Visualization*

Cells were incubated in 250 nM TubulinTracker Green (Life Technologies) for 30 min at 37 °C, then rinsed three times, and placed in recording buffer for fluorescence imaging with a standard FITC filter. For actin visualization, cells were fixed with 4 % paraformaldehyde (Electron Microscopy Sciences, Hatfield, PA; diluted with HBSS from a 16 % stock solution) for 15 min, then permeabilized for 3 min with 0.1 % Triton, followed by 20 min of dye-loading with 100 nM Acti-stain 488 phalloidin (Cytoskeleton, Inc, Denver, CO) taken from a 14 μM stock prepared according to the manufacturer's instructions. Images were taken on an Olympus IX 51 with a 40× oil-immersion objective.

24.2.3 *Elastic Modulus Measurement at Single Approach Velocity*

Cell stiffness was measured using force spectroscopy via an atomic force microscope (MFP 3D, Asylum Research, Oxford Instruments) as shown schematically in Fig. 24.1a. The technique includes indentation of an AFM tip into the cell and retraction of the tip resulting in the deflection of the AFM cantilever (Fig. 24.1c). Proper calibration of the AFM cantilever stiffness relates the deflection of the cantilever to force. The force-indentation curve in addition to a nanomechanical model appropriate for the material and tip shape is used to extract an elastic modulus. For our work, the Hertz model best represents the tip-sample interaction. The application of the Hertz model to extract an elastic modulus for biological materials is common, but some specific requirements must be met. The Hertz model requires a small strain assumption, which is satisfied when the AFM tip radius is sufficiently large and depth of indentation is relatively small. We modify an AFM cantilever with a silica colloid with radii of ~4 μm (Fig. 24.1b) and limit the depth of indentation (365 nm on average) thus satisfying this requirement. Another requirement is to consider substrate effects on the measurement due to the spread nature of attached cells. Because measurements are taken perinuclear where cell thickness is near its maximum, we limit AFM tip indentations to 10 % of the expected cell height for each cell type. After calibrating the AFM cantilever using the thermal tuning method

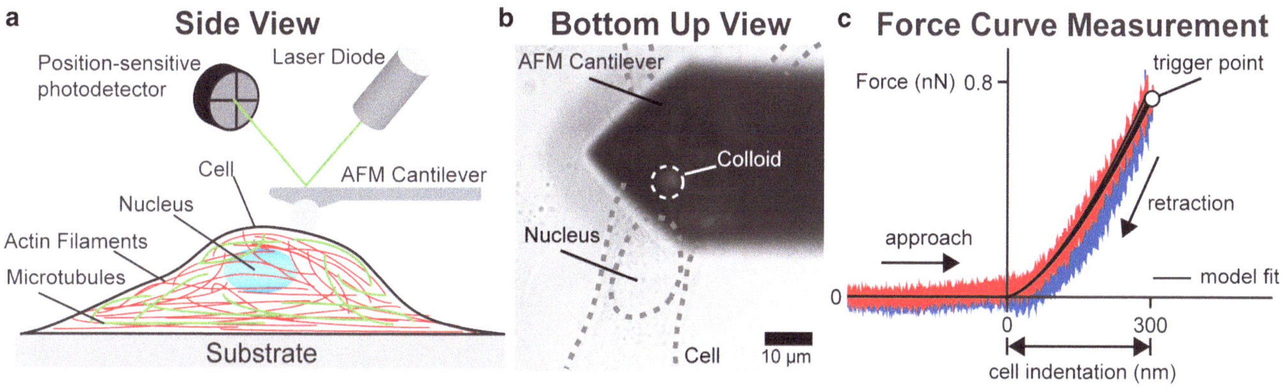

Fig. 24.1 (**a**) *Side view* schematic of an AFM cantilever over a cell, highlighting the substrate effects from measurements taken in the cell periphery. (**b**) Five successive force curve measurements taken on a fibroblast cell with Hertz model fit in black, approach curve in *red*, retraction curve in blue. Using Eqs. (24.2) and (24.3), average modulus value for the five curves is 1.97 ± 0.07 kPa. (**c**) Image from the inverted optical microscope (Nikon oil-immersion 100× objective) of a human dermal fibroblast cell underneath the AFM cantilever with colloid positioned perinuclear (outlines of cell and nucleus, and colloid for visualization). Adapted from [17]

to obtain the cantilever stiffness (~0.05 N/m), subsequent measurements are set to trigger on cantilever deflection. Approach velocity of the AFM cantilever during indentation is set at 500 nm/s and kept constant. Because cells can exhibit frequency dependence as explored in Sect. 24.3.2, we limit the indentation depth and set the same approach velocity so that the same probing frequency is maintained for all cells regardless of stiffness. All measurements were conducted in recording buffer solution.

24.2.4 Approach Velocity Modulation

Following Caporizzo et al. [7], who pioneered the Variable Indentation-rate Viscoelastic Analysis (VIVA) method, the apparent elastic modulus is measured on the same cell, in the same location, with varying approach velocity. By varying the approach velocity while maintaining the same indentation depth, the corresponding probing frequency is therefore varied by,

$$f = \mathrm{V}/\delta, \tag{24.1}$$

where f is the frequency (s^{-1}), V is the approach velocity (m/s), and δ is the indentation depth (m). Six approach velocities were chosen: 0.1, 0.2, 0.5, 1, 2, and 5 (μm/s). At each approach velocity, 3–5 force indentation curves were recorded, and the frequency was calculated from Eq. (24.1) and ranged 0.3–24 s^{-1}. The elastic modulus at each approach velocity was extracted using the force curve analysis discussed in Sect. 24.2.5. The geometric mean of the elastic modulus values was calculated for each approach velocity. The calculated frequencies were averaged at each approach velocity. This procedure gives rise to an apparent elastic modulus over a range of frequencies in order to identify the viscoelastic properties of living cells.

24.2.5 Force Curve Analysis

Stiffness measurements were acquired on multiple cultured substrates for a given test condition and were repeated. One perinuclear location on each cell was indented for elasticity measurements. Five successive measurements were taken and the moduli values from the Hertz model were averaged with good agreement (Fig. 24.1c). The trigger point was set to cantilever deflection indicated in Fig. 24.1b, which ranged 5–20 nm. The amount of resulting indentation within the cell will depend on the stiffness of the cell; therefore the cantilever deflection trigger set point was varied to maintain similar indentation depths for all cell lines. The average indentation depth across all measurements was 360 ± 200 nm. The number of cells tested in each condition ranged from 33 to 49, an overall average of 36 cells per condition, resulting in 542 total number of cells tested.

Elastic parameters are calculated by using the Hertz nanomechanical model for an infinitely stiff colloid tip and a soft, flat sample. Material properties are extracted by applying the model to force-indentation data obtained from AFM experiments. The relationship between force, F, cell indentation, δ, probe tip radius, R, and E^*, the complex elastic modulus, is described by the following,

$$F = \frac{4}{3}E^* \sqrt{R}\delta^{\frac{3}{2}}, \tag{24.2}$$

where E^* is given by,

$$\frac{1}{E^*} = \frac{1 - \upsilon_1^2}{E_1} + \frac{1 - \upsilon_2^2}{E_2}. \tag{24.3}$$

The material parameters E_1 (Young's modulus) and υ_1 (Poisson's ratio) of the colloid (silica glass) are 68 GPa and 0.19, respectively. The cell modulus, E_2 is an outcome of the model fit using 0.37 as υ_2, the cellular Poisson's ratio [18]. Young's moduli were obtained by a least square analysis of the force-indentation curve using routines in the Igor Pro (WaveMetrics, Inc.) software package and the material properties listed above.

Sufficient cleaning of the AFM probe tip prior to testing with a methanol rinse and ozone treatment for 15 min resulted in little to no adhesion between the probe tip and cell surface. Adhesion between the tip and cell surface would result in a retraction curve that significantly deviates from the approach curve at the end of indentation. Significant adhesion would render the Hertz model invalid and a more appropriate model such as JKR or DMT should be applied.

24.2.6 Statistical Analysis

Moduli values exhibited a log-normal distribution as indicated by high r-squared values in quantile–quantile plots discussed below. Since a non-normative distribution exists, an arithmetic average and standard deviation would be insufficient parameters to perform tests for statistical significance. Instead, a Wilcoxon rank sum test was performed to determine statistical significance using commercially available software (MATLAB, MathWorks).

24.3 Results and Discussion

Multiple different methodologies have been used to assess and compare mechanical properties of live cells. Each type of measurement has technical advantages as well as methodological disadvantages in terms of both performance of the measurement and robustness of the measured value as a true and representative reflection of cell elasticity. There being no standardized technique, this imposes significant challenges in interpreting and comparing values of mechanical properties of live cells obtained using different methods. For example, some techniques deform the entire cell (micropipette aspiration) as opposed to relatively small, localized indentations as in AFM. Even within AFM, depending on tip shape, values for elastic moduli can be severely overestimated without a model correction. Our work advances previous studies by providing a head to head comparison across 15 conditions: five cell lines under control, actin destabilization, and microtubule destabilization, using the same AFM method, plating protocol, and nanomechanical analysis. Our report also includes a relatively large number of experiments, for which we demonstrate the lognormal distribution of values utilizing quantile–quantile plots.

24.3.1 Cytoarchitectural Effects of Cytochalasin D and Nocodazole

Imaging of the cells indicated a well spread morphology as expected 2 days after plating, the same time point as elastic stiffness measurements. Cancer cells tended to form colonies, especially in the case of hepatocellular carcinoma. Fluorescent staining of tubulin (Fig. 24.2, left columns) and actin (Fig. 24.2, right columns) was performed on each of the five cell lines (Fig. 24.2 contains HUVECs see [17] for other images) including control cells and drug treated cells. Actin filaments in

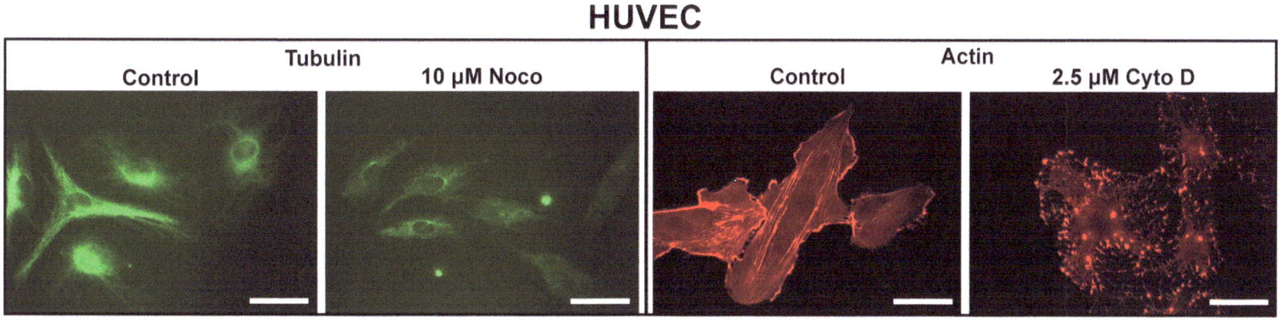

Fig. 24.2 Effect of cytoskeletal destabilizers, 10 μM nocodazole (Noco) or 2.5 μM cytochalasin D (Cyto D), on HUVECs. Tubulin Tracker™ was used to label microtubules. Phalloidin was used to label actin filaments. Images taken at 40× magnification. Scale bars are 50 μm. Adapted from [17]

control cells showed some qualitative degree of orientation or alignment as well as clearly visible fibers across the length of the cell (tens of microns). Actin staining of cytochalasin D-treated cells demonstrated the effect of the drug by the lack of long stress fibers. Residual patches of actin left behind by depolymerized stress fibers were located with high concentration at the cell perimeter, likely indicating focal adhesion sites. Tubulin staining in control cells revealed a dense microtubule network across all cell lines. Tubulin staining in nocodazole-treated cells resulted in a much less dense network of microtubules, where often the microtubule spindle is visible.

24.3.2 Cells Exhibit Probing Frequency Dependence

The strain-rate dependence of viscoelastic materials such as living cells has been established for a number of different cell types. Figure 24.3 includes the elastic moduli extracted using the Hertz model at increasing approach velocity and therefore increasing frequency in untreated fibroblasts in (a), fibroblasts treated with cytochalasin D (b), and fibroblasts treated with nocodazole in (c). The average calculated frequency from Eq. (24.1) and the corresponding geometric mean of the elastic modulus measurement is plotted in black squares on each plot. Figure 24.3d includes an overlay of the geometric mean of the elastic modulus for each treatment condition. Figure 24.3 includes the measurement of ten untreated fibroblasts, 10 fibroblasts treated with cytochalasin D, and 11 fibroblasts that were treated with nocodazole. The comparison between the elastic modulus probed at lower frequencies and the elastic modulus of the same treatment at higher frequencies demonstrates the frequency dependence of the elastic modulus of living cells. The dependence on probing frequency is exacerbated by the application of cytochalasin D. For control cells, the geometric mean increases from 1.7 kPa at a probing frequency of 0.4 s^{-1} to 2 kPa at a probing frequency of 24 s^{-1}. Nocodazole treated cells had a similar disparity between lower and higher probing frequencies increasing from 1.5 kPa at 0.4 s^{-1} to 2 kPa at 23 s^{-1}. Cytochalasin D treated cells exhibited the largest frequency dependence increasing from 0.8 kPa at 0.3 s^{-1} to 1.4 kPa at 22 s^{-1}. It is important that elastic moduli measurements control for differences in probing frequency by maintaining the same indentation depth regardless of cell type or drug treatment. For our single frequency studies, we chose the approach velocity of 500 nm/s and given an average indentation depth of approximately 300 nm, corresponds to a frequency of about 2 s^{-1}, about midway through the range of frequencies probed.

24.3.3 Elastic Stiffness Measurements Exhibit a log-Normal Distribution

The values for elastic stiffness were compared to a standard normal distribution and a log-normal distribution. Of those two, the log-normal distribution compared more favourably with our discrete measurements. To quantify the appropriateness of a log-normal distribution to describe our elastic stiffness measurements, a quantile–quantile plot (q–q plot) was produced with a corresponding r-squared value to assess linearity. Histograms and q–q plots for the HUVEC cell line are included in Fig. 24.4. The highly linear q–q plots demonstrate that cellular elastic moduli follow a log-normal distribution. The r-squared value reported is between a line generated between the first and third quantiles and the measurement quantiles. R-squared values generated from q–q plots comparing log(modulus) values to a standard normal curve ranged from 0.87 to

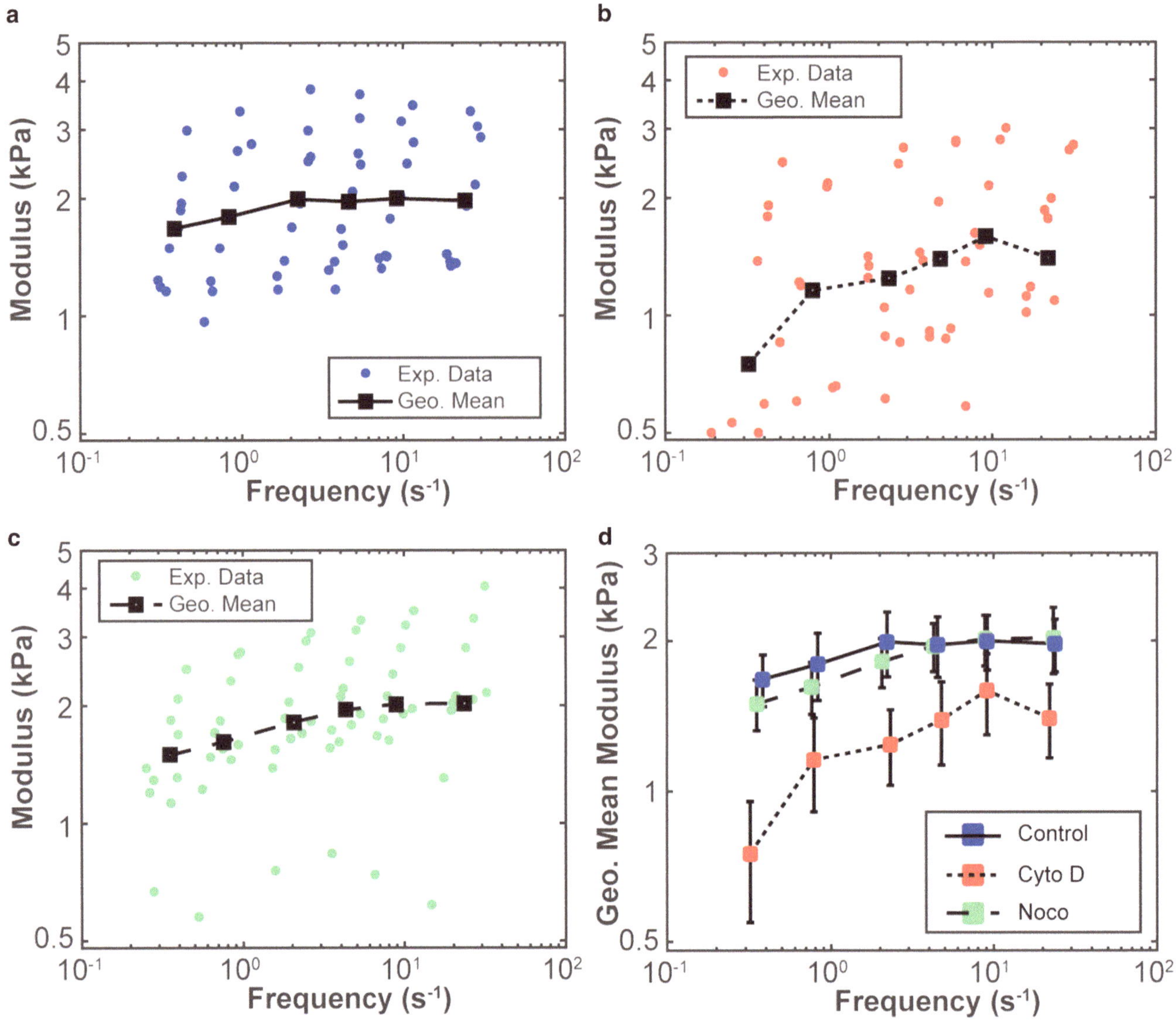

Fig. 24.3 Frequency dependence of fibroblasts. Measured elastic moduli for (**a**) fibroblasts under control conditions (*blue dots*), (**b**) fibroblasts treated with 2.5 μM cytochalasin D (Cyto D, *orange dots*), and (**c**) fibroblasts treated with 10 μM nocodazole (Noco, *green dots*), with the geometric mean at each approach velocity (*black squares*). (**d**) Geometric mean of the measured elastic modulus at each approach velocity for control (*blue square*), cytochalasin D treatment (*orange square*), and nocodazole treatment (*green square*). Error bars are standard error

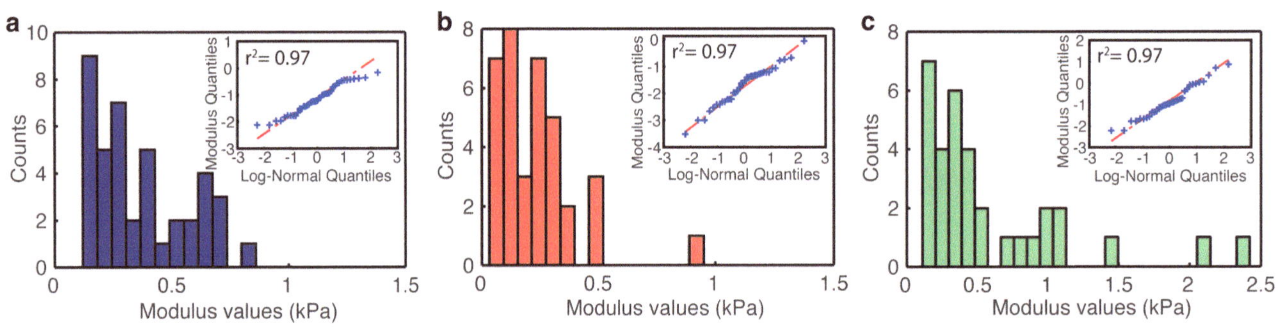

Fig. 24.4 Histogram of measured modulus values for: (**a**) HCCs under control conditions, (**b**) HCCs with 2.5 μM cytochalasin D, and (**c**) HCCs with 10 μM nocodazole, note the change in x-axis scale. *Inset*: Quantile–quantile plots (q–q plots), which are a graphical method of comparing the distribution of measured modulus values to a log-normal distribution. A highly linear q–q plot indicates similarity in the two distributions. Adapted from [17]

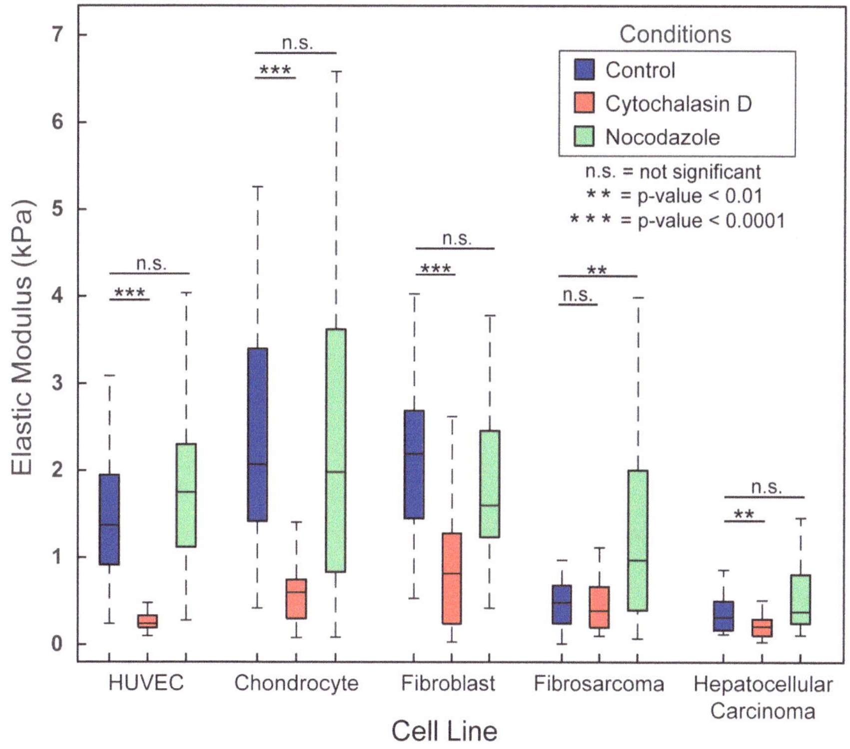

Fig. 24.5 Standard box plot of modulus values for HUVECs, chondrocytes, fibroblasts, fibrosarcomas, and hepatocellular carcinoma under control conditions (*blue*), 2.5 μM cytochalasin D treatment (*orange*), or 10 μM nocodazole treatment (*green*). Cancer cell measurements are highlighted in the shaded region. Wilcoxon rank sum test conducted for significance (**p-value < 0.01, ***p-value < 0.0001). Adapted from [17]

0.98, thereby indicating strong agreement with a log-normal distribution. Histograms and q–q plots for the other four cell lines are included in [17].

24.3.4 Actin Filaments Contribute More Than Microtubules to Elastic Stiffness in Normal Cells

All elastic stiffness measurements for the five cell lines are included in Fig. 24.5, which includes a standard box plot. Statistical significance is indicated by stars over the compared data sets. Since a non-normative distribution exists, a Wilcoxon rank sum test was performed to determine statistical significance. There is a dramatic decrease in cell elasticity when actin filaments are targeted via cytochalasin D for HUVECs, chondrocytes, and fibroblasts. The median values drop to between 18 and 38 % of the control value when cytochalasin D treated cells are compared to control cells for these three cell lines. Specifically, the median value of HUVEC control cells, 1.37 kPa, drops to 0.24 kPa upon cytochalasin D treatment ($p = 5e\text{-}11$), the median value of chondrocyte moduli at 2.07 kPa for control cells drops to 0.60 kPa ($p = 8e\text{-}11$), and the median value of fibroblast moduli at 2.19 kPa for control cells drops to 0.82 kPa ($p = 1e\text{-}7$). Considering the wide distribution of values for nocodazole treatment differed only slightly from controls: HUVEC median moduli value increased to 1.75 kPa ($p = 0.08$), chondrocyte median moduli value decreased to 1.98 kPa ($p = 0.47$), and fibroblast median moduli value decreased to 1.60 kPa ($p = 0.19$). Thus, destabilizing the microtubule network had little effect on the modulus for HUVECs, chondrocytes and fibroblasts; whereas destabilizing the actin filament network, decreased the measured modulus by half or more.

24.3.5 Destabilizing Microtubule Dynamics has a Greater Stiffness Effect in Cancer Cell Lines than Normal Cell Lines

Moduli values for the cancer cells were less than half that of the three other cell lines. Median moduli values for fibrosarcoma and hepatocellular carcinoma were 0.48 and 0.30 kPa, respectively. The response to cytochalasin D resulted in a decrease in the median modulus value for both the fibrosarcoma and hepatocellular carcinoma cells. The median modulus value decreased to 0.39 kPa for fibrosarcomas ($p = 0.70$, not significant) and to 0.21 kPa for the hepatocellular carcinomas ($p = 9e\text{-}4$). The nocodazole treatment *increased* the median modulus value of both cancer lines. Fibrosarcoma and hepatocellular carcinoma increased to 0.97 kPa ($p = 3e\text{-}3$) and 0.38 kPa ($p = 0.13$, not significant), respectively. While hepatocellular carcinoma line did not show a statistically significant increase after nocodazole treatment, the data is trending in that direction. The median value for control HCC is 0.30 kPa and for HCC after nocodazole treatment it increases to 0.38 kPa. Additionally, the geometric mean is 33 % higher after nocodazole treatment. Both median and geometric mean indicate increased stiffness. While use of Student's t-test to compare HCC and HCC after nocodazole treatment results in a statistically significant metric at $p < 0.05$, we have chosen the Wilcoxon rank sum test as a very robust and distribution-independent measure of significance, which may as a result not highlight all trends. Thus, it is evident that nocodazole treatment altered cell stiffness within the two cancer lines more than in the three normal cell lines for which measurements were made.

24.4 Conclusions

The comparison between the elastic modulus probed at lower frequencies and higher frequencies demonstrates the frequency dependence of the elastic modulus of living cells. The dependence on probing frequency was exacerbated by the depolymerisation of actin filaments using cytochalasin D. It is important that elastic moduli measurements control for differences in probing frequency by maintaining the same indentation depth during AFM measurements regardless of cell type or drug treatment. We confirm across multiple cell lines that actin filaments, and not microtubules, contribute most to elastic stiffness in healthy cell lines. The contribution of microtubules toward overall cell elasticity is highly contested, however, we have no evidence that suggests destabilized microtubules leads to a decrease in stiffness, to the contrary, that it may in fact *increase* cell stiffness in cells with disorganized cytoskeletal architectures such as cancer cells. Thus microtubule dynamics contributes more to elasticity in cancer cells than normal healthy cell lines. Our results provide a comprehensive platform upon which to build experimentally toward an understanding of the mechanisms by which microtubules or modifications thereof contribute to cell mechanical state. This, in turn, yields a foundation for linking mechanical and functional conditions, including cell bioenergetics in disease, susceptibility to toxins or therapeutic molecules in treatment regimens, or even metastatic potential for cancer phenotyping.

Acknowledgements The authors gratefully acknowledge our funding sources: ONR Grant N000141612100 (DME), the Provost's Postdoctoral Fellowship for Academic Diversity (MEG), NSF-NSEC Grant DMR08-32802 (RJC), and URF 4-000002-4820 (DME), which made this work possible. The work was performed at and supported by the Nano Bio Interface Center at the University of Pennsylvania through an instrumentation grant, DBI-0721913, and DMR-0425780. We also thank Judith Kandel for cell culture training and Dr. Matt Brukman and Dr. Matt Caporizzo for instrument support. We thank the following for their generous donations: Dr. Robert Mauck (chondrocytes), Dr. Ben Stanger (HUH-7), and Dr. Bruce Malkowicz (HT-1080).

References

1. Discher, D.E., Janmey, P., Wang, Y.L.: Tissue cells feel and respond to the stiffness of their substrate. Science **310**(5751), 1139–1143 (2005)
2. Bao, G., Suresh, S.: Cell and molecular mechanics of biological materials. Nat. Mater. **2**(11), 715–725 (2003)
3. Lee, G.Y.H., Lim, C.T.: Biomechanics approaches to studying human diseases. Trends Biotechnol. **25**(3), 111–118 (2007)
4. Rother, J., Noding, H., Mey, I., Janshoff, A.: Atomic force microscopy-based microrheology reveals significant differences in the viscoelastic response between malign and benign cell lines. Open Biol. **4**(5), 140046 (2014)
5. Nalam, P.C., Gosvami, N.N., Caporizzo, M.A., Composto, R.J., Carpick, R.W.: Nano-rheology of hydrogels using direct drive force modulation atomic force microscopy. Soft Matter **11**(41), 8165–8178 (2015)
6. Raman, A., Trigueros, S., Cartagena, A., Stevenson, A.P.Z., Susilo, M., Nauman, E., Contera, S.A.: Mapping nanomechanical properties of live cells using multi-harmonic atomic force microscopy. Nat. Nanotechnol. **6**(12), 809–814 (2011)
7. Caporizzo, M.A., Roco, C.M., Ferrer, M.C.C., Grady, M.E., Parrish, E., Eckmann, D.M., Composto, R.J.: Strain-rate dependence of elastic modulus reveals silver nanoparticle induced cytotoxicity. Nanobiomedicine **2** (2015). pii: 9
8. Ketene, A.N., Schmelz, E.M., Roberts, P.C., Agah, M.: The effects of cancer progression on the viscoelasticity of ovarian cell cytoskeleton structures. Nanomedicine **8**(1), 93–102 (2012)
9. Corbin, E.A., Kong, F., Lim, C.T., King, W.P., Bashir, R.: Biophysical properties of human breast cancer cells measured using silicon MEMS resonators and atomic force microscopy. Lab Chip **15**(3), 839–847 (2014)
10. Moreno-Flores, S., Benitez, R., Vivanco, M.D., Toca-Herrera, J.L.: Stress relaxation and creep on living cells with the atomic force microscope: a means to calculate elastic moduli and viscosities of cell components. Nanotechnology **21**(44), 445101 (2010)

11. Kuznetsova, T.G., Starodubtseva, M.N., Yegorenkov, N.I., Chizhik, S.A., Zhdanov, R.I.: Atomic force microscopy probing of cell elasticity. Micron **38**(8), 824–833 (2007)
12. Rebelo, L.M., de Sousa, J.S., Mendes, J., Radmacher, M.: Comparison of the viscoelastic properties of cells from different kidney cancer phenotypes measured with atomic force microscopy. Nanotechnology **24**(5), 055102 (2013)
13. Park, S., Koch, D., Cardenas, R., Kas, J., Shih, C.K.: Cell motility and local viscoelasticity of fibroblasts. Biophys. J. **89**(6), 4330–4342 (2005)
14. Ribeiro, A.S., Khanna, P., Sukumar, A., Dong, C., Dahl, K.: Nuclear stiffening inhibits migration of invasive melanoma cells. Cell. Mol. Bioeng. **7**(4), 544–551 (2014)
15. Buda, A., Pignatelli, M.: Cytoskeletal network in colon cancer: from genes to clinical application. Int. J. Biochem. Cell Biol. **36**(5), 759–765 (2004)
16. Lindberg, U., Karlsson, R., Lassing, I., Schutt, C.E., Höglund, A.-S.: The microfilament system and malignancy. Semin. Cancer Biol. **18**(1), 2–11 (2008)
17. Grady, M.E., Composto, R.J., Eckmann, D.M.: Cell elasticity with altered cytoskeletal architectures across multiple cell types. J. Mech. Behav. Biomed. Mater. **61**, 197–207 (2016)
18. Trickey, W.R., Baaijens, F.P.T., Laursen, T.A., Alexopoulos, L.G., Guilak, F.: Determination of the Poisson's ratio of the cell: recovery properties of chondrocytes after release from complete micropipette aspiration. J. Biomech. **39**(1), 78–87 (2006)

MIX
Papier aus verantwortungsvollen Quellen
Paper from responsible sources
FSC® C105338

If you have any concerns about our products,
you can contact us on
ProductSafety@springernature.com

In case Publisher is established outside the EU,
the EU authorized representative is:
Springer Nature Customer Service Center GmbH
Europaplatz 3, 69115 Heidelberg, Germany

Printed by Libri Plureos GmbH
in Hamburg, Germany